T0337828

COMPUTATIONAL PARALINGUISTICS

COMPUTATIONAL PARALINGUISTICS

EMOTION, AFFECT AND PERSONALITY IN SPEECH AND LANGUAGE PROCESSING

Björn W. Schuller
Technische Universität München, Germany

Anton M. Batliner
Friedrich-Alexander-Universität Erlangen-Nürnberg, Germany

This edition first published 2014
© 2014 John Wiley & Sons, Ltd

Registered office
John Wiley & Sons Ltd, The Atrium, Southern Gate, Chichester, West Sussex, PO19 8SQ, United Kingdom

For details of our global editorial offices, for customer services and for information about how to apply for permission to reuse the copyright material in this book please see our website at www.wiley.com.

Library of Congress Cataloging-in-Publication Data

Schuller, Björn.
   Computational paralinguistics : emotion, affect and personality in speech and language
processing / Björn Schuller, Anton Batliner. – First Edition.
      pages cm
   Includes bibliographical references and index.
   ISBN 978-1-119-97136-8 (hardback)
   1. Psycholinguistics–Data processing.   2. Linguistic models–Data processing.   3. Paralinguistics.
4. Language and emotions.   5. Speech processing systems.   6. Human-computer interaction.
7. Emotive (Linguistics)   8. Computational linguistics.   I. Batliner, Anton.   II. Title.
   P37.5.D37S38 2014
   401′.90285–dc23

                                                                          2013019325

A catalogue record for this book is available from the British Library.

Typeset in 10/12pt Times by Aptara Inc., New Delhi, India

Printed and bound in Singapore by Markono Print Media Pte Ltd

1   2014

To Dagmar and Gisela

Contents

Preface

It might be safe to claim that 20 years ago, neither the term 'computational paralinguistics' nor the field it denotes existed. Some 10 years ago, the term did not yet exist either. However, in hindsight, the field had begun to exist if we think of the first steps towards the automatic processing of emotions in speech in the mid-1990s. For example, Picard's book on *Affective Computing* published in 1997, and the International Speech Communication Association (ISCA) workshop on emotion and speech in 2000, just to mention some of the many topics and events related to and belonging to computational paralinguistics. The term 'paralinguistics' had already been coined in the 1950s – with different broad or narrow denotations; we will try and sketch this history in Part I of this book. Yet, in the realm of 'hard core' automatic (i.e., computational) processing of speech, the topic was still not fully acknowledged; as one of our colleagues said: 'Emotion recognition, that's esoterics with HMMs.'

Today, it might be safe to claim that computational paralinguistics has been established as a discipline in its own right – although surprisingly, not the term itself. It is only natural that, as a new and still somewhat exotic field, it has to cope with prejudices on the one hand, and unrealistic promises on the other hand.

This book represents the first attempt towards a unified overview of the field, its extremely ramified and diverse 'genealogy', its methodology, and the state of the art. 'Computational paralinguistics' is not an established subject that can be studied, and this fact is mirrored in the 'scientific CVs' of both authors. B.S. studied electrical engineering and information technology. However, his doctoral thesis dealt with one aspect of computational paralinguistics: the automatic recognition of human emotion in speech. During his habilitation period he broadened the scope of his work to 'intelligent audio analysis' – dealing with quite a number of further paralinguistic aspects, including those found in sung language and many other audio processing problems such as emotion in music and general sound. At the time of finalisation of the manuscript, he was a professor in computer science. A.B. started within philology, came from diachronic phonology to phonetics in general to prosody in particular, and via prosody in the interface for and within syntax and semantics to the automatic processing of acted and very soon naturalistic emotions – realising, moreover, that he had been dealing with different topics for a long time that all can be located within (computational) paralinguistics.

Originally, the intended focus of this book was on the computational processing of emotions and affect in speech and language, taking into account personality as well. In the early conceptual stages, however, we realised that this would be sub-optimal and thus decided to deal with everything 'besides' linguistics – namely, the computational processing of 'para'-linguistics in a broad sense. However, we confine ourselves to the acoustic/phonetic/linguistic aspects, that is, we only deal with one modality, namely speech/language including non-verbal components, and disregard other modalities such as facial gestures or body posture. Moreover, we do not aim at a complete description of human–human or human–machine communication which would include the generation and production (synthesis) of speech, the interaction with other components within a multimodal system, the role within application systems, or real-life applications and their evaluation. Apart from the fact that most of these aspects would not be part of our core competence, we feel that it makes sense to try to establish computational paralinguistics as one building block amongst several others. Besides, there are already good overview and introductory books available on these other topics. And last but not least, it would be rather too complex for one book.

We wish to provide the reader with a sort of map presenting an overview of the field, and useful for finding one's own way through. The scale of this map is medium-size, and we can only display a few of the houses in this virtual paralinguistic 'city' with their interiors, on an exemplary basis. In so doing, we hope to provide guidelines for the novice and to present at least a few new insights and perspectives to the expert. Many studies are referred to and core results are summarised. For all of them, the caveat holds that basically all such studies are restricted – confined to a specific choice of subjects, research questions, operationalisations, and features employed, just to mention a few of the decisive factors. There are errors such as the famous erroneous decimal point that made spinach more healthy than anything else – note that reports of this error might be erroneous themselves. And of course, there is much more that can go wrong – and hopefully, we will find out: scrutiny of results and replications of studies will eventually converge to more stable claims.

We decided not to describe basic phonetic/linguistic knowledge such as vowel or consonant charts, details on pitch versus F0, loudness, morphological and grammatical systems, basics on production and perception of speech, and the like. Such information can easily be obtained in introductory and overview books from the respective fields, as well as from online sources. In a similar way, selections had to be made for the computation aspects. For example, many approaches to linguistic modelling exist, and the fields of machine learning and signal processing each deserve at least one book in its own right. Thus, we limited our choices to the methods most established and common in the field – serving as a solid basis and inspiration for the interested reader to look further. A connected resource of information is the book's homepage found at http://www.cp.openaudio.eu which includes features such as links to the openSMILE toolkit and (part of) the data described.

Björn W. Schuller and Anton M. Batliner
Munich, February 2013

Acknowledgements

We would like to thank in particular Florian Eyben and Felix Weninger at Technische Universität München in Munich, Germany, for their great help in finishing Chapters 7–13; we also extend our thanks to Martina Römpp for her help with the illustrations in this part, and to Tobias Bocklet and Florian Hönig at the Friedrich Alexander Universität in Erlangen-Nuremberg, Germany, for commenting on Chapter 5.

We stand on the proverbial shoulders of giants – and of many other 'normal' people who all contributed to our understanding. In the same way, we are grateful to colleagues who do not share our points of view: opinions are not only reinforced by encouragement but also by dissent which helps us to rethink or revise our own approaches and theoretical positions, or simply to stick with them.

There are too many people in the field(s) that supported us, discussed with us, and exchanged ideas with us, too many to list in detail, so we wish to express our gratitude to all of them in a generic way.

For the encouragement in the first place, excellent continued support and guidance as well as patience, we sincerely thank the editor and publisher – John Wiley & Sons.

Finally, we would like to thank the HUMAINE Association (henceforth the Association for the Advancement of Affective Computing) for providing an excellent network not only for affective computing, but also for the broader field of computational paralinguistics. The authors also acknowledge funding from the European Commission under grant agreement no. 289021 from the ASC Inclusion Project committed to provide interactive serious emotion games for children with autism spectrum condition.

List of Abbreviations

ACF	Autocorrelation Function
ACM	Association for Computing Machinery
AD(H)D	Attention Deficit (Hyperactivity) Disorder
AEC	(FAU) Aibo Emotion Corpus
AL	Active Learning
ALC	Alcohol Language Corpus
AM	Acoustic Model
ANN	Artificial Neural Network
API	Application Programming Interface
ARFF	Attribute Relation File Format (WEKA)
ARMA	Autoregressive Moving Average
ASC	Autism Spectrum Condition
ASCII	American Standard Code for Information Exchange
ASD	Autism Spectrum Disorder
ASR	Automatic Speech Recognition
AUC	Area Under Curve
AVEC	Audiovisual Emotion Challenge
AVIC	(TUM) Audiovisual Interest Corpus
BAC	Blood Alcohol Content
BES	Berlin Emotional Speech Database
BFI	Big Five Inventory
BLSTM	Bidirectional Long Short-Term Memory
BoCNG	Bag-of-Character-N-Grams
BoNG	Bag-of-N-Grams
BoW	Bag-of-words
BPTT	Back-Propagation Through Time
BRAC	Breath Alcohol Concentration Test
BRNN	Bidirectional Recurrent Neural Network
C-AuDiT	Computer-Assisted Pronunciation and Dialogue Training
CALL	Computer-Aided Language Learning
CAPT	Computer-Aided Pronunciation Training
CC	(Pearson) Correlation Coefficient
CD	Compact Disc

CEICES	Combining Efforts for Improving Automatic Classification of Emotion in Speech
CFS	Correlation based Feature Selection
CGN	Spoken Dutch Corpus from Centre for Genetic Resources, the Netherlands
CR	Compression Rate
CSV	Comma Separated Value (format)
DAT	Digital Audio Tape
DBN	Dynamic Bayesian Networks
DCT	Discrete Cosine Transformation
DES	Danish Emotional Speech (Database)
DET	Detection Error Trade-off (curve)
DFT	Discrete Fourier Transform
DLL	Dynamic Link Library
DT	Decision Tree (machine learning) / Determiner (linguistics)
EC	Emotion Challenge (Interspeech 2009)
ECA	Embodied Conversational Agent
EER	Equal Error Rate
EM	Expectation Maximisation
EMMA	Extensible Multi-Modal Annotation (markup language, XML-style)
ERB	Equivalent Rectangular Bandwidth
ETSI	European Telecommunications Standards Institute
EU	European Union
EWE	Evaluator Weighted Estimator
F0	Fundamental Frequency
FAU	Friedrich Alexander University
FFT	Fast Fourier Transformation
FN	False Negatives
FNN	Feed-forward Neural Network
FNR	False Negative Rate/Ratio
FP	False Positives
FPR	False Positive Rate/Ratio
GM	Gaussian Mixture
GMM	Gaussian Mixture Model
GPL	GNU/General Public Licence
GUI	Graphical User Interface
HMM	Hidden Markov Model
HNR	Harmonics-to-Noise Ratio
HTK	Hidden Markov (model) Toolkit
ICA	Independent Component Analysis
IDF	Inverse Document Frequency
IG	Information Gain
IGR	Information Gain Ratio
IIR	Infinite Impulse Response
IP	Interruption Point
ISCA	International Speech Communication Association
ISLE	Italian and German Spoken Learners English (corpus)

ITU	International Telecommunication Union
KL	Kullback–Leibler (divergence/distance)
L1	first language
L2	second language
LDA	Linear Discriminant Analysis
LDC	Linguistic Data Consortium
LIWC	Linguistic Inquiry and Word Count
LLD	Low-Level Descriptor
LM	Language Model
LOO	Leave One Out
LP	Linear Predictor / Linear Prediction
LPC	Linear Predictive Coding
LPCC	Linear Predictive Cepstral Coefficient
LSF	Line Spectral (pair) Frequency
LSP	Line Spectral Pair
LSTM	Long Short-Term Memory
LSTM-RNN	Long Short-Term Memory Recurrent Neural Network
LVCSR	Large Vocabulary Continuous Speech Recognition
MAE	Mean Absolute Error
MAPE	Mean Absolute Percentage Error
MFB	Mel-Frequency Band
MFCC	Mel-Frequency Cepstral Coefficient
MIML	Multimodal Interaction Markup Language
ML	Maximum Likelihood
MLE	Mean Linear Error
MLP	Multilayer Perceptron
MSE	Mean Square Error
NCSC	NKI CCRT Speech Corpus
NHD	Null Hypothesis Decision
NHR	Noise-to-Harmonics Ratio
NHT	Null Hypothesis Testing
NL	Non-likeable
NMF	Non-negative Matrix Factorisation
NN	Noun (linguistic)
NP	Noun Phrase (linguistic)
NPV	Negative Predictive Value
OCEAN	Openness, Conscientiousness, Extraversion, Agreeableness, Neuroticism
OOV	Out Of Vocabulary (words)
openEAR	open-source Emotion and Affect Recognition (toolkit)
openSMILE	open-source Speech and Music Interpretation by Large Scale Extraction (toolkit)
PC	Paralinguistic Challenge (Interspeech 2010)
PCA	Principal Component Analysis
PCM	Pulse Code Modulation
PDA	Pitch Detection Algorithm
PDF	Probability Density Function

PLP	Perceptual Linear Prediction
PLP-CC	Perceptual Linear Prediction Cepstral Coefficient
PLTT	Post Laryngectomy Telephone Test
POS	Part Of Speech
PP	Prepositional Phrase (linguistic)
PR	Precision
RASTA	RelAtive SpecTrA
RASTA-PLP	RelAtive SpecTrA Perceptual Linear Prediction (Coefficients)
RE	Recall
RIFF	Resource Interchange File Format
RMS	Root Mean Square
RMSE	Root Mean Squared Error
RNN	Recurrent Neural Network
ROC	Receiver Operating Characteristic
SAL	Sensitive Artificial Listener
SAMPA	Speech Assessment Methods Phonetic Alphabet
SEMAINE	Sustained Emotionally coloured Machine-human Interaction using Non-verbal Expression (project)
SFFS	Sequential Floating Forward Search
SFS	Speech Filing System
SHS	Sub-Harmonic Summation
SI	International System of Units (French: Système international d'unités)
SIFT	Simplified Inverse Filtering Technique
SIMIS	Speech In Minimal Invasive Surgery (database)
SLC	Sleepy Language Corpus
SLD	Speaker Likability Database
SNR	Signal-to-Noise Ratio
SP	SPecificity
SPC	Speaker Personality Corpus
SSC	Speaker State Challenge
STC	Speaker Trait Challenge
SVM	Support Vector Machine
SVQ	Split Vector Quantisation
SVR	Support Vector Regression
T-expression	Ternary expression
TF	Term Frequency
TFIDF	Term Frequency and Inverse Document Frequency
TN	True Negatives
TP	True Positives
TPR	True Positive Rate/Ratio
TUM	Technische Universität München
UA	Unweighted Accuracy
UAR	Unweighted Average Recall
VAM	Vera-Am-Mittag (German TV show, corpus)
VB	Verb
VP	Verb Phrase

VQ	Vector Quantisation
WA	Weighted Accuracy / Word Accuracy (ASR)
WAR	Weighted Average Recall
WYALFIWYG	What You Are Looking For Is What You Get
WYSIWYG	What You See Is What You Get
XML	eXtensible Mark-up Language
ZCR	Zero Crossing Rate

Part I

Foundations

1

Introduction

1.1 What is Computational Paralinguistics? A First Approximation

So difficult it is to show the various meanings and imperfections of words when we have nothing else but words to do it with.

(John Locke)

The term *computational paralinguistics* is not yet a well-established term, in contrast to *computational linguistics* or even *computational phonetics*; the reader might like to try comparing the hits for each of these terms – or for any other combination of 'computational' with the name of a scientific field such as psychology or sociology – in a web search. This terminological gap is a little puzzling given the fact that there is a plethora of studies on, for example, affective computing (Picard 1997) and speech – which can partly be conceived as a sub-field of computational paralinguistics (as far as speech and language are concerned). But let us first take a look at the coarse meanings of the two words this term consists of: 'computational' and 'paralinguistics'.

Here, 'computational' means roughly that something is done by a computer and not by a human being; this can mean analysing the phenomenon in question, or generating humanlike behaviour. Note that nowadays computers are used for practically all systematic and scientific work, even if it is only for listing data, detailed information on subjects, or annotations in an ASCII (American Standard Code for Information Exchange) file. In traditional phonetic or psychological approaches, this can go along with the use of highly sophisticated signal extraction and statistical programs. A borderline between the 'simple' use of computers for tedious work and the use of computers for actually modelling and performing human behaviour is of course difficult to define. Here, we simply mean both: doing the work with the help of computers, and letting computers do the work of analysing and processing.

'Paralinguistics' means 'alongside linguistics' (from the Greek preposition $\pi\alpha\rho\alpha$); thus the phenomena in question are not typical linguistic phenomena such as the structure of a language, its phonetics, its grammar (syntax, morphology), or its semantics. It is concerned with *how* you say something rather than *what* you say.

Computational Paralinguistics: Emotion, Affect and Personality in Speech and Language Processing, First Edition.
Björn W. Schuller and Anton M. Batliner. © 2014 John Wiley & Sons, Ltd. Published 2014 by John Wiley & Sons, Ltd.

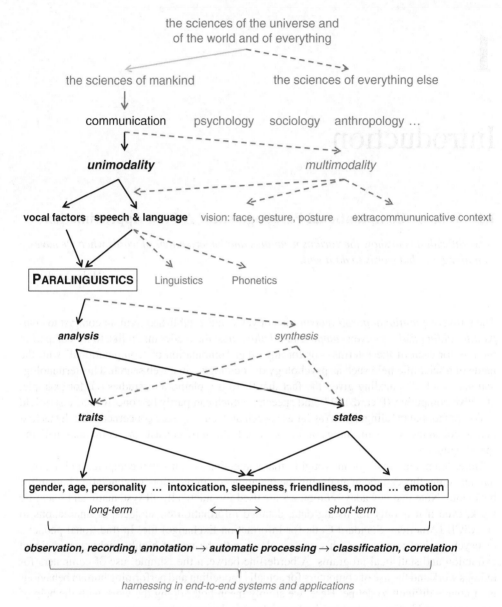

Figure 1.1 The realm of computational paralinguistics

In Figure 1.1 we try to narrow down the realm of paralinguistics in a reasonable way, as we conceive it and as we will deal with it in this book. Of course, there are other conceptualisations of paralinguistics, some broader, some narrower in scope. Figure 1.1 is a sort of flowchart that we will follow from top to bottom. A grey font indicates fields and topics that are not part of paralinguistics, for instance, the global science of mankind or of everything else that can be found in this world. Dashed lines lead to fields that are more or less disregarded in this book.

The first word shown in black is 'communication', denoting that interactions between human beings are focal. Paralinguistics deals with speech and language which both are primarily means of communication; even a soliloquy has to be overheard and eventually recorded and processed by the computer in order to be an object of investigation. The same holds for a private diary in its written form: it might not be intended as communication with others, but as soon as it is read by someone else, it is. Of course, human communication is an important part of related fields such as psychology, sociology, or anthropology. Thus, we have to follow the flowchart further down to point out what distinguishes paralinguistics from all these related fields.

In traditional linguistics, the term 'language' refers to the (innate and/or acquired) mental competence, and the term 'speech' to the performance, that is, to the ability to convert this competence into motor signals, acoustic waves, and percepts. In this book we adhere to a shallower definition of these two terms, based on their use in speech and language technology. Language is more or less synonymous with 'natural language' which is modelled and processed within computational linguistics; speech is the object of investigation within automatic speech processing, that is, 'spoken language', as opposed to written language.

We want to restrict paralinguistics to the unimodal processing of events primarily produced with the voice, or secondarily encoded in written language. 'Communication' is used in a broad sense: speech and language are primarily means of interaction between human beings; however, they can be decoupled from this function and analysed on their own, that is, when not used in a communicative setting. Note that this sort of secondary communication is natural for written language, because here the communication of sender and addressee is normally decoupled as for time and place.

We do not want to distinguish between *extralinguistics* and paralinguistics; Laver (1994, pp. 22f) attributes extralinguistics to *informative functions* denoting age, sex, and suchlike, and paralinguistics to *communicative functions*. Implicitly, extralinguistic functions are always communicated, as a sort of background; we will subsume these functions under *biological trait primitives*; see Section 5.1. Moreover, it would be simply too cumbersome to introduce 'computational extralinguistics' as an additional field.

There are alternative conceptualisations, the most important one arguably paralinguistics in the sense of *'multimodality'*; this holds practically for every aspect, whether it be emotion, or personality, or social signals (see Chapter 5). Undoubtedly, the most natural human–human interaction is face-to-face, with each partner employing all the means available to them: voice, linguistic message, face, gestures, and body. However, there are some common and natural (interaction) scenarios where only the acoustic channel is used, for instance, in telephone conversations or in radio plays. Moreover, it is simply natural that interaction/conversation partners sometimes speak, and sometimes only listen. In the latter case, there is no speech available that can be investigated; in the first case, when people are talking, analysing faces is more difficult because the movements of speech are superposed onto the facial gestures.

In this book, apart from vocal factors, speech, and language, everything else such as facial expressions, gestures, body posture, and any extracommunicative context is not part of paralinguistics. Needless to say, all these other aspects are very important within human–human and human–machine communication; we will return to such multimodal aspects throughout this book.

Moreover, paralinguistics is sort of defined *ex negativo*; it comprises everything which is *not* the object of investigation in phonetics or linguistics: It does not address the systematic aspects of speech and language which are dealt with in sub-fields such as phonology, morphology, syntax, or semantics. Note, however, that the *use* of specific phonological or grammatical structures within a specific context may very well be object of investigation within paralinguistics. All this will be illustrated more extensively below.

In this book we will focus on analysis, basically excluding generation and synthesis. At first sight, this might appear exotic: after all, analysis and generation/synthesis can be considered as two sides of the same coin. Both are necessary for a complete account. However, methodologies differ considerably; in Batliner and Möbius (2005) the methodological differences between analysis on the one hand and generation/synthesis on the other hand have been detailed for the modelling and processing of prosody. From a methodological point of view, the analysis of gestures has perhaps more in common with the analysis of speech than its synthesis. Moreover, analysis and not synthesis is the core competence of the authors. We therefore decided to treat synthesis the same way as vision and extracommunicative context, as a fringe phenomenon in this book.

So far, we have addressed broad *fields of science*, either including or excluding them from our definition of paralinguistics. We will now briefly sketch the *phenomena* we are dealing with, as well as the *processing chain*. All this will be dealt with in more depth in the chapters to follow. In simple terms, paralinguistics deals with *traits* and *states*; traits are long-term events, whereas states are short-term. Examples are given in Figure 1.1. Typical traits are gender, age, and personality, and typical states are emotions. Then, there are phenomena which are somehow in between: People can be friendly towards everybody, or, towards a specific person, only for a very short time. You can get tipsy, that is, intoxicated, for a short time, or you can be a regular heavy drinker. The title of this book mentions three exemplary phenomena:

> **personality** denoting long-term character traits which are specific to individuals or groups. In a broader sense, this encompasses everything that characterises a specific individual, including traits such as age, gender, race, and suchlike. In a narrower sense, this encompasses psychological traits such as neuroticism.

> **emotion** denoting short-term states: prototypical ones such as anger, fear, joy, or less prototypical ones such as surprise.

> **affect** as a broader term, encompassing all kinds of manifestations of personality such as mood, interpersonal stances, or attitudes as displayed in Table 2.1 – a very common term since Picard (1997).

The last terms to be commented upon in the title are *speech and language processing*: In basic research, the two fields of phonetics and linguistics deal with different data: phonetics with (the production/acoustics/perception of) speech, and linguistics with (written) language. Accordingly, there are two different lines of research traditions in paralinguistics: one dealing mostly with the acoustic signal (called, for instance, 'emotion/affect processing'), and one dealing exclusively with written language (called, for instance, 'sentiment analysis'). In automatic speech processing, the approach is different: acoustic and linguistic information are combined in a hybrid fashion. Following this tradition, we will address both acoustic and linguistic phenomena in this book.

With observations, recordings, and annotations, we decide which phenomena we are dealing with, and how long the single event takes. In *computational* paralinguistics, we then try to process these phenomena automatically. Ultimately, this means producing some performance measures which tell us how good we are at doing that. All this is the core topic of this book. Eventually, we of course want to evaluate our models not as single components but within end-to-end-systems and to harness them in applications; this will be touched upon and exemplified *passim*.

1.2   History and Subject Area

Language is not an abstract construction of the learned, or of dictionary makers, but is something arising out of the work, needs, ties, joys, affections, tastes, of long generations of humanity, and has its bases broad and low, close to the ground.

(Noah Webster)

So far, we have outlined the realm of computational paralinguistics. In this section, we want to sketch the history of paralinguistics and to narrow down its subject area.

Ever since the advent of structuralism (Saussure 1916), the study of (speech and) language has been more or less confined to the skeleton of language: phonetics/phonology, morphology, syntax, and grammar in general; there were only rather anecdotal remarks on functions of language which go beyond pure linguistics, for example, the following from Bloomfield (1933):

...pitch is the acoustic feature where gesture-like variations, non-distinctive but socially effective, border most closely upon genuine linguistic distinctions. The investigation of socially effective but non-distinctive patterns in speech, an investigation scarcely begun, concerns itself, accordingly, to a large extent with pitch.

Pike (1945) was amongst the few who noticed these additional functions of intonation:

Other intonation characteristics may be affected or caused by the individual's physiological state – anger, happiness, excitement, age, sex, and so on. These help one to identify people and to ascertain how they are feeling...

The basic neglect of paralinguistics holds for both European and American linguistics at that time – both displaying different varieties of structuralism. Thus, the central focus of linguistics in the last century was on structural, on genuine linguistic and, as far as speech is concerned, on formal aspects within phonetics and phonology. Language was conceived of as part of semiotics which deals with *denotation*, that is, with the core meaning of items.

This conviction was clearly expressed by Sapir (1921):

If speech, in its acoustic and articulatory aspect, is indeed a rigid system, how comes it, one may plausibly object, that no two people speak alike? The answer is simple. All that part of speech which falls out of the rigid articulatory framework is not speech in idea, but is merely a superadded, more or less instinctively determined vocal complication inseparable from speech in practice. All the individual color of speech – personal emphasis, speed, personal cadence, personal pitch – is a non-linguistic fact, just as the incidental expression of desire and emotion are, for the most part, alien to linguistic expression. Speech, like all elements of culture, demands conceptual selection, inhibition of the randomness of instinctive behavior.

On the other hand, in Sapir (1927) we can find an – albeit informal – conceptualisation of 'speech as a personality trait', giving a rough but fair enumeration of parameters which are relevant for characterising personality – and, by the way, emotion as well:

> *To summarize, we have the following materials to deal with in our attempt to get at the personality of an individual, in so far as it can be gathered from his speech. We have his voice. We have the dynamics of his voice, exemplified by such factors as intonation, rhythm, continuity, and speed. We have pronunciation, vocabulary, and style. Let us look at these materials as constituting so and so many levels on which expressive patterns are built.*

Such remarks were, however, normally anecdotal and somehow spurious. Generally, non-linguistic aspects were conceived as fringe phenomena, often taken care of by neighbouring disciplines such as anthropology, ethnology, or psychology. This attitude slowly changed in the middle of the last century; linguists and phoneticians began to be interested in all these phenomena mentioned by Bloomfield (1933) and Pike (1945), that is, in a broader conceptualisation of semiotics, dealing with *connotation* (e.g., affective/emotive aspects) as well.

According to Trager (1958), Laver (1994), and Rauch (2008), the term 'paralanguage' was first introduced by the American linguist Archibald Hill (1958).

Terms such as 'extralinguistic', 'paralanguage', and 'paralinguistics' were used by Trager (1958), and later elaborated on by Crystal (1963, 1966, 1971, 1974, 1975a,b). To start with, Crystal (1963) mentions the neglect of paralinguistics by linguistics:

> *The last decade has brought renewed study of this linguistic backwater, now called paralanguage; but there has been surprisingly little attempt to approach the subject in a sufficiently systematic and empirical way to satisfy the critical linguist.*

This critical attitude seems to have persisted during the decades to come (cf. Rauch 1999):

> *...paralinguistics is to linguistics, unfortunately, a neglected stepchild at most... (p. 165)*
>
> *...the seeds for obscuring the domain of paralanguage were inherent in its twentieth-century rebirth for linguists by linguists. (p. 166)*

One of the few who not only dealt with paralinguistic phenomena but also tried to really propagate this field was Fernando Poyatos (1991, 1993, 2002).

On the other hand, both within linguistics proper and especially with the advent of human–computer interaction, we can say that paralinguistics and neighbouring disciplines have been safely established. Yet, the subject areas are still defined differently. These are the definitions given in two renowned dictionaries:

> **paralanguage** (n.) *A term used in SUPRASEGMENTAL PHONOLOGY to refer to variations in TONE of voice which seem to be less systematic than PROSODIC features (especially INTONATION and STRESS). Examples of **paralinguistic features** would include the controlled use of BREATHY or CREAKY voice, spasmodic features (such as giggling while speaking), and the use of secondary ARTICULATION (such as lip-ROUNDING or NASALIZATION) to produce a tone of voice signalling attitude, social*

role, or some other language-specific meaning. Some analysts broaden the definition of paralanguage to include KINESIC features; some exclude paralinguistic features from LINGUISTIC analysis. (Crystal 2008)

paralanguage *... 1. Narrowly, non-segmental vocal features in speech, such as tone of voice, tempo, tut-tutting, sighing, grunts, and exclamations like* Whew! *2. Broadly, all of the above plus non-vocal signals such as gestures, postures and expressions – that is, all non-linguistic behaviour which is sufficiently coded to contribute to the overall communicative effect. ... (Trask 1996)*

Thus, since it first came into use in the middle of the last century, 'paralinguistics' has been confined to the realm of human–human communication, but with a broad and a narrow meaning. We follow Crystal (1974) who excludes visual communication and the like from the subject area and restricts the scope of the term to 'vocal factors involved in paralanguage'; cf. Abercrombie (1968) for a definition along similar lines. 'Vocal factor', however, in itself is not well-defined. Again, there can be a narrow meaning excluding linguistic/verbal factors, or a broad meaning including them. We use the last one, defining *paralinguistics* as the discipline dealing with those phenomena that are *modulated onto* or *embedded into* the verbal message, be this in acoustics (vocal, non-verbal phenomena) or in linguistics (connotations of single units or of bunches of units). This scope is mirrored and, at the same time, instantiated by the possibility of late fusion in multimodal ('non-verbal') processing and by the (relative) independence of computational paralinguistic approaches from other fields. Many tools and procedures have been developed specifically for dealing with the speech signal or with (written) language; many sites and researchers, specialising in speech and language, have extended their focus onto computational paralinguistics.

To give examples for acoustic phenomena: everybody would agree that coughs are not linguistic events, but they are somehow embedded in the linguistic message. The same holds for laughter and filled pauses (such as *uhm*) which display some of the characteristics of language, for example, as far as grammatical position or phonotactics is concerned. All these phenomena are embedded in the word chain and are often modelled the same way as words in automatic speech processing; they can denote (health) state, emotion/mood, speaker idiosyncrasies, and the like. In contrast, high pitch as an indicator of anxiety and breathy voice indicating attractiveness, for example, are modulated onto the verbal message. As for the linguistic level, paralinguistics also deals with everything beyond pure phonology/morphology/syntax/semantics. Let us give an example from semantics. The 'normal' word for a being that can be denoted with these classic semantic features [+human, +female, +adult] is *woman*. In contrast, *slut* has the same denotation but a very different connotation, indicating a strong negative valence and, at the same time, the social class and/or the character of the speaker. Bunches of units, for instance the use of many and/or specific adjectives or particles, can indicate personality traits or emotional states.

Whereas the 'garden-fencing' within linguistics, that is, the concentration on structural aspects, was mainly caused by theoretical considerations, a similar development can be observed within automatic speech (and language) processing which, however, was mainly caused by practical constraints. It began with concentrating on single words; then very constrained, read/acted speech, representing only one variety, that is, one rather canonical speech register, was addressed. Nowadays, different speech registers, dialects, and spontaneous speech in general are processed as well.

At least amongst linguists, language has always been seen as the principal mode of communication for human beings (Trager 1958) which is accompanied by other communication systems such as body posture, movement, facial expression, cf. (Crystal 1966) where the formal means of indicating communicative stances are listed: (1) vocalisations such as 'mhm', 'shhh', (2) hesitations, (3) 'non-segmental' prosodic features such as tension (slurred, lax, tense, precise), (4) voice qualifiers (whispery, breathy, . . .), (5) voice qualification (laugh, giggle, sob, cry), and (6) non-linguistic personal noises (coughs, sneezes, snores, heavy breathing, etc.).

The extensional differentiation between terms such as verbal/non-verbal or vocal/non-vocal is sometimes not easy to maintain and different usages do exist; as often, it might be favourable to employ a prototype concept with typical and fringe phenomena (Rosch 1975). A fringe phenomenon, for example, is filled pauses which often are conceived of as non-verbal, vocal phenomena; however, they normally follow the native phonotactics, cannot be placed everywhere, can be exchanged by filler words such as *well*, and are modelled in automatic speech recognition the same way as words.

We can observe that different strands of research – having much in common – evolved more or less independently of each other; thus what sometimes has been subsumed under 'paralinguistics' by linguists has been called *non-verbal behaviour* research by psychologists (cf. Harrigan *et al.* 2008): facial actions, vocal behaviour, and body movement. Jones and LeBaron (2002) mention that '. . . the study of "nonverbal communication" emerged in the 1960s, largely in reaction to the overwhelming emphasis placed upon verbal behavior in the field of communication.' They argue in favour of integrating verbal and non-verbal approaches. Non-verbal communication from a multi-disciplinary perspective is dealt with in Burgoon *et al.* (2010).

Interestingly, the terms used are normally rather *ex negativo* such as 'para-/extra-linguistics' or 'non-verbal/non-vocal' parameters – again indicating that from its very beginning, the field had to be delimited from the more established discipline of linguistics.

1.3   Form versus Function

Form follows function – that has been misunderstood. Form and function should be one, joined in a spiritual union.

(Frank Lloyd Wright)

The distinction between *form* and *function* is arguably constitutive for modern phonetics and linguistics – form roughly meaning 'what does it look like, and how does it relate to other elements?', function meaning 'what is it used for?'. We can compare this basic distinction with the distinction between knowledge about fabrics (the substance for clothing) and fashion (the form, the code of clothing) on the one hand, and the function of clothing (used for an evening in the opera, or used for mountaineering) on the other hand. There are specialists in each of these aspects.

A phonetic form is constituted by some higher-level, structural shape or type which can be described holistically and analysed/computed using between 1 and *n low-level descriptors (LLDs)* such as pitch or intensity values and *functionals* such as mean or maximum values over time. A simple example is a high rising final tone which very often denotes, that is, functions as indicating a question. This is a genuine linguistic function. In addition, there are paralinguistic functions encoded in speech or in other vocal activities. Examples include a slurred voice when the speaker is inebriated, or a loud and high-pitched voice when a person is angry. *Phonetics* deals with the acoustic, perceptual, and production aspects of spoken language (speech), and *linguistics* with all aspects of written language; this is the traditional point of view. From an engineering point of view, there is a slightly different partition: normally, the focus is on recognising and subsequent understanding of the content of spoken or written language; for speech, acoustic modelling is combined with linguistic modelling whereas, naturally enough, (written) language can only be modelled by linguistic means. Form is rather a means to handle the function of speech and language.

Laver (1994, p. 20) refers to the contrast between (phonological) *form* – how does an element relate to other elements? – and (phonetic) *substance* – for example, how does its acoustics look? Crystal (2008, pp. 194, 204) tells apart functions within and outside linguistics: linguistic and phonetic form and substance do have paralinguistic functions, for example, the word *somewhat* with its specific phonetic realisation in a specific syntactic position functioning as a hedge can characterise personality and/or communicative situations. In this book we will always contrast phonetic/linguistic form (consisting of form and substance) with paralinguistic function.

The distinction between form and function is also constitutive for discriminating paralinguistics as it typically is performed by linguists/phoneticians from paralinguistics as it typically is performed by engineers, psychologists, and other neighbouring disciplines. Linguists and phoneticians start with some formal element and try to find out which functions can be attributed to this specific form. Engineers and psychologists are primarily interested in modelling (manually or automatically) specific phenomena such as personality, emotion, or speech pathology, with the help of acoustic and/or linguistic parameter; that is, they are primarily interested in one specific (type of) function and want to find out which (form) features to use for modelling and classifying this function. A simple test whether the author of a study follows a formal or a functional approach is to estimate the number of pages dedicated to the one or the other aspect; of course, there are transitional forms in between.

Figure 1.2 illustrates the two different approaches. While the figure is straightforward, what is behind it can be extremely complicated. Conceptually, it is always a one-to-many mapping but the direction is reversed. To the left, there is the typical phonetic/linguistic approach. We start with one – more or less complex – formal element; this can be one word, one type of words *(part-of-speech)*, one syntactic construction; it can be one phoneme with its *allophonic* (free) variants, or one *supra-segmental* parameter, just to mention a few. Then we want to

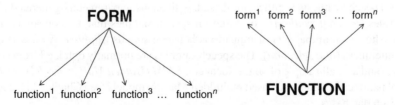

Figure 1.2   Form versus function: (left) linguistic/phonetic approach; (right) sociological, sociolinguistic, psychological, and psycholinguistic approach

find out which functions this formal element can be used for. This can be some intralinguistic function – for instance, a pronoun serves an *anaphoric* function if it refers to a noun that can be found earlier in the word chain; in this book, we are mostly interested in paralinguistic functions. To the right, the approach typical of psychology and other neighbouring fields is depicted. We start with one specific function – for instance, one emotion, one personality trait, or a specific non-native accent. Then we try to find out which formal elements denote this function and can be used for automatic modelling. In a few cases, this might be one form such as a high final pitch value, denoting questions or proneness towards questions. Nowadays, in brute-force approaches, we employ many formal elements, up to several thousand features.

In fact, it is mostly not a one-to-many but a many-to-many relationship because of the intrinsically multi-functional nature of acoustic-linguistic parameters. We will return to the distinction of form and function when presenting the different research strategies in Chapters 4 and 5.

1.4   Further Aspects

As pointed out above, we restrict the realm of paralinguistics to the analysis of vocal and verbal aspects. Of course, this is not the whole picture. There is *generation* and *synthesis* of paralinguistics as well, often embedded in a *multimodal* interaction. All this has to be modelled for human–computer interactions in prospective *application* scenarios. In order to be successful, *usability* has to be considered from the very beginning. Above all, and at a very early stage, *ethical* considerations have to be taken into account.

These aspects are not all relevant or pivotal for all subfields of paralinguistics: 'emotionally intelligent' virtual agents and robots might arguably be the main target group for generation and synthesis of adequate behaviour. In contrast, the synthesis of deviant speech (e.g., of a foreign accent or of some variety of pathological speech) most likely comes last, as far as meaningful applications are concerned. Of course, we can always imagine some application: there might be some place for a virtual agent in a computer game that impersonates a foreign language learner. Apart from being somehow exotic, such characters might be less attractive from a marketing point of view, and more difficult to implement.

In this section, we will first give a short account of synthesis, concentrating on emotion and personality. Then, both generation and analysis of multimodality are addressed. We will conclude with applications and usability, and ethical considerations.

1.4.1   The Synthesis of Emotion and Personality

Basically, speech synthesis is either rule-based, with acoustic parameters generated following specific rules (*formant synthesis*), or based on speech samples that are concatenated, which can be as short as minimal sets of transitions between sounds (*diphone synthesis*) or whole phrases/utterances (*unit selection*). The speech samples are normally obtained from controlled recordings and a small sample of single speakers. *HMM (Hidden Markov model) synthesis* is a statistical parametric synthesis, based on hidden Markov models (see Chapter 11), and trained from speech databases.

Formant synthesis allows for systematic manipulation but does not sound fully natural. Unit selection sounds most natural if the transitions between units can be smoothed correctly but necessitates too much pre-recorded information, especially if different paralinguistic phenomena – normally different emotional states – have to be modelled. Explicit modelling is flexible but not fully natural; in contrast, concatenative modelling sounds natural but is not flexible.

Of course, the building blocks are the same for synthesis and analysis, such as phones, words, phrases, and utterances. Methodologies, however, differ considerably. These differences did not show up very clearly in the early days of phonetic and emotion research, when only a few features were explicitly modelled, that is, manipulated or analysed. Nowadays, however, it seems difficult to bridge the gap between the thousands of features used for brute-force modelling of many speakers on the one hand, and the relatively few features or speakers modelled for rule-based or concatenative synthesis. The basically different procedural approaches towards analysis and synthesis of prosody – which is one of the main building blocks for emotional modelling, apart from voice quality – are elaborated on in (Batliner and Möbius 2005). *Embodied conversational agents (ECAs)* can be cartoon-like or very pronounced; they can be based on acted emotions produced by one single actor. It is conceivably not possible to manipulate and generate thousands of acoustic-prosodic features – which is no problem in a brute-force automatic classification. Thus, the perspective of paralinguistic synthesis differs considerably from that of paralinguistic analysis; it is more similar to that of traditional lab phonetics where specific hypotheses are proved with the help of carefully manipulated stimuli presented in *identification* or *discrimination* tests.

Let us give a short account of the synthesis of emotional speech. First attempts towards emotional, rule-based speech synthesis were reported in Murray and Arnott (1993, 1995). Schröder (2001) gave an overview of what had been done in the field; this was continued in Schröder (2004); see also Gobl and Ní Chasaide (2003) and Schröder *et al.* (2010). Black (2003) deals with unit selection and emotional speech. The synthesis of 'personality primitives' such as age and gender is straightforward. The synthesis of personality traits is not yet a fully established field. It is addressed in Trouvain *et al.* (2006); Schröder *et al.* (2012) describe a framework for generating and synthesising emotionally competent embodied conversational agents having four different personalities – aggressive, cheerful, gloomy, and pragmatic – within a prototype of a multimodal dialogue system, the Sensitive Artificial Listener (SAL) scenario. Schröder *et al.* (2011) present a conceptual view on the generic representation of emotions using an *Emotion Markup Language* (an agreed-upon computer-readable representation) and *ontologies* (formal specifications of shared conceptualisations such as paralinguistic states).

Of course, the divide between synthesis and analysis can be overcome, but this will take time. Indications are, on the one hand, the use of HMM synthesis based on multiple speakers, and on the other hand, the use of synthesised data for augmenting real-life databases used for training automatic classifications of paralinguistic phenomena.

1.4.2   Multimodality: Analysis and Generation

Evidently, it is not only speech and language that communicate personality, emotion, affect and the like. Darwin (1872) attributed a leading role to the face: 'Of all parts of the body, the face is most considered and regarded, as is natural from its being the chief seat of expression and the source of the voice.' In addition, there are gestures and body movements/posture (Kleinsmith and Bianchi-Berthouze 2012).

Although confined to a specific experimental condition, the results of Mehrabian and Wiener (1967) were often taken as proof that the verbal channel contributes only little (7%) to the communication of attitudes; this is called the 7%–38%–55% myth. However, already Ekman and Friesen (1980) state that the '... claims in the literature that the face is most important or that the nonverbal-visual channel is more important than the verbal-auditory channel have not been supported' in their experiments. O'Sullivan *et al.* (1985) elaborate further on the complex interrelationship between different types of messages and the relative importance of verbal compared to non-verbal factors. The answer is simply that 'no channel is always most important'. Further arguments can be found in Trimboli and Walker (1987), Lapakko (1997), and Krauss *et al.* (1981).

Generic statements on the relative importance of single modalities do not make any sense; we can only ask about the contribution of single modalities in specific communicative settings. Now, does it make sense to describe single modalities at all? Jorgensen (1998) claimed that researchers focusing only on one modality, for example, the verbal channel, 'are no longer studying valid communication processes, but rather disassociated parts of the whole'. A simple but important argument against this position can be found in Planalp and Knie (2002) where it is argued that even '... the simplest research on cue and channel combinations ... produces incredibly complicated results'.

An overview on affect and emotion recognition methods in multimodal human–computer systems is given in Zeng *et al.* (2009) and Pantic *et al.* (2011). The complex task of coordinating multiple modalities in an affective agent – this holds for analysis as well – is nicely illustrated in the following list quoted from Martin *et al.* (2011):

- *Equivalence/substitution: one modality conveys a meaning not borne by the other modalities (while it could be conveyed by these other modalities)*
- *Redundancy/repetition: the same meaning is conveyed at the same time via several modalities*
- *Complementarity:*
  - *Amplification accentuation/moderation: one modality is used to amplify or attenuate the meaning provided by another modality*
  - *Additive: one modality adds congruent information to another modality*
  - *Illustration/clarification: one modality is used to illustrate/clarify the meaning conveyed by another modality*
- *Conflict/contradiction: the meaning transmitted on one modality is incompatible or contrasting with the one conveyed by the other modalities; this cooperation occurs when the meaning of the individual modalities seems conflicting but indeed the meaning of their combination is not and emerges from the conflicting combination of the meanings of the individual modalities.*
- *Independence: the meanings conveyed by different modalities are independent and should not be merged.*

The claim of Planalp and Knie (2002) might be slightly exaggerated; however, there is surely a trade-off between the basic complexity of multimodality and the possibilities for investigating complex phenomena within one single modality. Moreover, there are of course constellations where only one channel is used and available for analysis. We will come back to this in Section 2.12.

1.4.3   Applications, Usability and Ethics

The computational processing of paralinguistics could be conceived of as encapsulated – data in, measures out; thus neither applications nor usability need to be addressed. Potential applications are often mentioned in the introduction and/or final remarks of articles; yet, it is often not clear how the approach presented really can be harnessed in these applications. However, they are, together with ethical considerations, decisive for success or failure of approaches if we do not confine ourselves to pure research. Applications are, as it were, at the lowest, practical level; mostly they are presented in the form of single examples. However, we will try and present a tentative taxonomy. Usability is on a methodologically higher level; pertinent considerations are based on psychological and theoretical theories. Ethics is, of course, at the highest level, and not a genuine topic of this technologically oriented book. However, we definitely want to stress its importance.

Applications

Examples of applications for affective computing are given in Picard (1997), Picard (2003), and Batliner *et al.* (2006), and for paralinguistics in a broader sense in Burkhardt *et al.* (2007) and Schuller *et al.* (2013). In the following, examples and presentation are inspired by the last three references.

Basic types of application approaches are (1) speech recognition in itself, (2) analysis, screening and monitoring of paralinguistic events or phenomena, and (3) interaction, normally of humans with an ECA on a computer or with a robot; all these aspects can be employed alone or in combination. Speech recognition can hopefully be improved when the speaker's paralinguistic peculiarities are modelled, for instance, by a preceding attribution to speaker classes. For all the other aspects, it is the other way round: speech recognition should be as good as possible, especially for employing linguistic features. Human–human communication can be analysed and monitored. We are interested in the type of communication. That is, is the communication symmetric or asymmetric? Which roles are taken up to what extent by which communication partner? Is the communication 'normal', 'as it should be', 'not as it should be', or even 'pathological'? We can assess conversations of married couples having problems (Lee *et al.* 2010), and we can monitor and summarise meetings (Kennedy and Ellis 2003; Laskowski 2009) or call centre interactions. All types of deviant, especially 'pathological', speech (see Section 5.6) can be the object of analysis and monitoring, and of periodic screening. Human–machine interaction is a wide field, encompassing all kinds of gaming, tutoring, information, and assistive and communicative robotics. Media retrieval is a genuine object of investigation for written data. Even cross-modal control is possible by paralinguistic rather than linguistic means, for example, in helping artists with upper limb disabilities to use the volume of their voice to control cursor movements to create drawings on the screen.

A tentative taxonomy of basic properties for emotion-oriented systems is given in Batliner *et al.* (2006); this was extended to speaker classification systems in Burkhardt *et al.* (2007). Table 1.1, adapted from Batliner *et al.* (2006) and slightly edited, summarises the criteria that can distinguish different (types of) applications. 'Meta-assessment' means that we are looking at success or failure from the outside. For instance, telling a call-centre customer during the interaction that she is angry (single instance decision, system design: on-line, mirroring) can

Table 1.1   Some basic features of applications

features	description
meta-assessment	
critical	application's aims are impaired if the paralinguistic phenomenon is processed erroneously (single instance decision)
non-critical	erroneous processing does not impair application's aims (cumulative evidence)
system design	
on-line	system reacts (immediately/delayed) while interacting with user
off-line	no system reaction, or delayed reaction after actual interaction
mirroring	user gets feedback as for his/her (expressive) behaviour
non-mirroring	system does not give any explicit feedback
emotional	system reacts in an emotional way
non-emotional	system does not behave emotionally but 'neutrally'

be detrimental if she is not. Collecting cumulative evidence, for instance, checking whether customers are, in the long run, more content with one or another automatic system, is non-critical as long as there is some correlation with the ground truth. Generally, on-line, mirroring, and emotional system (re)actions will be more critical than off-line, non-mirroring, and non-emotional (re)actions. Note that 'non-critical' as used here refers to the immediate context of an application; later decisions based on the processing within such applications can of course still be wrong, that is, economically unwise or unethical.

Further examples of applications will be touched upon *passim* in the following chapters.

Usability

Trivially, usability is largely irrelevant in the case of pure recording and monitoring or screening, and subsequent analysis and evaluation – all this constitutes the largest part of paralinguistic research. Of course, usability becomes relevant if the user is somehow interacting with the system (see Table 1.1).

Kaye *et al.* (2011) present the historical background and contrast the traditional goals of software design, such as utility, effectiveness and learnability, with the goals of user-centred design (user experience goals: being motivating, fun, enjoyable) which is not a linear but an iterative process involving users at any step of design and evaluation. Historically, the concept of usability evolved from a narrow focus on the computational expert (engineers) in the 1950s, passing through a focus on psychological and cognitive experts in the 1980s – the users of the personal computer were no longer experts but rather lay people; the interface was not only a green or amber screen with a text console but a graphical interface, and nowadays, it can be ECAs and even robots as well. Most recent is a focus on experience-focused human–computer interaction. An up-to-date account of the whole field is given in Rogers *et al.* (2011). The specific requirements for multimodal interfaces are addressed in Oviatt (1999, 2008) and Oviatt and Cohen (2000).

Ethical Considerations

The first question often asked – not by engineers but by people directly concerned – about technology at large is whether we want it that way, and, in particular, whether we mind if it takes over human work. Secretaries fear unemployment when dictation systems are used, and speech therapists fear unemployment when screening and assessment of speech pathologies are taken over by machines. Is technology harmful to those whose expertise is substituted, or will they then be free to do more expert work?

The next question is whether technology does what it is supposed to do and what it promises to do. Failure to do that might not be detrimental in the case of dictation systems: it is straightforward to find out. It requires more effort in the case of automatic screening: a counter-check introduces exactly that kind of manual work that should be avoided. In the case of the lie detector, such a counter-check is not possible in real-life situations, for instance in court, thus we have to rely on transferability from scientific studies. A chapter in Kreiman and Sidtis (2011) describes nicely how this definitely should not be done because it simply would violate the principle *in dubio pro reo*. This would be the case of an erroneous single instance decision (Table 1.1), with monstrous unethical consequences.

Even if technology can do what it is supposed to do, we have to ask whether this is acceptable: is the monitoring of call-centre agents ethically acceptable – even if it might be reliable if done off-line and in an accumulative way?

Basic research on computational paralinguistics might not be much concerned with such questions but it definitely is necessary to know about them. Ethical concerns about privacy, however, are of utmost relevance. How can we ensure that ethical principles are observed during recruitment of participants in experiments, during recording, storing, and during dissemination/displaying of recordings and other types of results? Here, we should follow the principle of informed consent (Sneddon *et al.* 2011): amongst others things, participants should be informed about the goals of the study and the experiment, and they should be given the possibility to withdraw during and after the experiment. It should go without saying that strict anonymity must be guaranteed later on. All these provisions might be cumbersome to maintain but universities, research organisations and legal provisions will safeguard them.

Several further aspects of ethical concern are discussed in Ragin and Amoroso (2011), Cowie (2011), Döring *et al.* (2011), and Goldie *et al.* (2011).

1.5 Summary and Structure of the Book

In Part I, we will lay the foundations for the computational processing that is dealt with in Part II. In this first chapter, we began by defining 'computational paralinguistics', and sketched the history of the term and the subject area. The opposition between form and function is a guiding principle not only of phonetics and linguistics but also of paralinguistics, and was therefore described next. In the rest of this chapter, we sketched all those aspects – generation, synthesis, multimodality, usability, applications, and ethical considerations that we do not focus on in the following. Chapter 2 presents a taxonomy of oppositions that can – but need not in every case – be relevant for the different sub-fields, topics and phenomena within paralinguistics. Chapter 3 tries to point out important aspects of modelling, that is, theories and methodologies that heavily influence the way how we see, approach, deal with, and evaluate paralinguistic phenomena. Chapter 4 presents examples for formal elements – segmental and supra-segmental, phonetic and linguistic, verbal and non-verbal – that constitute the building blocks for the marking of all those functions that are described in Chapter 5.

Corpus Engineering is dealt with in Chapter 6, especially annotations and exemplars of paralinguistics corpora; this constitutes the transition of Part I into Part II.

In Part II, we first give an overview of the chain of processing within computational paralinguistics in Chapter 7. Acoustic features and their extraction on a 'low' frame-by-frame level are described in Chapter 8. Linguistic features are then described in Chapter 9. Both types of features are used for feature generation on a supra-segmental level as described in Chapter 10. In Chapter 11 we deal with the field's most common approaches to modelling from a machine learning point of view. This also includes a statement on feature relevance analysis and testing protocols. In Chapter 12 insight is given into how best to embed computational paralinguistics in a running system's working context. The selected aspects cover distribution in a client–server architecture, weakly surpervised learning and confidence measure calculation. To provide the chance of experiencing what the book describes, Chapter 13 provides a 'hands-on' tutorial alongside a description of the 'usual suspects' when it comes to toolkits in the field.

Outside of these two main parts, Chapter 14 provides a short general epilogue.

Last but not least, the Appendix contains the description of standardised feature sets and a feature encoding scheme.

References

Abercrombie, D. (1968). Paralanguage. *International Journal of Language & Communication Disorders*, **3**, 55–59.

Batliner, A. and Möbius, B. (2005). Prosodic models, automatic speech understanding, and speech synthesis: Towards the common ground? In W. Barry and W. Dommelen (eds), *The Integration of Phonetic Knowledge in Speech Technology*, pp. 21–44. Springer, Dordrecht.

Batliner, A., Burkhardt, F., Ballegooy, M. v., and Nöth, E. (2006). A taxonomy of applications that utilize emotional awareness. In *Proc. of IS-LTC 2006*, pp. 246–250, Ljubljana.

Black, A. (2003). Unit selection and emotional speech. In *Proc. of Eurospeech*, pp. 1649–1652, Geneva.

Bloomfield, L. (1933). *Language*. Holt, Rinehart and Winston, New York.

Burgoon, J. K., Guerrero, L., and Floyd, K. (2010). *Nonverbal Communication*. Allyn & Bacon, Boston.

Burkhardt, F., Huber, R., and Batliner, A. (2007). Application of speaker classification in human machine dialog systems. In C. Müller (ed.), *Speaker classification I: Fundamentals, Features, and Methods*, pp. 174–179. Springer, Berlin.

Cowie, R. (2011). Editorial: 'Ethics and good practice' – Computers and forbidden places: Where machines may and may not go. In P. Petta, C. Pelachaud, and R. Cowie (eds), *Emotion-Oriented Systems: The Humaine Handbook*, Cognitive Technologies, pp. 707–712. Springer, Berlin.

Crystal, D. (1963). A perspective for paralanguage. *Le Maître Phonétique*, **120**, 25–29.

Crystal, D. (1966). The linguistic status of prosodic and paralinguistic features. *Proc. of the University of Newcastle-upon Tyne Philosophical Society*, **1**, 93–108.

Crystal, D. (1971). Prosodic and paralinguistic correlates of social categories. In E. Ardener (ed.), *Social Anthropology*, pp. 185–206. Tavistock, London.

Crystal, D. (1974). Paralinguistics. In T. Sebeok (ed.), *Current Trends in Linguistics 12*, pp. 265–295. Mouton de Gruyter, The Hague.

Crystal, D. (1975a). Paralinguistic behaviour as continuity between animal and human communication. In W. McCormack and S. Wurm (eds), *Language and Man: Anthropological Issues*, pp. 13–27. Mouton de Gruyter, The Hague.

Crystal, D. (1975b). Paralinguistics. In J. Benthall and T. Polhemus (eds), *The Body as a Medium of Expression*, pp. 162–174. Institute of Contemporary Arts, London.

Crystal, D. (2008). *A Dictionary of Linguistics and Phonetics*. Blackwell, Oxford, 6th edition.

Darwin, C. (1872). *The Expression of the Emotions in Man and Animals*. John Murray, London.

Döring, S., Goldie, P., and McGuinness, S. (2011). Principalism: A method for the ethics of emotion-oriented machines. In P. Petta, C. Pelachaud, and R. Cowie (eds), *Emotion-Oriented Systems: The Humaine Handbook*, Cognitive Technologies, pp. 713–724. Springer, Berlin.

Ekman, P. and Friesen, W. V. (1980). Relative importance of face, body, and speech in judgments of personality and affect. *Journal of Personality and Social Psychology*, **38**, 270–277.

Gobl, C. and Ní Chasaide, A. (2003). The role of voice quality in communicating emotion, mood and attitude. *Speech Communication*, **40**, 189–212.

Goldie, P., Döring, S., and Cowie, R. (2011). The ethical distinctiveness of emotion-oriented technology: Four long-term issues. In P. Petta, C. Pelachaud, and R. Cowie (eds), *Emotion-Oriented Systems: The Humaine Handbook*, Cognitive Technologies, pp. 725–734. Springer, Berlin.

Harrigan, J. A., Rosenthal, R., and Scherer, K. R. (eds) (2008). *New Handbook of Methods in Nonverbal Behavior Research*. Oxford University Press, Oxford.

Hill, A. (1958). *Introduction to Linguistic Structures: From Sound to Sentence in English*. Harcourt Brace, New York.

Jones, S. E. and LeBaron, C. D. (2002). Research on the relationship between verbal and nonverbal communication: Emerging integrations. *Journal of Communication*, **52**, 499–521.

Jorgensen, P. (1998). Affect, persuasion, and communication processes. In P. Andersen and L. Guerrero (eds), *Handbook of Communication and Emotion*, pp. 403–422. Academic Press, San Diego.

Kaye, J. J., Laaksolahti, J., Höök, K., and Isbister, K. (2011). The design and evaluation process. In P. Petta, C. Pelachaud, and R. Cowie (eds), *Emotion-Oriented Systems: The Humaine Handbook*, Cognitive Technologies, pp. 641–656. Springer, Berlin.

Kennedy, L. and Ellis, D. (2003). Pitch-based emphasis detection for characterization of meeting recordings. In *Proc. of ASRU*, pp. 243–248, Virgin Islands.

Kleinsmith, A. and Bianchi-Berthouze, N. (2012). Affective body expression perception and recognition: A survey. *IEEE Transactions on Affective Computing*, **99**.

Krauss, R. M., Apple, W., Morency, N., Wenzel, C., and Winton, W. (1981). Verbal, vocal, and visible factors in judgments of another's affect. *Journal of Personality and Social Psychology*, **40**, 312–320.

Kreiman, J. and Sidtis, D. (2011). *Foundations of Voice Studies: An Interdisciplinary Approach to Voice Production and Perception*. Wiley-Blackwell, Malden, MA.

Lapakko, D. (1997). Three cheers for language: A closer examination of a widely cited study of nonverbal communication. *Communication Education*, **46**, 63–67.

Laskowski, K. (2009). Contrasting emotion-bearing laughter types in multiparticipant vocal activity detection for meetings. In *Proc. of ICASSP*, pp. 4765–4768, Taipei, Taiwan.

Laver, J. (1994). *Principles of Phonetics*. Cambridge University Press, Cambridge.

Lee, C.-C., Black, M., Katsamanis, A., Lammert, A., Baucom, B., Christensen, A., Georgiou, P., and Narayanan, S. (2010). Quantification of prosodic entrainment in affective spontaneous spoken interactions of married couples. In *Proc. of Interspeech*, pp. 793–796, Makuhari, Japan.

Martin, J.-C., Devillers, L., Raouzaiou, A., Caridakis, G., Ruttkay, Z., Pelachaud, C., Mancini, M., Niewiadomski, R., Pirker, H., Krenn, B., Poggi, I., Caldognetto, E. M., Cavicchio, F., Merola, G., Rojas, A. G., Vexo, F., Thalmann, D., Egges, A., and Magnenat-Thalmann, N. (2011). Coordinating the generation of signs in multiple modalities in an affective agent. In P. Petta, C. Pelachaud, and R. Cowie (eds), *Emotion-Oriented Systems: The Humaine Handbook*, Cognitive Technologies, pp. 349–367. Springer, Berlin.

Mehrabian, A. and Wiener, M. (1967). Decoding of inconsistent communications. *Journal of Personality and Social Psychology*, **6**, 109–114.

Murray, I. and Arnott, J. (1993). Toward the simulation of emotion in synthetic speech: A review of the literature on human vocal emotion. *Journal of the Acoustical Society of America*, **93**, 1097–1108.

Murray, I. R. and Arnott, J. L. (1995). Implementation and testing of a system for producing emotion by rule in synthetic speech. *Speech Communication*, **16**, 369–390.

O'Sullivan, M., Ekman, P., Friesen, W. V., and Scherer, K. R. (1985). What you say and how you say it: The contribution of speech content and voice quality to judgment of others. *Journal of Personality and Social Psychology*, **48**, 54–62.

Oviatt, S. (1999). Ten myths of multimodal interaction. *Communications of the ACM*, **42**, 74–81.

Oviatt, S. (2008). Multimodal Interfaces. In A. Sears and J. J. A. (eds), *The Human-Computer Interaction Handbook. Fundamentals, Evolving Technologies, and Emerging Applications*, pp. 413–432. Lawrence Erlbaum, New York.

Oviatt, S. and Cohen, P. (2000). Multimodal interfaces that process what comes naturally. *Communications of the ACM*, **43**, 45–53.

Pantic, M., Caridakis, G., André, E., Kim, J., Karpouzis, K., and Kollias, S. (2011). Multimodal emotion recognition from low-level cues. In P. Petta, C. Pelachaud, and R. Cowie (eds), *Emotion-Oriented Systems: The Humaine Handbook*, Cognitive Technologies, pp. 115–132. Springer, Berlin.

Picard, R. (1997). *Affective Computing*. MIT Press, Cambridge, MA.

Picard, R. (2003). Affective Computing: Challenges. *Journal of Human-Computer Studies*, **59**, 55–64.

Pike, K. L. (1945). *The Intonation of American English*. University of Michigan Press, Ann Arbor.

Planalp, S. and Knie, K. (2002). Integrating verbal and nonverbal emotion(al) messages. In S. R. Fussell (ed.), *The Verbal Communication of Emotions: Interdisciplinary Perspectives*, pp. 55–77. Lawrence Erlbaum.

Poyatos, F. (1991). Paralinguistic qualifiers: Our many voices. *Language & Communication*, **11**, 181–195.

Poyatos, F. (1993). *Paralanguage: A Linguistic and Interdisciplinary Approach to Interactive Speech and Sounds*. John Benjamins, Amsterdam.

Poyatos, F. (2002). *Nonverbal Communication across Disciplines, Volume II*. John Benjamins, Amsterdam.

Ragin, C. C. and Amoroso, L. M. (2011). *Constructing Social Research: The Unity and Diversity of Methods*. Sage Publications, Thousand Oaks, CA.

Rauch, I. (1999). *Semiotic Insights: The Data Do the Talking*. University of Toronto Press, Toronto.

Rauch, I. (2008). *The Phonology/Paraphonology Interface and the Sounds of German across Time*. Peter Lang, New York.

Rogers, Y., Sharp, H., and Preece, J. (2011). *Interaction Design: Beyond Human - Computer Interaction*. Wiley, Chichester.

Rosch, E. (1975). Cognitive representations of semantic categories. *Journal of Experimental Psychology: General*, **104**(3), 192–233.

Sapir, E. (1921). *Language: An Introduction to the Study of Speech*. Harcourt Brace, New York.

Sapir, E. (1927). Speech as a personality trait. *American Journal of Sociology*, **32**, 892–905.

Saussure, F. de (1916). *Cours de linguistique générale*. Payot, Paris.

Schröder, M. (2001). Emotional speech synthesis: a review. In *Proc. of Eurospeech*, pp. 561–564, Aalborg.

Schröder, M. (2004). *Speech and Emotion Research. An Overview of Research Frameworks and a Dimensional Approach to Emotional Speech Synthesis*, volume 7 of *Reports in Phonetics, University of the Saarland*. Institute for Phonetics, University of Saarbrücken.

Schröder, M., Burkhardt, F., and Krstulovic, S. (2010). Synthesis of emotional speech. In K. R. Scherer, T. Bänziger, and E. Roesch (eds), *Blueprint for Affective Computing*, pp. 222–231. Oxford University Press, Oxford.

Schröder, M., Pirker, H., Lamolle, M., Burkhardt, F., Peter, C., and Zovato, E. (2011). Representing emotions and related states in technological systems. In P. Petta, C. Pelachaud, and R. Cowie (eds), *Emotion-Oriented Systems: The Humaine Handbook*, Cognitive Technologies, pp. 369–388. Springer, Berlin.

Schröder, M., Bevacqua, E., Cowie, R., Eyben, F., Gunes, H., Heylen, D., Maat, M. t., McKeown, G., Pammi, S., Pantic, M., Pelachaud, C., Schuller, B., Sevin, E. d., Valstar, M., and Wöllmer, M. (2012). Building autonomous sensitive artificial listeners. *IEEE Transactions on Affective Computing*, **3**(2), 165–183.

Schuller, B., Steidl, S., Batliner, A., Burkhardt, F., Devillers, L., Müller, C., and Narayanan, S. (2013). Paralinguistics in speech and language – state-of-the-art and the challenge. *Computer Speech and Language, Special Issue on Paralinguistics in Naturalistic Speech and Language*, **27**(1), 4–39.

Sneddon, I., Goldie, P., and Petta, P. (2011). Ethics in emotion-oriented systems: The challenges for an ethics committee. In P. Petta, C. Pelachaud, and R. Cowie (eds), *Emotion-Oriented Systems: The Humaine Handbook*, Cognitive Technologies, pp. 753–768. Springer, Berlin.

Trager, G. L. (1958). Paralanguage: A First Approximation. *Studies in Linguistics*, **13**, 1–12.

Trask, R. (1996). *A Dictionary of Phonetics and Phonology*. Routledge, London.

Trimboli, A. and Walker, M. B. (1987). Nonverbal dominance in the communication of affect: A myth? *Journal of Nonverbal Behavior*, **11**, 180–190.

Trouvain, J., Schmidt, S., Schröder, M., Schmitz, M., and Barry, W. (2006). Modelling personality features by changing prosody in synthetic speech. In *Proc. of Speech Prosody*, Dresden.

Zeng, Z., Pantic, M., Roisman, G. I., and Huang, T. S. (2009). A survey of affect recognition methods: Audio, visual, and spontaneous expressions. *IEEE Transactions on Pattern Analysis and Machine Intelligence*, **31**(1), 39–58.

2

Taxonomies

Taxonomy is described sometimes as a science and sometimes as an art, but really it's a battleground.

(Bill Bryson)

We now want to present a 'taxonomic skeleton' for paralinguistic events. Taxonomies deal with the same-different problem which we could call the 'which box(es) game'. There are between 2 and *n* boxes into which the cases should be put. If *n* is a large number, it is a matter of ordered or unordered classes.

Normally, the single taxonomies presented in this chapter do not describe distinct classes but rather continua; we will not call them dimensions because this term is reserved for the dimensional representation of phenomena such as emotional states and personality traits in theories and models which will be dealt with especially in Sections 5.3 and 5.4.

What we are dealing with in this section are inherent characteristics of paralinguistic phenomena. We want to tell these apart from aspects of modelling which are dealt with in Chapter 3. Admittedly, this is not always a sharp distinction. The idea behind it is not necessarily to establish the ultimate taxonomy for paralinguistics but to make it possible to answer to the question as to what phenomena, such as emotions, personality traits, or types of deviant speech, can be described within which taxonomy, and how relevant is this for the different stages of computational paralinguistics? Examples will be given below.

In this chapter, we often have more to say on emotion and personality than on biological or cultural traits primitives, or on the other phenomena. This is simply due to the nature of these different paralinguistic phenomena.

2.1 Traits versus States

Time is the measurer of all things, but is itself immeasurable, and the grand discloser of all things, but is itself undisclosed.

(Charles Caleb Colton)

In the Merriam-Webster online dictionary, a *trait* is (1) 'a distinguishing quality (as of personal character)' or (2) 'an inherited characteristic'. This characterises nicely the two

Computational Paralinguistics: Emotion, Affect and Personality in Speech and Language Processing, First Edition.
Björn W. Schuller and Anton M. Batliner. © 2014 John Wiley & Sons, Ltd. Published 2014 by John Wiley & Sons, Ltd.

prevailing uses of the term: The narrow one (1) is reserved for personality characteristics, and used widely in psychological theories. The broader one (2) is more general, denoting a quality characterising persons or things which can be inherited or acquired. We will use the broader meaning, subsuming under 'trait' characteristics such as age and gender which we want to call 'biological trait primitives', (see Section 5.1), as well as characteristics such as being a native speaker of a first language and its regional or social variety or varieties, and suchlike, which we will call 'cultural trait primitives' (see Section 5.2). All these constitute different layers of a person; even if personality traits in the narrow sense might not necessarily be influenced by biological or cultural traits, this certainly can happen: in specific situations, males can feel more or less comfortable than women, and vice versa; moreover, their perception and evaluation by other people might differ, because of these differences in biological or cultural traits. To give another example, Lev-Ari and Keysar (2010) show that non-native speech – which is deviant and more or less influenced by the first language – not only is harder to understand, but also this causes non-native speakers to sound less credible. In other words, a cultural trait primitive influences the perception of a personality trait.

While traits are at least longer-lasting or fixed characteristics, *states* are short-term; this is the first out of many definitions of 'state' in Merriam-Webster: 'mode or condition of being'. Although a state can be self-induced – you can put yourself into some mood – normally it is caused by someone or something else. It takes a much shorter period of time to be put into some state than to be put into some trait and to 'get out of it' again. Early approaches towards taxonomies of traits and states can be found in Allport and Odbert (1936) and Norman (1967); see Section 3.1.

Between long-term traits and short-term states is something we want to call *medium-term between traits and states*. Such states can be self-induced – a typical example is alcoholisation – or may be interpersonal attitudes (interpersonal stances, (see Table 2.1) towards other persons, or social roles such as leadership taken over in specific conversational situations (dyadic or meetings). Table 2.1, after Scherer (2003), displays some design feature delimitations

Table 2.1 Design feature delimitation of different affective states, after Scherer (2003)

Type of affective state	Intensity	Duration	Synchron-isation	Event focus	Appraisal elicitation	Rapidity of change	Behav. impact
Emotion	++-+++	+	+++	+++	+++	++++	+++
Mood	+-++	++	+	+	+	++	+
Interpers. stances	+-++	+-++	+	++	+	+++	++
Attitudes	0-++	++-+++	0	0	+	0-+	+
Personality traits	0-+	+++	0	0	0	0	+

Emotion: brief episode, response to an event (angry, sad, joyful, . . .)
Mood: diffuse affect state, low intensity, long duration (cheerful, gloomy, buoyant, . . .)
Interpersonal stances: toward another person (distant, cold, warm, . . .)
Attitudes: enduring, preferences, predispositions (liking, loving, hating, . . .)
Personality traits: dispositions and tendencies (nervous, anxious, morose, . . .)

0 : low, + : medium, ++ : high, +++ : very high,-: indicates a range; Behav.: Behavioural.

of affective states – which constitute a substantial part of traits and states dealt with in paralinguistics, but not the whole story. We can see that, implicitly, traits and states and something in between are modelled in the 'Duration' column as 'low', 'medium', and 'very high'; this is partly correlated with rapidity of change – the shorter events are, the faster their characteristics can change (cf. the 'Rapidity of change column' in Table 2.1).

However, can we always know what is what and how long it takes, and do we have to know at all? Words are often inherently vague and do not care whether they are used for denoting long-term traits or short-term states. Let us take 'interest', which does not belong to the 'classic' emotions; nowadays, however, it is normally included in the list of emotions (see Silvia 2005, 2008). Interest is conceived of as a rather short-term state and modelled the same way as other emotions (Schuller *et al.* 2009, 2010); we think of someone being interested in something, be this a display in a shop window, or a conversation partner in a speed dating scenario, or a specific remark or information given by such a partner. But what about this description of a guy on a dating website who remarks 'Interests: I'm a very interested person'? This is a generic statement, and, as such, it cannot denote a short-term state. It could mean that he is eager to take interest in everything, and/or that interest is a trait characterising his personality. We can try to solve these ambiguities by establishing two different readings/meanings of 'interest' (cf. the two readings of 'trait' in Merriam-Webster above), or we can simply leave them unspecified. This might not be conceivable in psychological theories and models on emotion and personality, but it is a feasible way for empirical paralinguistics. We mainly have to know how long a unit is that we have to establish for processing; this holds both for formal – phonetic and linguistic – characteristics and for functional traits and states as well. Formal and functional units have to be time-aligned, which means that they first have to be segmented on the time axis (speech) or in the chain of written words (text).

This is arguably the basic paralinguistic taxonomy, from long-term traits to short-term states along the time axis. Again, the following listing is neither complete nor do we mention all possible varieties. For instance, we assume only two genders, even if there exist more varieties in between. These traits and states can have different intensity, of course, apart from the ones that are binary or can be measured on an interval scale such as age or height. The following list is adapted from Schuller *et al.* (2013b); we substitute references to studies with references to the sections in this book that deal with the specific phenomena in Chapters 4 and 5.

- **Long-term traits**:
    - *biological trait primitives* such as height, weight, age, and gender, cf. Section 5.1;
    - *cultural trait primitives*, such as group/ethnicity membership: race/culture/social class with a weak borderline towards linguistic concepts such as speech registers, dialect, or first language (see Section 5.2);
    - *personality traits* (see Sections 5.3 and 5.5) such as personality in general (the 'big five'), and single, more specific traits such as likeability;
    - a *bundle of traits* constitutes speaker idiosyncrasies, that is, speaker-ID; speaker traits can be used to mutually improve classification performance, by assigning speakers to distinct classes. For instance, subjects can be automatically assigned to either males or females, or to children or adults, and later on be processed separately in emotion or personality classification.

- **Medium-term between traits and states**:
  - *(partly) self-induced more or less temporary states*: sleepiness, intoxication (e.g., alco-holisation), health state, mood (e.g., depression) (see Section 5.6);
  - *structural (behavioural, interactional, social) signals*: role in dyads and groups, friendship and identity, positive/negative attitude, intimacy, interest, politeness; (non-verbal) social signals (see Section 5.7);
  - *discrepant signals*: irony, sarcasm, lying (see Section 5.8);
  - *mode*: speaking style and voice quality (see Section 4.2), which can also be long-term or sometimes short-term.
- **Short-term states**:
  - *emotions* (full-blown, prototypical), (see Sections 5.4 and 5.5);
  - *emotion-related states or affects,* such as stress, confidence, uncertainty, frustration, pain (cf. Sections 5.4 and 5.5).

Interdependencies with Other Taxonomies

We assume that traits and states are mostly orthogonal to all other taxonomies; it is the most basic distinction, laying the foundation for all other taxonomies. The expression of biological trait primitives is mostly not intentional; it is not possible to change one's age – but it is possible to influence others' perception of one's age. Normally, sex is either/or – but there are stages in between, and the choice amongst them can be traced back to some mixture of inheritance and intention. Traits are long-term, states are short-term; medium-term states (which could be named 'traits' as well) are in between. To find out more about the duration of traits and states, we have to resort to measured versus perceived. To give a somewhat exotic example, namely displaying non-native traits in a second language: we can simply measure the time of formal education in the second language, or the duration of residence in the foreign country, but these are most likely not the best measures; or we can use the assessment of native annotators (experts or naïve). In the case of alcoholisation, we can measure blood alcohol content and use this as reliable reference (ground truth), see Section 5.6.2; or we can employ human assessments which normally are, however, less reliable.

Main Relevance for Computational Processing

So far, segmentation of traits and states into most appropriate lengths of units of analysis has been a stepchild of computational approaches. Normally, whole recordings – either whole dialogues, or single wave files representing single utterances or dialogue moves – are taken as units, or a pause detection algorithm is employed. Sometimes, unit length is addressed by segmenting into equally spaced parts – for instance, an utterance is split into units of 1 second each, or into three parts of equal length (initial part, middle part, final part). We do not feel that this is an optimal strategy because it is simply 'dumb', ignoring any genuine segment boundaries in the data; however, it can be used without any linguistic modelling. Segmentation (cf. Section 3.3) gets even more complicated when it comes to multimodal processing, when different modalities have to be time-aligned. It is not clear yet whether and when a 'most appropriate' segmentation really pays off in terms of performance measures; there are some indications that it does (see Batliner *et al.* 2010). Pure performance is not the only criterion, though: in the long run, a closer dovetailing of, for instance, acoustic and linguistic analysis will necessitate a segmentation into linguistically meaningful units as well. As traits normally change not at all or only very slowly, experiments are possible only between speakers and not within speakers; for short-term states, both are feasible, and for medium-term states, within-speaker designs might be possible if the speakers are available for a longer period of time. Modelling medium-term states such as alcoholisation or sleepiness by fusion of utterance-level decisions can improve classification performance considerably (Schuller *et al.* 2013a).

2.2   Acted versus Spontaneous

The most important lesson for acting: Be yourself. Be authentic. Be honest.
(Arnold Schwarzenegger)

There might be some common-sense agreement on the difference between *acted* and *spontaneous*. When acting, you learn by heart what you have to say, you dress up in some way, you pretend you are in love when in fact you are not; when being spontaneous, you are just your usual self, you do not pretend to be someone else, you show the emotions that you really have, no more, no less, you speak as you normally do.

In the same vein, there might be consensus that in principle we would like to aim at 'spontaneous data' when collecting speech or language data for paralinguistic research. However, this is rather difficult to accomplish, and moreover, at second glance, the term is not very precise. Obtaining acted data is easier but has its drawbacks, too.

The prototypical actor went to drama school and acts in a dramatic production, on stage or in a movie. Besides, anybody can act in a community theatre. Acting styles will differ, between amateurs and professionals, between cultures, and depending on time and generation – just compare the style of movie actors from the 1950s with those of the present day. Konijn (2000) deals with different professional styles of acting emotions; although she is not concerned with acted emotions used for the automatic classification of emotions, her taxonomy is of interest to us. Her starting point is Diderot's *Paradoxe* in Diderot (1883):

> *A good actor feels nothing at all and can therefore evoke the strongest of feelings in the audience – this was Diderot's proposal in 'Paradoxe sur le Comédien'. The actor should act emotions on stage without feeling.*

This leads to 'the actor's dilemma': should the actor be involved or detached? Should the emotions – or any other attitudes and feelings – of the portrayed character be identical with the momentary private emotions, attitudes and feelings, or not? In contrast to Diderot and to other schools – for instance, the 'epic theatre' of Bertold Brecht, the Russian director Stanislavsky and the acting teacher Lee Strasberg propagate the style of *involvement*: the actor should identify himself with the character. It is often reported that great actors project themselves into the roles they have to impersonate, and change their appearance and their own speech for the time being. This can be a real impersonation and/or the portrayal of a stereotype – excellently acted by good actors but often badly acted in daytime TV soaps. But what about participants in experiments who are told to act as if they had specific emotions, feelings, or attitudes?

Table 2.2 attempts to summarise the most frequent strategies employed for collecting data, displaying different types on an exemplary basis. Such speech data can be analysed

Table 2.2 Type of speech data and methods

type	subjects	method	experimental control
read	specific target groups	predefined, written stimuli	high
acted: stage, movie	professionals	screenplay	high
acted: prompted	professionals, laymen	scripted	high
acted: non-prompted	laymen	elicited	medium
induced explicitly	professionals, laymen	emotion/mood-ind. techniques	high
induced implicitly	professionals, laymen	emotion/mood-ind. context	medium
naturalistic, scripted	preselected groups	TV recordings	medium
naturalistic, unscripted	any	role attribution	medium
naturalistic, free	any	(hidden) rec. in the wild	low

acoustically, or transcribed orthographically and then serve for grammatical or sentiment analyses. There are genuinely written data besides, but for these the question whether they are acted or spontaneous or something in between is not immediately relevant; of course, in writing, we can also pretend, lie, or be honest. There are many types and subtypes in between, and especially mixtures of types, but we hope that this table is useful for clarifying the approaches mostly used in research.

The most prototypical types shown in Table 2.2 are those on the first and last lines. For many decades, the main object of investigation was *read*, *scripted*, *prompted* speech obtained in the lab. The speaker should be representative of a language, a dialect, or some other variety, or of specific target groups such as females or children, or have some speech pathologies such as stuttering, or be non-native in the language. Recordings are done in the lab, the stimuli to be produced are normally predefined and presented in written form, the experimental control is very high and both audio quality and acoustic conditions of the recordings are excellent. Such data are still believed to be indispensable for research (see Xu 2010).

The opposite can be found at the bottom of Table 2.2: *naturalistic*, free, *spontaneous* productions, by any speaker or speaker group, or by specific target groups, recorded in real-life settings. However, speech quality and experimental control are low, and we might not find the specific phenomena we are interested in – there can be a severe sparse-data problem. Moreover, we are confronted with Labov's *observer's paradox* (Labov 1972, p. 209):

> *The aim of linguistic research in the community must be to find out how people talk when they are not being systematically observed; yet we can only obtain these data by systematic observation.*

There may be some ways out: people forget that they are being recorded, either by force of habit (they are connected to audio recording devices for hours and weeks) or because they are distracted by some tasks. Other methods are described in Llamas (2007). However, there can be severe privacy issues (cf. Section 1.4.3) if people are being observed in their daily life.

Now let us go down the table, starting with three types of acted data. 'Acted: stage, movie' is normally based on a screenplay and involves professional actors, and the recording quality is high. Such data can be taken as prototypical performances; a genuine use can be, in media retrieval, to look for specific combinations of factors, for instance, for love scenes in American versus Italian movies from the 1940s. Control is high because, basically, meta-information should be available, for instance, the screenplay with instructions.

'Acted: prompted' is like doing a (very) short movie in the lab, employing either professional actors or laymen; the actors are told what they should do, which emotion they should express, either by always using the same carrier sentence, or by choosing their own words. This can be called 'prompting' (Schiel 1999) the acting of emotions. This has been the usual strategy in earlier studies, for instance in the cases of the well-known Berlin emotional database (Burkhardt *et al.* 2005) and the Danish emotional database (Engberg *et al.* 1997). Experimental control is high.

The next type, 'Acted: non-prompted', is rather special. Imagine that someone is told to act in a simulated call-centre scenario. Her task is to get some information and to buy a flight ticket. She is not told in any way that she should get angry when the system does not work properly, or to stay calm even if the systems does not work properly. In such a scenario, we can at least assume that any emotion triggered by system malfunctions are not acted. This is sometimes called 'inducing' emotion; however, the explicit task is scripted in some way even if the wording can be freely chosen. It is a 'slight' type of acting, close to the roles the participants can take on in daily life as well. Experimental control is medium: the design of the task makes it likely that participants react in specific ways but they do not have to.

A closely related task is given in the row called 'induced implicitly'. We have to distinguish the 'as if' task of booking a ticket – it is evident that this is not the real act of booking a ticket and having to pay for it later – from a task such as telling children to direct a robot from A to B (Batliner *et al.* 2008) because this is a real task that can be accomplished or not. Here, the context (the task and difficulties introduced by the experimental design) can implicitly induce mood or emotions.

Induction can be explicit as well, such as in the approach put forward by Velten (1968). A film or story is presented which can induce both positive and negative states. This is claimed to be a reliable method by Westermann *et al.* (1996) and Scherer (2013).

The next two types are called 'naturalistic' – not because they necessarily are, but because that epithet is often used especially for the first type, 'naturalistic, scripted'. Scherer (2013) calls this type 'convenience sampling' because TV recordings such as reality shows or YouTube videos are convenient to obtain. He cautions against this type because participants mostly will be scripted in some way, thus playing predefined roles and not behaving 'naturally'. Experimental control is medium, albeit *post hoc*, because the scripted TV recordings can be selected.

The typical scenario for 'naturalistic, unscripted' is a multi-party interaction where more than two participants play some roles, for instance, in a project group. This type is more or less close to 'acted, non-prompted', and to 'induced implicitly', depending on the experimental setting. In many of these scenarios, the role taken over by the participant is not necessarily exclusive; for instance, they can speak aside to a third party present, and thus leave their role for a moment. The distinction between the types in Table 2.2 has of cause something to do with the lab versus life distinction dealt with in Section 3.6; however, lab speech is not necessarily acted, and vice versa. The researcher decides whether to use the lab or to record in some real-life setting, or to use existing recordings.

The opposition of acted versus spontaneous (real life) is weakened by the fact that, more or less intentionally (cf. Section 2.7), the expression of traits and states in speech is controlled by *display rules* (Ekman 1972) and by control and regulation. This makes it possible that the divide between the two varieties is not too large. However, there is no easy means – or no means at all – to assess these differences and similarities.

There is a long tradition within psychology of preferring scenarios with a high experimental control. Scherer (2013) strongly advocates emotion/mood-inducing techniques or even portrayal or enacting of different emotions because such techniques guarantee a proper balance of classes, claiming that 'there is theoretical and empirical evidence that strategic emotion regulation, particularly for expression, is extremely frequent in day-to-day "natural" expressions'. He finds 'a rather strong similarity between the vocal changes produced by an established mood induction procedure and a scenario-based acting or portrayal procedure'. This does not yet prove, however, that both methods really yield 'natural emotions'. A decisive experiment would be to compare really 'naturally produced' emotions with induced ones for the same speakers, not to compare two elicitation methods. We do not know of any such study and doubt their feasibility in general. Appropriate 'auxiliary' measures are surely post-experimental questionnaires to check the impression of the participants, and careful observations of edge conditions and participants' behaviour – not just recording and extracting of information.

Thus we are left with a problem for which we cannot find the ultimate answer: what is a real (natural/spontaneous) expression of emotion (or of attitudes, personality traits, and suchlike)? It might be easier to decide upon our own research interest: as Bühler (1934) assumes for speech signs in general, Scherer (2013) attributes a three-part function to nonverbal vocal expressions as well: *symptoms* of speakers, *appeals* to listeners, and *symbols* for concepts which are iconically represented. Now what are the interests of computational paralinguistics that normally – but not exclusively – has the intention to employ the results in some application? They are consistency, performance and function in the application – criteria close to but not fully identical with reliability and validity. We can largely disregard symbols; appeals to listeners are closest to human–human or human–machine interaction and constitute the main interest. Symptoms will be in focus, for instance, when investigating personality traits, especially pathologies.

The disadvantages set out by Scherer, as far as 'naturalistic' data are concerned, can be seen as advantages for applications which need training data as close as possible to test data; this pertains not only to the phenomena one wants to model, but also to everything else that influences modelling, even sparsity of data, unbalanced data, or reverberation and noise. From this point of view, there are (only) a few constellations where acted databases are fully appropriate; we have to distinguish acted data that are congruent with and practically the same as the data we want to model, from acted data that only intend to be close to those we want to model. When we conduct experiments with movie speech aiming at multimedia retrieval applications such as searching for love scenes in a movie database, acted data are fully appropriate. The same holds for the use of acted emotion data for teaching children with autism spectrum conditions to understand and produce emotions 'correctly' – in the latter case, what we want the children to do is first to act and subsequently to be able to produce emotions less voluntarily. This can be compared with the acquisition of the prosody of a foreign language. *Wizard of Oz* studies (the role of the machine is taken over by a human supervisor) for call-centre scenarios are basically a workaround for real data because the users do not really need the information they want to obtain. However, it is a familiar task and chances are high that the speech produced is very similar to that in real-life situations. Another constellation where acting is tolerable is, for instance, cross-cultural studies; here, we have to keep constant as many factors as possible because the one factor we are interested in – cross-cultural differences, similarities or transfer – is complex enough. We have to keep in mind, however, that with such a design, we might first of all model *stereotypes* and not necessarily and exactly the production of realistic, non-prompted emotions; however, such stereotypes are interesting enough to warrant such studies.

Interestingly, acted emotions are fairly common in research. In contrast, acted traits – be it personality traits, non-native/dialectal/vernacular traits, or pathological speech – are not employed for investigating their basic – and thus, their non-acted – characteristics. The traditional means of investigating personality has been self-assessment or others' assessment with questionnaires; for non-native or pathological traits, it has been read speech or sometimes less controlled narratives. We know, however, that good actors can impersonate all these different traits.

Matters are a little different in engineering approaches where we can find a few studies that use acted data for the automatic detection/classification of personality (Polzehl *et al.* 2010) or of focus of attention (Hacker *et al.* 2006); see Section 5.8.3. Scripted mood-induction procedures (for moods such as aggressive, intoxicated, tired, cheerful, nervous) are used in Schuller *et al.* (2007). Realistic medium-term state induction procedures are, for example, employed in Krajewski *et al.* (2009) for sleepiness by keeping the subjects awake at night and in Schiel *et al.* (2012) for intoxication by providing the subjects with controlled amounts of alcoholic drinks.

Interdependencies with Other Taxonomies

The different ways of acting dealt with above have something to do with 'felt versus perceived' and 'intentional versus instinctual'. Different cultures have different styles of acting; this holds especially for different cultural spheres, for example, Western versus Japanese ways of acting; it holds, however, even for different nationalities within Western society and for different periods and generations – we mentioned different styles of movie actors above. Thus, cross-cultural considerations should play a role (cf. 'universal versus culture-specific'). Acted and stereotypical traits and states might tend more towards prototypes than towards peripheral types of expression.

Main Relevance for Computational Processing

A promising maxim for the building of databases aimed at training and subsequent testing is that the training data should be as close as possible to the test data; this is one of the basic requirements in automatic speech recognition. It is never clear whether the claim of universality or at least transferability from training data onto test data is warranted until it has been proven. What has been proven so far is that transferability is at least rather weak (Batliner *et al.* 2000; Eyben *et al.* 2010), as far as prompted versus non-prompted, or cross-database studies are concerned. Thus, if possible, the best way to ensure transferability is, for instance, to use the same call centre and the same (type of) clients for training and testing.

We always make certain assumptions – such as 'naturalistic data are really natural/ realistic/spontaneous' or 'acting with well-induced emotions really yields realistic emotions' – and question or criticise the opposite assumption. For such discussions, a shift towards the question why we are doing so and what our aim is might be beneficial; in computational processing, we normally aim at usability in applications – and chances are higher if performance is better. Thus, we normally do not want to use read speech and standard varieties when trying to model (and test) spontaneous, vernacular speech. In the same way, it most often will be sub-optimal to use prompted, acted emotions for the training and modelling of spontaneous emotions, even if these spontaneous emotions are regulated and altered by display rules.

2.3   Complex versus Simple

'When I use a word,' Humpty Dumpty said, in rather a scornful tone, 'it means just what
I choose it to mean – neither more nor less.'
              (Lewis Carroll, *Through the Looking Glass and What Alice Found There*)

Imagine two different situations. In the one, you have just learned that the one you are madly in love with reciprocates your feelings – which are now plain and utter joy and elation. In the other, you learn that the one you live with and still love does not love you any longer: your feelings may well be a mixture of love, hate, disappointment and despair.

Less dramatic is the situation where you wanted to buy something on eBay, forgot the last bid and failed – only to learn that the final price was pretty low. This might cause plain, straightforward anger, or just slight disappointment, depending on your hopes and expectations. More complex situations can yield more complex emotions and feelings; see the study of Scherer and Ceschi (2000) on baggage loss at an airport where customers reported experiencing several different emotions at the same time. However, it might not be necessary for specific applications to model such a complexity: it might suffice to realise that a customer is 'really angry', and thus that the system should take some action.

Ben-Ze'ev (2000) points out the complexity of emotions due to their 'great sensitivity to personal and contextual circumstances' and because '. . . they often consist of a cluster of emotions and not merely a single one'. Moreover, there are many different types and shades.

Emotions are good exemplars for demonstrating complex versus simple; however, every other paralinguistic phenomenon, for instance, a speech variety, can be as complex as feelings, emotions, or bundles of personality traits – or it can display only a few 'simple' formal traits. Fossilised remains of non-native traits in a foreign language can be a few incorrect word accent (stress) placements while everything else seems to be fine. For example, Swedish speakers who master English perfectly can sometimes still be recognised by the quality of their /a/ sounds. Similar phenomena can be observed in non-typical, pathological speech.

Interdependencies with Other Taxonomies

Complexity might be directly connected to one of the other taxonomies; they all can be more or less complex. Mostly, it might rather be in the eye of the observer, that is, in the approach chosen, whether phenomena are conceived of and treated as complex or not.

Main Relevance for Computational Processing

It is a basic approach of any kind of science, thus also of paralinguistic research, to focus on specific aspects only in order to make the task manageable. We will present several exemplars in Chapter 4, as far as formal aspects are concerned. As for functional aspects, we normally concentrate on one phenomenon only, be it (some specific) emotion or (some specific type of) deviant speech. Decisions can be based on theoretical considerations; this is mostly the case in basic research, for instance, when investigating the 'big five' personality traits, (see Section 3.1). It can be based on data and on requirements of applications we are having in mind; a typical approach is to record call-centre interactions and look only for angry users, that is, for only one specific emotion. On the other hand, Scherer (2013) advocates the precise modelling of several predefined emotions; this is only possible when inducing these emotions in the lab, getting rid of the sparse-data problem, of noise, reverberation, and suchlike.

Thus, irrespective of the corpora and phenomena to be processed, we can reduce complexity from the very beginning, or later. *Mixed* or *blended* emotions are addressed in the recordings and annotations of Sobol-Shikler and Robinson (2010), and in Devillers *et al.* (2005) who model *major* and *minor* emotions for the same segments. The inevitable reduction of complexity comes later: Sobol-Shikler and Robinson (2010) resort to separate classifications of pairs of emotions, and Devillers *et al.* (2005) to a coarser-grained modelling of emotions, and thus have fewer classes to classify. *Majority voting*, that is, assigning the label chosen by most of the annotators (Steidl 2009), is one of many other strategies to reduce complexity. Another is dimensional modelling where the problem of arranging many more or less related categories is boiled down to assigning values on a few dimensions, for instance, the one or the other dimension in the traditional emotion model with arousal and valence. As for personality, the two prevailing approaches are to model either main traits such as the 'big five' *big five* – disregarding the complexity below this highest level – or specific traits that seem to be interesting and promising within applications, such as leadership, attractiveness or aggressiveness.

2.4   Measured versus Assessed

If it can't be expressed in figures, it is not science; it is opinion.

(Robert Heinlein)

For some paralinguistic phenomena, objective measures are applicable that can be taken as 'ground truth', or at least as being highly correlated with the ground truth. Measures can be mapped one-to-one onto the variable/factor in question: height in centimetres/inches, weight in pounds, or age in years/days. Blood alcohol content is a reliable measure, and specific values often serve as legal thresholds. Acoustic parameters can be measured: F0, intensity, formants, and the like. However, there is no simple one-to-one relation to their perceptual equivalent (see Section 4.2). Physiological signals (biosignals) can be measured more or less 'directly' with biomedical sensors (Kaniusas 2012), or indirectly – see Skopin and Baglikov (2009) and Mesleh *et al.* (2012) who extract heart rate from voiced parts of the speech signal. It is still debated to what extent physiological measures – for example, skin conductance as measure of stress – can be taken as ground truth, or whether they should not rather be modelled as yet another feature set (Knapp *et al.* 2011).

On the other hand, there are phenomena that can only be assessed (annotated/labelled) perceptually, such as (degree of) interest, likeability, attractiveness, or politeness; this holds especially for the valence dimension but even for arousal, where at least some correlation with 'lower, smaller' versus 'higher, louder, greater' can been assumed. Basically, this holds for deviant speech as well: of course, the deviation of some acoustic-phonetic measure from typical, 'normal' speech can be measured but the combined impact of different (types of) deviations on the degree of, for example, non-nativeness or pathology has to be assessed with perceptive judgements. Perceptual assessment is more or less subjective, and subjects can have different 'perceptual reference points' serving as 'anchor'; however, chances are high that employing several labellers reveals central tendencies (see Section 3.10). Ranking-based measures (Han *et al.* 2012) with relative, not absolute, judgements can be favourable as well.

Statistical *levels of measurement* (Stevens 1946) go along with the order from measured to assessed. To start with, the *ratio* level and the *interval* level can be measured, the *ordinal* level is normally be obtained via human assessment/annotation, and the *nominal* (categorical) level is either annotated or given (such as sex or nationality). In reality it is more complicated, and the levels are often mapped onto each other: parametric procedures suited for the ratio/interval level are often used for annotation measures, especially mean values, and ordinal data are mapped onto categories.

An important aspect of measured versus assessed is the problem of how 'real' the object of our investigation is. The terms implicitly used to refer to the certainty of 'realness' are *ground truth*, *gold standard*, and *reference*.

The term *ground truth* comes from cartography and satellite imagery. All information taken from distance can be verified with reference to the ground; someone literally can be sent there to check. This is a very strong verification: the object in question really can be measured. Thus, the ground truth lies in the characteristics of the object.

The term *gold standard* is taken from the monetary gold standard – see Claassen (2005) and Morris (2003), who describes its origin:

> As the monetary gold standard was waning, the gold standard in clinical research was gaining currency. Under a monetary gold standard, the value of money is anchored to measurable reserves of the precious metal gold. By analogy, the gold standard in clinical research anchors a patient's diagnosis to another objective measure – results of unequivocal tests.

In speech and language research, we normally have to do with some manual annotation which has to be scrutinised to determine whether it can serve as such a standard – see Carbone *et al.* (2004) who present '. . . several techniques for unifying multiple discourse annotations into a single hierarchy, deemed a "gold standard" – the segmentation that best captures the underlying linguistic structure of the discourse'. Thus, such a gold standard lies in the human agreement.

Reference is a more neutral term; its history is sketched by Hájek (2007) who attributes to Reichenbach (1949, p. 374) the standard formulation of the reference class problem:

> If we are asked to find the probability holding for an individual future event, we must first incorporate the case in a suitable reference class. An individual thing or event may be incorporated in many reference classes, from which different probabilities will result. This ambiguity has been called the problem of the reference class.

Thus, the reference lies in the perspective of the observer – which (types of) classes she is interested in for the moment. The term is somewhat neutral and used as an extensional definition: reference is what belongs to the cases/items belonging, for instance, to our training set.

If we aim to distinguish the different types of references, that is, 'ground truth', 'gold standard', and 'reference', we should reserve 'ground truth' for characteristics that can be measured (age, weight), 'gold standard' for agreed-upon human annotation procedures, and use 'reference' in a technical sense, taking a 'conservative' (cautious) stance, for instance, as denoting the class(es) we have available in our training data and want to assign automatically in our test data.

Interdependencies with Other Taxonomies

There does not seem to be a clear relationship to other taxonomies in the sense that there is some high contingency – sometimes we can measure, sometimes we have to assess.

Main Relevance for Computational Processing

The relevance for computational processing is straightforward. According to the levels of measurement, we have to choose appropriate procedures. This is straightforward for parameters that 'really can be measured', such as time in milliseconds, or any acoustic measure. For practical reasons, such as data sparseness or needs imposed by applications, we often map 'precise' measures onto coarser-grained levels.

2.5   Categorical versus Continuous

PICTURE, n. *A representation in two dimensions of something wearisome in three.*
(Ambrose Bierce, *The Devil's Dictionary*)

The notion of categories and categorisations, the nature of categories (discrete or vague), and their status (universal and/or innate) is dealt with extensively in Cohen and Lefebre (2005). The paradigm of *categorical perception* (Harnad 1987) was one of the prevailing phonetic models in the 1980s (Repp 1981): it was claimed that consonants especially are perceived in a categorical way, with sharp boundaries between the categories, for instance, between /p/ and /b/ (IPA/SAMPA notation). This was investigated with psychoacoustic experiments (identification or discriminations tasks, among others). However, the strong paradigm gave way to more 'less categorical opinions' – interestingly, in a voluminous reader on hearing and speech, it was not even mentioned (Summerfield 1991). It seems to be more difficult to make such strict hypotheses – which can be proved right or wrong – when it comes to paralinguistics.

In paralinguistics, the formal characteristics can normally be measured somehow, and this means a continuous representation on some scale; different formal characteristics aligned with different scales represent some paralinguistic function which can be conceived of holistically, as phenomenological *gestalt* – this holds for straightforward phenomena such as filled pauses or laughter, and for complex phenomena such as specific emotions or personality traits. Whether we perceive or conceive something as categorical, that is, as belonging to mutually exclusive classes, or as continuous, that is, as more or less representing classes that might not be mutually exclusive, almost always seems to be a matter of perspective, even of taste. We will come back to these different perspectives in Section 3.9. Biological trait primitives such as sex/gender are normally taken as two categories, that is, either male or female, when we try to optimise performance by telling apart these two classes that display formal characteristics in slightly different ways (e.g. the pitch register is higher for females than for males). On the other hand, there are varieties of sex/gender that definitely are not either/or, and within these classes there are degrees of femininity and masculinity which in turn are closely interrelated with degrees of attractiveness and sexiness (see Section 5.3). The same holds for categories such as nationality, first language versus second language, or for different varieties of the same language. We sometimes speak of diglossia if two language varieties are distinct and used in different situational settings. Yet, this might as well be only a very strong tendency and not a fully clear-cut difference between categories. There are clear-cut cases of lying, but we can also lie just a little bit.

As for personality and emotion, there is a long-standing discussion on whether they should be modelled as classes or as continua. In personality research, types are contrasted with traits; in emotion research, categories are contrasted with dimensions. "*Personality types* are discrete categories that differ qualitatively in kind rather than in degree. Personality types are categories that can involve a constellation of personality characteristics that are present in an all-or-none fashion' (Flett 2007, p. 28). Nowadays, *personality trait* models prevail with a dimensional approach, and within such models the so-called five-factor model is predominant; it is described in Section 3.1. In emotion research, categories seem to be more common than in personality research; however, both categorical and dimensional models coexist. Emotion categories have a long-standing tradition and are, for example, referred to in chapter headings in Darwin (1872): 'Low spirits, anxiety, grief, dejection, despair' (Chapter VII) and 'Joy, high spirits, love, tender feelings, devotion' (Chapter VIII). The first (modern) study using emotional dimensions seems to be by Schlosberg (1941) who proposed two scales: pleasantness/unpleasantness and attention/rejection; note that Wundt (1896) had already spoken of 'affective directions' such as 'pleasurable and unpleasurable' (see Section 3.2). Schlosberg (1954) further introduced the activation dimension, referring to physiological variables such as galvanic skin response.

The seminal work of Osgood *et al.* (1957) and Osgood (1964) introduced the *semantic differential* method. Scales with contrasting adjectives such as 'good–bad', 'strong–weak', 'loud–soft', or 'pleasant–unpleasant' were given to subjects who had to use these scales for rating the semantics of concepts. A subsequent factor analysis resulted in three most important factors: *evaluation, potency* and *activity*. These dimensions were reviewed and evaluated in Mehrabian and Russell (1974) for determining the emotional space. Evaluation/valence and arousal/activity remain the best-established and most widely used emotional dimensions; studies dealing with emotion dimensions and speech are referred to in Section 5.4. However, for a complete modelling of emotions, two dimensions will hardly suffice. Fontaine *et al.* (2007) eventually establish four dimensions: 'In order of importance, these dimensions are evaluation-pleasantness, potency-control, activation-arousal, and unpredictability.' Cochrane (2009) proposes eight dimensions: '(1) attracted-repulsed, (2) powerful-weak, (3) free-constrained, (4) certain-uncertain, (5) generalized-focused, (6) future directed-past directed, (7) enduring-sudden, (8) socially connected-disconnected.'

Plutchik (1980) proposed a circular structure of emotion. Complex emotions are modelled as mixtures of primary emotions. This is elaborated on in Plutchik and Conte (1997) where the 'circumplex model' is described: personality traits and emotions are conceived as being structurally similar and can be described in a circular/circumplex order. Another attempt to relate categories to dimension can be found in Scherer (2000) where (categorical) emotion terms are placed within a two-dimensional, circular representation of activity and valence.

Obviously, all has not yet been said about the number and the characteristics of continua and dimensions and their relationship with categories: which dimensions we should eventually assume, and where we can assume a continuum, where an ordered relationship, and where clear-cut categories. We definitely should distinguish coarse processing from in-depth processing. Sex and gender can serve as examples: a coarse partition into two different sex classes for a subsequent processing of other phenomena, simply to get better performance, will mostly do. However, this will not suffice to do justice for all the different manifestations of gender differences.

Interdependencies with Other Taxonomies

Perhaps 'categorical versus continuous' is, besides 'traits versus states', another global and universal distinction, orthogonal to the other taxonomies: for each of the other taxonomies, we can imagine more or less categorical or continuous representations.

Main Relevance for Computational Processing

The relevance is straightforward: in the case of categories, the normal way of processing is classification, based on n classes, n being at least 2 but not too high. In the case of continua, these are normally modelled with regression and correlation. However, there is always the possibility of mapping categories onto continua and vice versa, albeit normally not without loss of information.

2.6   Felt versus Perceived

All that we see or seem
Is but a dream within a dream.

(Edgar Allan Poe)

Let us take up the father–son example from Batliner *et al.* (2010): a young boy is messing around, and his father gets angry and shouts at him. Now the father may be angry because he has told his son several times to stop, because he is a person inclined to get angry (personality trait), or for many other reasons. However, the father may only be pretending to be angry (see Section 2.7) – while in fact he is amused – just because he thinks as a father he should, because he has been told by his wife to act in this way in such a situation, or for many other reasons. The boy – especially if he is very young and not yet sophisticated enough – may take his father's anger at face value, and may stop or continue, running the risk of facing even greater anger; however, if older and cleverer, he may figure out that the anger is only a pretense, and react accordingly. Of course, both such felt and perceived states can be straightforward or mixed/complex (see Section 2.3). In a real father–child interaction, the boy might be well advised to obey – even if he knows that his father is only pretending.

Normally, what we feel is induced – by oneself or by others – and described and annotated afterwards, by the person herself who felt that way (internal perspective). In contrast, we perceive from the outside (external perspective). We can imagine more complicated scenarios, for instance, trying to tell apart felt and perceived states or traits by the person herself or by others. We can try and look at ourself from the outside ('externalised perspective') and when observing other people, we might not only label their feelings but feel the same way as they do (we might call this 'internalised perspective'). However, all this might be far too complicated to be employed in any computational processing (see also Section 2.8).

Naturally enough, as far as type of annotation is concerned, self-annotation for 'felt' can only be done by one person describing her own states, whereas annotation of perceived by others normally is conducted by several labellers.

Emotion is exemplary for 'felt versus perceived'; to a lesser extent, we might 'feel' person-ality traits or even cultural trait primitives, and we can adopt speech characteristics in order to play a role. Personality scales can be obtained based on self-assessment or assessment by others, both normally with the help of questionnaires (see Section 5.3). As personality traits are long-term, they are easier to assess by the subjects themselves than emotions which have to be time-aligned: in the case of emotions, the subjects have to remember when they were in which state. However, in both constellations, self-observations and others' observations do not necessarily coincide. When we are interested in the actual communication with others and the impact of personality traits and emotions on others, perceptual annotation (i.e., others' obser-vations) might be appropriate. However, when we are interested in the self, self-observation (in combination with correspondences and discrepancies with respect to others' observations) might be more adequate.

Self-assessment is somewhat controversial. Some researchers claim that this is the most reliable assessment – after all, people themselves should know what they are feeling. Others argue against this, especially if self-assessment is done some time after the event, for instance, when the subjects listen to their own recordings, or observe themselves video-taped. Again, studies on convergence of different methodologies are needed.

Interdependencies with Other Taxonomies

There are many interdependencies with other taxonomies. One type of acting, recommended by Scherer (2013), for example, implies the induction of specific felt emotions. Of course, there is a relationship to 'measured versus perceived'. Here, the perspective is pivotal: seen from one's own perspective or seen from the outside. 'Measured versus perceived' relates to possibilities of constituting references and modelling them. When aiming at 'felt', we for instance can try to measure physiological signals; 'perceived' can normally only be obtained by 'external' annotation. Another relationship is that with 'intentional versus instinctual'. If the father in our example above pretends to be angry but is not – or only a little – then he does so intentionally. 'Instinctual' can be congruent with 'felt' but need not be so. 'Per-ceived' is within the 'receiver' who, at the same time, can reason about any intention of the 'sender'. Moreover, we can speculate whether discrepant communication can be based on consistent feelings.

Main Relevance for Computational Processing

Both for 'felt' and 'perceived', only 'vague' reference values – at best constituting some gold standard but not any ground truth in its strict sense – are available, based on human assessment.

2.7   Intentional versus Instinctual

Good intentions aren't good enough!

(George W. Bush)

We will follow Tomasello *et al.* (2005, p. 675) who establish a close link between intentionality and *cognition* in both phylogeny and ontogeny of mankind – shared intentionality as a species-unique feature of cultural cognition: 'Participation in such activities requires not only especially powerful forms of intention reading and cultural learning, but also a unique motivation to share psychological states with others and unique forms of cognitive representation for doing so.' Lacewing (2004, p. 175) summarises a recent framework which claims that '... emotions are a form of evaluative response to their intentional objects, centrally involving cognition or something akin to cognition, in which the evaluation of the object relates to the concerns, interests, or well-being of the subject.' See also Roberts (1988).

A well-known example is the snake that appears all of a sudden: we do need cognition to be able to recognise that it is a snake, and that it is potentially dangerous. However, especially if we have a snake phobia, our reaction is very instinctual. If not, it might depend on the circumstances: how big the snake is, how far away, whether we have been told that there are poisonous snakes around – in the latter case, there will be a specific cognitive (and most likely, affective) priming. In cognition the pre-frontal cortex is involved, in affect the amygdala (see Roesch *et al.* 2011).

Now this example looks very conclusive; however, it might not be very representative for human–human or human–machine communication, or other scenarios representative of computational processing. A better example might be that of the father getting angry with his child depicted in Section 2.6: in such a situation we can imagine everything from fully intentional/cognitive/acted, via any stage in between, to fully instinctual/automatic/spontaneous.

Social interaction in general – and this can pertain all paralinguistic phenomena – and emotion in particular are constantly regulated, and past and future actions of the communication partner are taken into account (Marinetti *et al.* 2011). This can mean following socially agreed-upon display rules that are more or less culture-specific (Matsumoto 1990). There seems to be no agreement on whether phenomena such as emotions should be dealt with in their 'pure' form, or when altered by such display rules. Scherer (2013) obviously aims at the pure form, arguing against employing 'convenience' databases from TV recordings which are heavily influenced by display rules. On the other hand, for many scenarios in human–human or human–machine interaction, display rules are an integral part because they are applied all the time, and are constitutive for the interaction partners' reactions.

Interdependencies with Other Taxonomies

There are many interactions and interdependencies with other taxonomies and types of paralinguistic phenomena. Acting is normally intentional. We can reason about whether 'felt' is more 'instinctual' and 'perceived' transformed by more or less intentionally applied display rules. Biological/cultural trait primitives and deviant signals might normally not be changed intentionally, but they can be, in specific situations. Discrepant speech is very often intentional, for instance, when someone is lying or being sarcastic. 'Intentional versus instinctual' might constitute a continuum or two categories, or two continua with a mutual, complex

interrelationship; admittedly, such interrelationships can be imagined but are far too complicated to be adequately processed automatically.

Main Relevance for Computational Processing

In general, we always have to consider the general scenario and the specific situation where
we will obtain our data, and especially our research interest. There is a high relevance, as
far as biosignals as reference are concerned. It might be possible to establish biosignals
as a real ground truth in the case of very instinctual events, such as being confronted
with a dangerous snake. There might be a possibility that instinctual emotions yield more
pronounced acoustic-prosodic feature characteristics. However, this will not be always the
case (see Section 2.2): a good actor might – but need not – produce higher arousal than in the
case of spontaneous emotions – and vice versa.

2.8   Consistent versus Discrepant

Life is full of misery, loneliness, and suffering – and it's all over much too soon.
(Woody Allen)

The ideal interaction is consistent and honest (Pentland 2008), and follows the maxims established by Grice (1975) for felicitous communication (see Section 5.8). Inconsistent, discrepant – because unexpected – information can be in the eye of the communication partner
or observer only; examples are unusual high-pitched voices in males, or low-pitched voices in
females. Or imagine a Caucasian ('white') boy in an African village speaking a strong local
dialect of Swahili. In such cases, the context triggering the expectation of the communication
partner is responsible for the discrepant impression. A discrepancy between the literal meaning
of a message and the context can point towards irony, which is defined in Merriam-Webster
as 'the use of words to express something other than and especially the opposite of the literal
meaning'. And a discrepancy between situational context and what is being said can point
towards the fact that the speaker is not telling the truth but lying.

Irony and sarcasm are rather rhetorical figures: the speaker normally wants to be understood
so that *Oh, that's really wonderful*, produced in a discordant context, is understood in a negative
way. In contrast, pretending and lying normally intend to be taken at face value.

Consistency or discrepancy can exist within and across modalities. In a fully consistent
behaviour, speech, language, gestures, facial gestures and posture all signal the same intention.
In contrast, the father described in Section 2.6 can signal pure anger with his voice but his
smile might indicate that he does not mean it really seriously.

Multimodal discrepancy might be more common and more 'normal': it might be easier –
and more 'natural' – to signal different messages with different modalities. This can make
multimodal processing more difficult and error-prone; in turn, it can make unimodal processing
more straightforward.

Interdependencies with Other Taxonomies

In discrepant communication, there might often be some type of acting involved. Measurement – if possible – might reveal other characteristics than in the case of consistent, honest
communication.

Main Relevance for Computational Processing

Discrepant signals can constitute 'rare birds' that normally are classified wrongly because of their low frequency which makes modelling cumbersome or impossible. Normally, we are assuming consistent, honest signals when trying to analyse and recognise paralinguistic phenomena. For a successful processing of discrepant signals, we have to know the difference with respect to consistent signals; thus, we do need data sets with both discrepant and consistent items. This normally requires some sophisticated experimental set-up or specific constellations with some external evidence: for instance, we should know that a speaker is lying or telling the truth.

2.9   Private versus Social

The only thing you owe the public is a good performance.

(Humphrey Bogart)

Parkinson (1996) claims that emotions are not primarily individual reactions but should be viewed as social phenomena because they usually have consequences for other people. When embedded in a communicative setting, the partners communicate all their states and traits, sometimes explicitly, if expressing plain anger, sometimes implicitly, when using social and regional varieties, or expressing personality traits. However, the opposite is not necessarily true. Imagine a person wearing a recorder all day long and being recorded; even when she is alone, she can express her emotions vocally and verbally, for instance, she can sigh, she can swear, she can talk to herself (private speech; see Section 5.8.3). All this is fairly 'normal'. However, when we hear her reprimanding or using motherese (Batliner *et al.* 2008) while alone, this is less 'normal' because there is no communication partner around. Thus the only explanation would be that this person takes on a role in an imagined dialogue – as if some addressee were present. Even interest or love need an object, to a greater extent than depression. Hareli and Parkinson (2008) and Manstead (2005) address different aspects of the social character of emotions.

There is a narrow and a broad conceptualisation of 'communication'. The narrow one is what we normally think of, typically a dialogue between two partners – which is a special case of multi-party communications. The broad conceptualisation includes every human production that can be observed and recorded. In the narrow conceptualisation, non-verbal factors, and thus multimodality, are genuinely involved. In the broad one, we can imagine many situations where this is not the case: for the screening of pathological speech, we only need speech data. For call-centre interactions, multimodality is non-existent because normally it cannot be recorded – even if, of course, the face displays emotions continuously.

A clear contrast exists between *expressing* the speaker's state (rather private) versus *directing* the listener (definitely social). However, even while 'only' expressing your own state, you act socially if someone else is present. Even if no one is present – as in the case of 'private speech' or when you write a diary – what you are saying or writing becomes social when someone else listens to the recordings or reads what you have written: one cannot not communicate (Watzlawick *et al.* 1967).

Basic aspects of speech and language – which variety you speak, whether you tend towards fluent or hesitant speech, which words you prefer, all this is not social *per se* but contributes to the social impression you are making on an interaction partner.

Interdependencies with Other Taxonomies

'Private versus social' is often orthogonal to the other taxonomies. However, you can act for yourself – albeit more often you want an audience. 'Perceived' is of course social, in contrast to 'felt'. 'Instinctual' might be more private than 'intentional'.

Main Relevance for Computational Processing

The main relevance of this opposition for computational processing might lie in the greater impact of the situational and acoustic/linguistic context in the case of phenomena that are more social than private.

2.10   Prototypical versus Peripheral

From the time of Aristotle to the later work of Wittgenstein, categories were thought be [sic] well understood and unproblematic. They were assumed to be abstract containers, with things either inside or outside the category. Things were assumed to be in the same category if and only if they had certain properties in common. And the properties they had in common were taken as defining the categories.

(George Lakoff)

This quotation from Lakoff (1987, p. 6) nicely describes the traditional concept of categories. This classic concept was questioned by Ludwig Wittgenstein in *Philosophische Untersuchungen*, posthumously published in 1953 (see Wittgenstein 2009), introducing *family resemblances* which can be described and defined by using language (*Sprachspiel*, language game). The concept of prototypes was developed by E. Rosch (see Rosch 1975; Rosch and Mervis 1975): a *prototype* is a salient, central member of a category and typically most often associated with this category (graded categorisation). Fuzzy boundaries between categories were introduced in (Zadeh 1965).

Ultimately, all alternatives to clear-cut categories such as mixed/blended emotions, stereotypes, and many more can be traced back to these works, or be described within these frameworks. Fehr and Russel (1984) applied the prototype theory to emotions and showed that the layman's conceptualisation of emotion is prototypical, with fuzzy boundaries and graded degrees of family resemblances.

Shaver *et al.* (1987) collected prototypicality ratings for 213 emotion words: love, anger, hate, depression, fear, jealousy, and happiness had the highest, and interest, self-control, alertness, carefulness, practicality, deliberateness, and intelligence had the lowest scores. Russell (1991) defends this conceptualisation against critics; see also Niedenthal and Cantor (1984) and Fuhrmann (1988). Encyclopaedic accounts of category concepts are Harnad (1987) and Cohen and Lefebre (2005).

Nowadays, the concept of prototypes is still appealing; however, it is not really established in theoretical or computational approaches. The reason might be that the instruments of science seem to be more inclined towards categorical or clear continuous approaches which are easier to manage than less marked-off classes. However, even if not named that way, all these concepts sneak, as it were, into the scientific discourse if researchers go from clear-cut, predefined categories onto real-life data. If no external criterion is available, real-life data have to be annotated manually to provide a reference for automatic processing. Thus a straightforward operationalisation is to speak of 'prototypical' cases if the labellers agree. Non-prototypical – weak and/or mixed – emotions can be found when labellers annotate more than one emotion per item, or when we preserve the disagreement of several labellers in some sort of graded/mixed annotation. Note that irrespective of the type of annotation, we can always generate either categorical labels representing pure or mixed cases, or a continuous representation by placing each case on some dimension scale, whether it be for valence or arousal. Basically, it is always possible to convert a continuous representation into a categorical one, and vice versa. Still, the relationship between prototypicality and dimensional or categorical approaches seems rather complicated.

A sequel of the prototype theory is the exemplar-based theory introduced in psychology; a key assumption is that categories are represented not (only) by a prototype but by clouds of exemplars (Nosofsky 1986; Nosofsky and Zaki 2002; Skousen *et al.* 2002). As far as we can see, it has not yet really been applied to paralinguistics but rather to linguistics proper, that is, phonetics, phonology and grammar (Johnson 2006; Lacerda 1995; Walsh *et al.* 2010).

Interdependencies with Other Taxonomies

Prototypes might be universal rather than culture-specific; for instance, Ekman's six basic (and thus prototypical) emotions are claimed to be universal; see, however, Kövecses (2000). Less prototypical cases might be more complex and not represent clear-cut categories but rather be found somewhere on a continuum.

Main Relevance for Computational Processing

Prototypes might be produced more unambiguously and thus easier to process; therefore, they might be good candidates for demonstrating and teaching. However, they might not be as frequent as less prototypical members of a category.

2.11 Universal versus Culture-Specific

If we spoke a different language, we would perceive a somewhat different world.
(Ludwig Wittgenstein)

So far, at several points, we have pointed out that for application-minded computational processing, we should aim at data used for modelling and training that are as close as possible to the data used for testing and found within the application. This governing principle will exclude a situation where we are faced with the question whether the phenomena we are dealing with are universal or culture-specific. Cross-corpora studies where the corpora used are cross-linguistic and cross-cultural – even if not generated for this purpose – are of course interesting *per se*. So far, there are only a few studies using non-acted emotions, and, due to the availability of different acted emotion corpora, a few more using acted emotions. Studies that, from the start, are designed as cross-cultural studies, normally use tightly controlled data. On the basis of such data, Sauter (2006) and Sauter *et al.* (2010) find that non-verbal emotional vocalisations (screams, laughs) are perceived similarly across two strikingly different culture groups, as far as the signalling of basic emotions is concerned. However, it seems that other, positive emotions are communicated by using culture-specific means. As for facial expressions and emotion, the strong hypothesis that has prevailed so far – that the 'big six' basic emotions are culturally universal – has recently been disproved by Jack *et al.* (2012) in a cross-cultural experiment with individuals from Western and Eastern cultures. Culturally specific 'display rules' for emotions have been addressed in the work of Paul Ekman and colleagues (Ekman 1972). There are certainly similar display rules for other states and traits, such as intoxication, sleepiness, openness, neuroticism, female versus male traits, or age.

As far as linguistic means are concerned, the discussion in modern times can be traced back to the old dispute on *linguistic relativity* as to whether language influences our way of conceptualising and thinking. Is this relativity a strong, weak, or non-existent tendency? Great battles were fought between supporters of the so-called Sapir–Whorf hypothesis (Whorf 1956) and of the universalist theories with Chomsky as protagonist. Some cognitive linguists, for example, Lakoff (1987), took an intermediate stance.

For instance, the term *Schadenfreude* for a specific emotion is a 'synthetic' term in German, not existing as such in English and thus an English loanword, but of course its semantics can be expressed 'analytically' in English (for instance, as 'pleasure in the misfortune of others') and other languages as well. It is still an empirical question whether the means made or not made available by a specific language predefine the ways of thinking and acting, and/or whether the means available are used differently by different cultures; as for culture-specific semantics of emotions, see Wierzbicka (1986) and Kövecses (2000); culture-specific appraisal biases are addressed in Scherer and Brosch (2009). A weak form of linguistic relativity, such as advocated by Boroditsky (2003), seems to be a reasonable stance to take. It is a chicken-and-egg problem whether language users shape their culture and create words denoting specific phenomena or whether the language partly influences the way of thinking by providing specific words.

The characteristics of deviant speech should by and large be universal: the speech of children with cleft palate, or that of laryngectomised elderly people, will basically show the same characteristics across languages. The same holds for biological trait primitives. Thus for these phenomena, cross-linguistic studies, or simply employing data from another language for modelling, seem to be more promising and frequent. There are exceptions, though. For instance, the pitch register of women can be higher or lower than expected, due to culture-specific influences; see Section 4.2 and the display rules mentioned above.

Interdependencies with Other Taxonomies

Acting might be different in different cultures, and due to weak relativity constraints, reports on felt states – using the different linguistic means available in different languages – might differ between cultures.

Main Relevance for Computational Processing

Although it might be possible simply to transfer methodologies and feature sets onto data obtained from other cultures, this is still a research topic; the few cross-cultural emotion studies so far (see above) show that there will be some loss of performance. However, we do not know yet whether this is due to cross-cultural factors or simply to differences in databases used for training and testing. As for deviant speech, cross-cultural differences might turn out not to be very relevant – apart from differences due to phonetic and linguistic systems, of course.

The data for training, validation, and testing are normally taken from a consistent population, for example, culture-specific when utmost performance is targeted.

2.12   Unimodal versus Multimodal

The computer can't tell you the emotional story. It can give you the exact mathematical design, but what's missing is the eyebrows.

(Frank Zappa)

We have briefly addressed multimodality in Section 1.4. On the one hand, it is evident that humans are 'multimodal beings' employing all modalities at hand for communicating and expressing themselves. However, sometimes not every modality is available for the sender and/or receiver of a communicative message; it is such situations that we want to address in this section.

Vision or hearing might not be employed either because people are handicapped (hard of hearing or vision-impaired) or because the situational context does not allow it. 'Speech-only' situations are: over the phone, in the dark, over long distances (facial gestures are not visible but perhaps posture and gestures are), or addressing from behind (a popular but outdated strategy for addressing hard-of-hearing children, to find out whether they can hear or not); the driver of a car can speak to his passenger but should not face her – however, the driver's face can be seen by the passenger and monitored by a camera.

A deaf father who normally uses sign language cannot possibly 'speak' – meaning communicate with his child in the normal way by signing – while carrying a heavy box with both hands. This is no severe handicap – he simply has to do the one after the other. The same holds for the normal way of communicating via sign language with eye contact – looking simultaneously at a picture book and communicating with signing is not possible.

'Vision-only' situations are: temporary loss of voice due to illness, 'conversation' between two people who do not understand each other's language at all (normally with some vocalisations, though), or interactions in very noisy surroundings.

Written language is by its very nature unimodal but can of course be enriched with drawings, icons, and the like – think of emoticons in email messages. Conversations with the help of written messages (with or without visual cues) do occur, for instance, in the case of a temporary or permanent loss of voice.

Telephone calls are speech-only (note that call-centre interaction is a standard use case for computational paralinguistics), as are pure audio media retrieval scenarios, audio documents in general (e.g., the famous report on the *Hindenburg* disaster when a German airship caught fire and was destroyed in 1937), radio plays and radio news. A prototypical multimodal setting is face-to-face interaction. From a processing point of view, such an interaction can be partly unimodal, for instance, when one partner is only listening while the other is talking. Moreover, there might be some reason for only monitoring the speech of the talking partner because the affect expressed in his/her face is distorted by the movements of the speaking organs.

In any case, we have to consider that the functional load for, on the one hand, linguistic-semantic messages, and on the other hand, paralinguistic messages, can be different. When deaf people use sign language, then not only hand gestures but also facial gestures are often used to convey a semantic message (e.g., raised eyebrows indicating a question). In such situations, this sign is encoded with a specific semantic function and might not function well any longer as 'free variant' – free for transmitting paralinguistic messages. The same restriction can apply for spoken language: In noisy environments, we have to speak very loudly and might thus not be able to express something like irony with the help of voice quality, intonation, and the like.

In such cases, a full synchrony of functional means might not be possible, but of course they can be employed sequentially – the same way as in written messages, the emoticon comes after the word sequence to which it pertains.

Even deviant speech can be claimed to be a multimodal phenomenon, and strictly speaking it certainly is because speaking is caused by the speech organs that can be tracked with video recordings. Speech therapy is genuinely multimodal. So far, the automatic processing of pathological speech has been exclusively unimodal.

Interdependencies with Other Taxonomies

There are cultural differences as to the importance of multimodality; people from southern Europe use gestures to a larger extent than people from northern Europe. Yet, we do not know of any study that has addressed the question whether acoustic-prosodic information is employed to a larger or lesser extent, or differently, in any of these cultures. A straightforward relation exists with 'complex versus simple': of course, multimodal communication is, other things being equal, more complex than unimodal communication. However, it can be that in the case of unimodal communication, subtler means are employed than in the case of multimodal communication, where one modality is fully free to indicate some specific paralinguistic (or extralinguistic) function.

Main Relevance for Computational Processing

The question is always whether one modality – in our case, speech/language – signals every-thing or at least the most important aspects of the relevant information. That is, do we face a loss of performance in the case of unimodal modelling?

Some of the examples above might be rather peripheral for the computational processing of paralinguistics but they are well suited for exemplifying the complex interdependency between modalities.

2.13   All These Taxonomies – So What?

> *There are known knowns. These are things we know that we know. There are known unknowns. That is to say, there are things that we know we don't know. But there are also unknown unknowns. There are things we don't know we don't know.*
>
> (Donald Rumsfeld)

We do not claim that the taxonomies presented in this chapter are on the same level as biological taxonomies, or as the taxonomies of personality and emotion research that have evolved over many decades. In a practical sense, all of them can be taken as a catalogue of questions we should ask when addressing any of the paralinguistic phenomena we want to deal with: whether they apply, whether they can be employed in any useful way, and what our stance is towards any of them. An experienced traveller does not necessarily need a checklist of what to pack – but it might come in handy because it is easy to forget one single but important item. An experienced researcher in computational paralinguistics might not need a checklist of the phenomenon she wants to address – but here as well, it might be useful to go through a list of all possible taxonomies and to consider what to do about them. A checked box can inspire us to get more acquainted with this taxonomy; for instance, when we check 'social' in 'private versus social', we might read more about the social constructivist perspective in emotion research, or about social signals (see Sections 3.2 and 5.7). It might not be that important to know the 'correct' answer, that is, to tick the 'correct' box; sometimes we simply do not know exactly, but even this is important.

We now want to demonstrate how this checklist can be used for databases; for this purpose, we use two databases that we are very familiar with, the FAU (Friedrich Alexander University) Aibo Emotion Corpus (AEC) used in the first Interspeech Emotion Challenge 2009 (Batliner *et al.* 2008; Steidl 2009), and the C-AuDiT database (Hönig *et al.* 2012).

2.13.1   Emotion Data: The FAU AEC

German children were recorded while communicating with Sony's Aibo pet robot. They were led to believe that Aibo was responding to their commands, whereas the robot was actually being controlled by a human operator who caused Aibo to perform a fixed, predetermined sequence of actions; sometimes Aibo behaved disobediently, thereby provoking emotional reactions. The data were collected at two different schools from 51 children (aged 10–13, 21 male, 30 female; about 8.9 hours of speech without pauses, sampling rate 16 bits at 16 kHz). The recordings were segmented automatically into 'turns' using a pause threshold of 1000 ms. Five labellers listened to the turns in sequential order and annotated each word as neutral (default) or as belonging to one of ten other classes. If three or more labellers agreed, the label was attributed to the word (majority voting). The number of cases with majority votes is given in parentheses: *joyful* (101), *surprised* (0), *emphatic* (2528) as a pre-stage of 'angry', *helpless* (3), *touchy*, that is, irritated (225), *angry* (84), *motherese* (1260), *bored* (11), *reprimanding* (310), *rest*, that is, non-neutral but not belonging to the other categories (3), *neutral* (39 169); 4707 words had no majority votes; all in all, there were 48 401 words. As some of the labels are very sparse, they are mapped onto cover classes (Steidl 2009): *touchy* and *reprimanding*, together with *angry*, are mapped onto *Angry* as representing different but closely related kinds of negative attitude. (*Angry* can consist, for instance, of two *touchy* and one *reprimanding* label; thus the number of *Angry* cases is far higher than the sum of *touchy*, *reprimanding*, and *angry* majority voting cases.) Some other classes, like *joyful*, *surprised*, *helpless*, *bored* and *rest*, do not appear in this subset.

Table 2.3 displays the FAU AEC checklist. on a five-point either/or scale. We have to decide between X (e.g., trait) and Y (e.g., state), with the following possibilities: I am *sure* that it is X; it is X *rather* than Y; it is something *in between* X and Y; it is Y *rather* than X; I am *sure* that it is Y. The phenomena we were annotating were clearly *states* and not traits; it was evident, though, that the children's personality determined which states they employed

Table 2.3 Checklist for FAU AEC

either	sure	rather	in between	rather	sure	or
trait					x	state
acted					x	spontaneous
complex			x				simple
measured					x	assessed
categorical			x				continuous
felt					x	perceived
intentional		x		x			instinctual
consistent	x						discrepant
private		x				x	social
prototypical		x					peripheral
universal			?				culture-specific
unimodal				x			multimodal

'x', 'applicable'; '?', 'don't know for sure'; '0', 'not applicable'; more than one box can be ticked.

how often – there were very active children, and very shy children. The children's behaviour was not prompted in any way; however, there is no clear-cut criterion for deciding whether your subjects are acting or behaving spontaneously: we simply have to design a non-prompted experiment and have to observe our subjects. In this case, there was no indication that the children were not *spontaneous*. We could observe both *simple*, that is, clear-cut states, and *complex*, mixed cases – the annotators agreed on some cases, and disagreed on others; this is the usual outcome. We did not measure but only annotated, thus, the data are fully *assessed*. Here, complex/simple and categorical/continuous are mutually dependent, thus we decide not to decide between *categorical* and *continuous*. There is no self-annotation, thus our labels are *perceived* by others. The children had to give commands to the robot, thus we can assume *intentional* behaviour which is mirrored in motherese/reprimanding and emphatic; on the other hand, there are cases of *instinctual* behaviour, for instance, 'real' anger. There is no indication that the children do not behave in a *consistent* way. Motherese and reprimanding are clearly *social*, anger at least partly *private*. The states that were annotated are rather *prototypical* and not exotic; seen from the point of view of emotion theory, only anger – being a member of the 'big six' (Section 3.2) – is a prototypical emotional state. As our subjects are from one specific culture, we cannot say anything about universal versus culture-specific characteristics. The children were not told that the robot could observe their behaviour or not; thus, we cannot decide from start whether they employed any multimodal behaviour in communicating with the robot. The video recordings revealed that some of the children were very static, some very lively, and that some of them pointed towards the target the robot should go to.

Of course, some of the boxes were 'ticked in advance'; the experimental setting resulted in states and not in traits to be annotated, and made it very likely that we really could elicit spontaneous speech. Yet, we did not envisage exactly this combination of rather 'private' emotions and 'interactional' states such as motheresing and reprimanding. The combination inspired us to use non-metrical dimensional scaling – a visualisation method in n-dimensional space which led to two dimensions, the first being valence, and the second being not arousal but social interaction (Batliner *et al.* 2008). Inspection of the video recordings – which were not intended to be analysed but to constitute a sort of fall-back documentation – inspired an analysis of the interplay between body movements, hand gestures (especially pointing), and affective states; in (Batliner *et al.* 2011) we demonstrate a correlation between body movements and hand gestures on the one hand, and the use of specific emotions on the other hand.

2.13.2 Non-native Data: The C-AuDiT corpus

For the C-AuDiT (Computer-Assisted Pronunciation and Dialogue Training) corpus (Hönig *et al.* 2012), 55 speakers of English as a second language (L2) were recorded: 25 German, 10 French, 10 Spanish, and 10 Italian speakers. They had to read aloud 329 utterances shown on the screen display of an automated recording program. The data to be recorded consisted of two short stories (broken down into sentences to be displayed on the screen), sentences containing, *inter alia*, different types of phenomena such as question intonation or position of phrase accent (*This is a house.* versus *Is this really a house?*), or tongue-twisters, and words/phrases such as *Arabic/Arabia/The Arab World/In Saudi Arabia,...* ; pairs such as *SUBject* versus *subJECT* had to be repeated after the pre-recorded production of a tutor. Some sentences were taken from the Italian and German Spoken Learners' English (ISLE) corpus (Menzel *et al.* 2000). Based on annotations of three experienced labellers (Hönig *et al.* 2009), a subset was defined consisting of those five sentences that were judged as 'prosodically most error-prone for L2 speakers of English' (Hönig *et al.* 2010).

For annotation, a perception experiment was conducted for scoring intelligibility, non-native accent, perceived first language (L1), melody and rhythm, using the PEAKS tool (Maier *et al.* 2009). Twenty native American English, 19 native British English, and 21 native Scottish English speakers with normal hearing abilities judged each sentence in random order. These were the possible answers to the melody question: This sentence's melody sounds: (1) normal; (2) acceptable, but not perfectly normal; (3) slightly unusual; (4) unusual; (5) very unusual. The labels on the Likert scales were averaged over all sentences of a speaker to get a single score for each criterion.

Table 2.4 displays the C-AuDiT checklist. Such recordings of read non-native speech typically display *traits*, that is, cultural trait primitives; one can imagine testing again, after an intensive learning period of a few weeks, therefore we entered '?' under 'rather' to indicate that this trait can change in the foreseeable future, and within-speaker experiments might be possible. 'Acted' versus 'spontaneous' does not really apply for read speech, especially if the data to be read are not embedded in a dialogue. The formal characteristics of this type of speech are rather *complex* – wrong word accent position, substitution or deletion of segments, non-native rhythm, and so on. The data were *assessed* by human annotators. Non-native speech is typically more or less non-native, thus a somewhat continuous phenomenon; the assessment was done on an ordinal scale resulting in a sort of *continuous* scale after averaging across annotators. The annotation was *perceived*; note that additionally, the subjects assessed their

Table 2.4   Checklist for C-AuDiT

either	sure	rather	in between	rather	sure	or
trait	x	?				state
acted			0			spontaneous
complex		x				simple
measured					x	assessed
categorical			x			continuous
felt					x	perceived
intentional			0			instinctual
consistent			?			discrepant
private			0			social
prototypical		x		x		peripheral
universal				x		culture-specific
unimodal	x					multimodal

'x', 'applicable'; "?", 'don't know for sure'; '0', 'not applicable'; more than one box can be ticked.

own L2 proficiency on a coarse scale. The next three taxonomies do not really apply; sometimes the formal characteristics might be discrepant, when, for instance, a strong non-native rhythm went together with a close-to-perfect segmental pronunciation. However, this happens rarely. There are L2 speakers who are *prototypical* for a specific L1, and there are L2 speakers who cannot easily be assigned to any L1. The non-native traits are *culture-specific* in the sense that the L1 can be seen as cultural trait primitive (see Section 5.2). These types of read data are typically *unimodal* – no video recordings, no interactions with peers or teachers. Of course, video recordings of the speakers could be taken to reveal their states, for instance, 'flow' when they do not have any problems in pronouncing the items, or 'stuck' when it is getting difficult for them.

Obviously, our taxonomies are more useful for emotion modelling than for non-native speech – most likely because non-native speech is less influenced by personality traits or by complex interactions of other traits or states (cf. the different levels displayed in Figure 5.1).

References

Allport, G. W. and Odbert, H. S. (1936). Trait-names: A psycho-lexical study. *Psychological Monographs*, **47**, i–171.
Batliner, A., Fischer, K., Huber, R., Spilker, J., and Nöth, E. (2000). Desperately seeking emotions: Actors, wizards, and human beings. In *Proc. of the ISCA Workshop on Speech and Emotion*, pp. 195–200, Newcastle, Co. Down.
Batliner, A., Steidl, S., Hacker, C., and Nöth, E. (2008). Private emotions vs. social interaction — a data-driven approach towards analysing emotions in speech. *User Modeling and User-Adapted Interaction*, **18**, 175–206.

Batliner, A., Seppi, D., Steidl, S., and Schuller, B. (2010). Segmenting into adequate units for automatic recognition of emotion-related episodes: a speech-based approach. *Advances in Human-Computer Interaction*. 15 p.

Batliner, A., Steidl, S., and Nöth, E. (2011). Associating children's non-verbal and verbal behaviour: Body movements, emotions, and laughter in a human-robot interaction. In *Proc. of ICASSP*, pp. 5828–5831, Prague.

Ben-Ze'ev, A. (2000). *The Subtlety of Emotions*. MIT Press, Cambridge, MA.

Boroditsky, L. (2003). Linguistic relativity. In L. Nadel (ed.), *Encyclopedia of Cognitive Science*, pp. 917–921. Macmillan Press, London.

Bühler, K. (1934). *Sprachtheorie. Die Darstellungsfunktion der Sprache*. G. Fischer, Jena.

Burkhardt, F., Paeschke, A., Rolfes, M., Sendlmeier, W., and Weiss, B. (2005). A database of German emotional speech. In *Proc. of Interspeech*, pp. 1517–1520, Lisbon.

Carbone, M., Gal, K., Shieber, S. M., and Grosz, B. (2004). Unifying annotated discourse hierarchies to create a gold standard. In *Proc. of the Fifth SIGdial Workshop on Discourse and Dialogue*, Boston.

Claassen, J. A. H. R. (2005). The gold standard: not a golden standard. *British Medical Journal (BMJ)*, **330**, 1121.

Cochrane, T. (2009). Eight dimensions for the emotions. *Social Science Information*, **48**, 379–420.

Cohen, H. and Lefebre, C. (eds) (2005). *Handbook of Categorization in Cognitive Science*. Elsevier, Amsterdam.

Darwin, C. (1872). *The Expression of the Emotions in Man and Animals*. John Murray, London.

Devillers, L., Vidrascu, L., and Lamel, L. (2005). Challenges in real-life emotion annotation and machine learning based detection. *Neural Networks*, **18**, 407–422.

Diderot, D. (1883). *The Paradox of Acting. Translated with Annotations from Diderot's 'Paradoxe sur le Comédien' by Walter Herries Pollock. With a Preface by Henry Irving*. Chatto & Windus, London.

Ekman, P. (1972). Universals and cultural differences in facial expressions of emotion. In J. R. Cole (ed.), *Nebraska Symposium on Motivation*, pp. 207–283, Lincoln. University of Nebraska Press.

Engberg, I. S., Hansen, A. V., Andersen, O., and Dalsgaard, P. (1997). Design, recording and verification of a Danish emotional speech database. In *Proc. of Eurospeech*, pp. 1695–1698, Rhodes.

Eyben, F., Batliner, A., Schuller, B., Seppi, D., and Steidl, S. (2010). Cross-corpus classification of realistic emotions – some pilot experiments. In L. Devillers, B. Schuller, R. Cowie, E. Douglas-Cowie, and A. Batliner (eds), *Proc. of the 3rd International Workshop on EMOTION: Corpora for Research on Emotion and Affect, satellite of LREC 2010*, pp. 77–82, Valletta, Malta.

Fehr, B. and Russel, J. A. (1984). Concept of emotion viewed from a prototype perspective. *Journal of Experimental Psychology: General*, **113**, 464–486.

Flett, G. L. (2007). *Personality Theory & Research*. Wiley, Mississauga, ON.

Fontaine, J. R., Scherer, K. R., Roesch, E. B., and Ellsworth, P. C. (2007). The world of emotions is not two-dimensional. *Psychological Science*, **18**, 1050–1057.

Fuhrmann, G. (1988). 'Prototypes' and 'fuzziness' in the logic of concepts. *Synthese*, **75**, 317–347.

Grice, H. (1975). Logic and conversation. In P. Cole and J. Morgan (eds), *Syntax and Semantics, Volume 3, Speech Acts*, pp. 41–58. Academic Press, New York.

Hacker, C., Batliner, A., and Nöth, E. (2006). Are you looking at me, are you talking with me – multimodal classification of the focus of attention. In P. Sojka, I. Kopecek, and K. Pala (eds), *Proc. of Text, Speech and Dialogue (TSD)*, Brno. Springer, New York.

Hájek, A. (2007). The reference class problem is your problem too. *Synthese*, **156**, 563–585.

Han, W., Li, H., Ma, L., Zhang, X., and Schuller, B. (2012). A ranking-based emotion annotation scheme and real-life speech database. In *Proceedings 4th International Workshop on Emotion Sentiment & Social Signals 2012 (ES' 2012) – Corpora for Research on Emotion, Sentiment & Social Signals, held in conjunction with LREC 2012*, pp. 67–71, Istanbul, Turkey.

Hareli, S. and Parkinson, B. (2008). What's social about social emotions? *Journal for the Theory of Social Behaviour*, **38**, 131–156.

Harnad, S. (1987). *Categorical Perception. The Groundwork of Cognition*. Cambridge University Press, Cambridge.

Hönig, F., Batliner, A., Weilhammer, K., and Nöth, E. (2009). Islands of failure: Employing word accent information for pronunciation quality assessment of English L2 learners. In *Proc. of SLATE*, Wroxall Abbey.

Hönig, F., Batliner, A., Weilhammer, K., and Nöth, E. (2010). Automatic assessment of non-native prosody for English as L2. In *Proc. of Speech Prosody, Chicago IL*.

Hönig, F., Batliner, A., and Nöth, E. (2012). Automatic assessment of non-native prosody annotation, modelling and evaluation. In *Proc. of the International Symposium on Automatic Detection of Errors in Pronunciation Training (ISADEPT)*, pp. 21–30, Stockholm.

Jack, R. E., Garrod, O. G. B., Yu, H., Caldara, R., and Schyns, P. G. (2012). Facial expressions of emotion are not culturally universal. *Proceedings of the National Academy of Sciences of the United States of America*, **109**, 7241–7244.

Johnson, K. (2006). Resonance in an exemplar-based lexicon: The emergence of social identity and phonology. *Journal of Phonetics*, **34**, 485–499.

Kaniusas, E. (2012). *Biomedical Signals and Sensors I: Linking Physiological Phenomena and Biosignals*. Springer, Berlin.

Knapp, R. B., Kim, J., and André, E. (2011). Physiological signals and their use in augmenting emotion recognition for human-machine interaction. In P. Petta, C. Pelachaud, and R. Cowie (eds), *Emotion-Oriented Systems: The Humaine Handbook*, Cognitive Technologies, pp. 133–159. Springer, Berlin.

Konijn, E. A. (2000). *Acting Emotions*. Amsterdam University Press, Amsterdam.

Kövecses, Z. (2000). The concept of anger: Universal or culture specific? *Psychopathology*, **33**, 159–170.

Krajewski, J., Batliner, A., and Golz, M. (2009). Acoustic sleepiness detection: Framework and validation of a speech-adapted pattern recognition approach. *Behavior Research Methods*, **41**, 795–804.

Labov, W. (1972). *Sociolinguistic Patterns*. University of Pennsylvania Press, Philadelphia.

Lacerda, F. (1995). The perceptual-magnet effect: An emergent consequence of exemplar-based phonetic memory. In *Proc. of ICPhS*, pp. 140–147, Stockholm.

Lacewing, M. (2004). Emotion and cognition: Recent developments and therapeutic practice. *Philosophy, Psychiatry and Psychology*, **11**, 175–186.

Lakoff, G. (1987). *Women, Fire, and Dangerous Things. What Categories Reveal about the Mind*. University of Chicago Press, Chicago.

Lev-Ari, S. and Keysar, B. (2010). Why don't we believe non-native speakers? The influence of accent on credibility. *Journal of Experimental Social Psychology*, **46**, 1093–1096.

Llamas, C. (2007). A new methodology: DATA elicitation for regional and social language variation studies. *York Papers in Linguistics*, pp. 138–163.

Maier, A., Haderlein, T., Eysholdt, U., Rosanowski, F., Batliner, A., Schuster, M., and Nöth, E. (2009). PEAKS: A system for the automatic evaluation of voice and speech disorders. *Speech Communication*, **51**, 425–437.

Manstead, T. (2005). The social dimension of emotion. *The Psychologist*, **18**, 484–487.

Marinetti, C., Moore, P., Lucas, P., and Parkinson, B. (2011). Emotions in social interactions: Unfolding emotional experience. In P. Petta, C. Pelachaud, and R. Cowie (eds), *Emotion-Oriented Systems: The Humaine Handbook*, Cognitive Technologies, pp. 31–46. Springer, Berlin.

Matsumoto, D. (1990). Cultural similarities and differences in display rules. *Motivation and Emotion*, **14**, 195–214.

Mehrabian, A. and Russell, J. (1974). *An Approach to Environmental Psychology*. MIT Press, Cambridge, MA.

Menzel, W., Atwell, E., Bonaventura, P., Herron, D., Howarth, P., Morton, R., and Souter, C. (2000). The ISLE corpus of non-native spoken English. In *Proc. of LREC*, pp. 957–964, Athens.

Mesleh, A., Skopin, D., Baglikov, S., and Quteishat, A. (2012). Heart rate extraction from vowel speech signals. *Journal of Computer Science and Technology*, **27**, 1243–1251.

Morris, D. G. (2003). Gold, silver, and bronze. Metals, medals, and standards in hypersensitivity pneumonitis. *American Journal of Respiratory and Critical Medicine*, **168**, 909–910.

Niedenthal, P. M. and Cantor, N. (1984). Making use of social prototypes: From fuzzy concepts to firm decisions. *Fuzzy Sets and Systems*, **14**, 5–27.

Norman, W. T. (1967). 2800 personality trait descriptors: Normative operating characteristics for a university population. The University of Michigan. College of Literature, Science, and the Arts, Department of Psychology.

Nosofsky, R. M. (1986). Attention, similarity, and the identification-categorization relationship. *Journal of Experimental Psychology: General*, **115**, 39–57.

Nosofsky, R. M. and Zaki, S. R. (2002). Exemplar and prototype models revisited: Response strategies, selective attention, and stimulus generalization. *Journal of Experimental Psychology: Learning, Memory, and Cognition*, **28**, 924–940.

Osgood, C., Suci, G., and Tannenbaum, P. (1957). *The Measurement of Meaning*. University of Illinois Press, Urbana.

Osgood, C. E. (1964). Semantic differential technique in the comparative study of cultures. *American Anthropologist*, **66**, 171–200.

Parkinson, B. (1996). Emotions are social. *British Journal of Psychology*, **87**, 663–683.

Pentland, A. (2008). *Honest Signals. How They Shape Our World*. MIT Press, Cambridge, MA.

Plutchik, R. (1980). *Emotion: A Psychoevolutionary Synthesis*. Harper & Row, New York.

Plutchik, R. and Conte, H. R. (eds) (1997). *Circumplex Models of Personality and Emotions.* American Psychological Association, Washington D.C.

Polzehl, T., Möller, S., and Metze, F. (2010). Automatically assessing personality from speech. In *Proc. of the IEEE Fourth International Conference on Semantic Computing (ICSC '10)*, pp. 134–140, Washington, DC.

Reichenbach, H. (1949). *The Theory of Probability.* University of California Press, Berkeley.

Repp, B. (1981). Phonetic trading relations and context effects: New experimental evidence for a speech mode of perception. Haskins Laborotories: Status Report on Speech Research SR-67/68.

Roberts, R. C. (1988). What an emotion is: A sketch. *Philosophical Review*, **97**, 183–209.

Roesch, E. B., Korsten, N., Fragopanagos, N. F., Taylor, J. G., Grandjean, D., and Sander, D. (2011). Biological and computational constraints to psychological modelling of emotion. In P. Petta, C. Pelachaud, and R. Cowie (eds), *Emotion-Oriented Systems: The Humaine Handbook*, Cognitive Technologies, pp. 47–62. Springer, Berlin.

Rosch, E. (1975). Cognitive representations of semantic categories. *Journal of Experimental Psychology: General*, **104**(3), 192–233.

Rosch, E. and Mervis, C. B. (1975). Family resemblances: Studies in the internal structure of categories. *Cognitive Psychology*, **7**, 573–605.

Russell, J. A. (1991). In defense of a prototype approach to emotion concepts. *Journal of Personality and Social Psychology*, **60**, 37–47.

Sauter, D. (2006). *An investigation into vocal expressions of emotions: The roles of valence, culture, and acoustic factors.* Ph.D. thesis, University College London.

Sauter, D. A., Eisner, F., Ekman, P., and Scott, S. K. (2010). Cross-cultural recognition of basic emotions through nonverbal emotional vocalizations. *Proceedings of the National Academy of Sciences of the United States of America*, **107**, 2408–2412.

Scherer, K. and Ceschi, G. (2000). Criteria for emotion recognition from verbal and nonverbal expression: Studying baggage loss in the airport. *Personality and Social Psychology Bulletin*, **26**, 327–339.

Scherer, K. R. (2000). Emotion. In M. Hewstone and W. Stroebe (eds), *Introduction to Social Psychology: A European Perspective*, pp. 151–191. Blackwell, Oxford.

Scherer, K. R. (2003). Vocal communication of emotion: A review of research paradigms. *Speech Communication*, **40**, 227–256.

Scherer, K. R. (2013). Vocal markers of emotion: Comparing induction and acting elicitation. *Computer Speech and Language*, **27**, 40–58.

Scherer, K. R. and Brosch, T. (2009). Culture-specific appraisal biases contribute to emotion dispositions. *European Journal of Personality*, **23**, 265–288.

Schiel, F. (1999). Automatic phonetic transcription of non-prompted speech. In *Proc. of ICPhS*, pp. 607–610, San Francisco.

Schiel, F., Heinrich, C., and Barfuesser, S. (2012). Alcohol Language Corpus – The first public corpus of alcoholized German speech. *Language Resources and Evaluation*, **46**, 503–521.

Schlosberg, H. (1941). A scale for judgment of facial expressions. *Journal of Experimental Psychology*, **29**, 497–510.

Schlosberg, H. (1954). Three dimensions of emotion. *Psychology Review*, **61**, 81–88.

Schuller, B., Wimmer, M., Arsić, D., Rigoll, G., and Radig, B. (2007). Audiovisual behavior modeling by combined feature spaces. In *Proc. of ICASSP*, pp. 733–736, Honolulu.

Schuller, B., Müller, R., Eyben, F., Gast, J., Hörnler, B., Wöllmer, M., Rigoll, G., Höthker, A., and Konosu, H. (2009). Being bored? Recognising natural interest by extensive audiovisual integration for real-life application. *Image and Vision Computing*, **27**(12), 1760–1774.

Schuller, B., Steidl, S., Batliner, A., Burkhardt, F., Devillers, L., Müller, C., and Narayanan, S. (2010). The Interspeech 2010 Paralinguistic Challenge. In *Proc. of Interspeech*, pp. 2794–2797, Makuhari, Japan.

Schuller, B., Steidl, S., Batliner, A., Schiel, F., Krajewski, J., Weninger, F., and Eyben, F. (2013a). Medium-term speaker states – a review on intoxication, sleepiness and the first challenge. *Computer Speech and Language*.

Schuller, B., Steidl, S., Batliner, A., Burkhardt, F., Devillers, L., Müller, C., and Narayanan, S. (2013b). Paralinguistics in speech and language – state-of-the-art and the challenge. *Computer Speech and Language, Special Issue on Paralinguistics in Naturalistic Speech and Language*, **27**(1), 4–39.

Shaver, P., Schwartz, J., Kirson, D., and O'Connor, C. (1987). Emotion knowledge: Further exploration of a prototype approach. *Journal of Personality and Social Psychology*, **52**, 1061–1086.

Silvia, P. J. (2005). What is interesting? Exploring the appraisal structure of interest. *Emotion*, **5**, 89–102.

Silvia, P. J. (2008). Interest - the curious emotion. *Current Directions in Psychological Science*, **17**, 57–60.

Skopin, D. and Baglikov, S. (2009). Heartbeat feature extraction from vowel speech signal using 2D spectrum representation. In *Proc. of 4th International Conference on Information Technology (ICIT)*, Amman, Jordan.

Skousen, R., Lonsdale, D., and Parkinson, D. B. (eds) (2002). *Analogic Modeling: An Exemplar-Based Approach to Language*. John Benjamins, Amsterdam.

Sobol-Shikler, T. and Robinson, P. (2010). Classification of complex information: Inference of co-occurring affective states from their expressions in speech. *IEEE Transactions on Pattern Analysis and Machine Intelligence*, **32**, 1284–1297.

Steidl, S. (2009). *Automatic Classification of Emotion-Related User States in Spontaneous Children's Speech*. Logos Verlag, Berlin.

Stevens, S. S. (1946). On the theory of scales of measurement. *Science, New Series*, **103**, 677–680.

Summerfield, Q. (ed.) (1991). *Hearing and Speech. A Special Issue of The Quaterly Journal of Experimental Psychology - Section A - Human Experimental Psychology*. Lawrence Erlbaum, Hove and London.

Tomasello, M., Carpenter, M., Call, J., Behne, T., and Moll, H. (2005). Understanding and sharing intentions: The origins of cultural cognition. *Behavioral and Brain Sciences*, **28**, 675–735.

Velten Jr., E. (1968). A laboratory task for induction of mood states. *Behavior Research & Therapy*, **6**(4), 473–482.

Walsh, M., Möbius, B., Wade, T., and Schütze, H. (2010). Multilevel exemplar theory. *Cognitive Science*, **34**, 537–582.

Watzlawick, P., Beavin, J., and Jackson, D. D. (1967). *Pragmatics of Human Communications*. W.W. Norton, New York.

Westermann, R., Spies, K., Stahl, G., and Hesse, F. W. (1996). Relative effectiveness and validity of mood induction procedures: a meta-analysis. *European Journal of Social Psychology*, **26**, 557–580.

Whorf, B. L. (1956). *Language, Thought, and Reality*. Technology Press of Massachusetts Institute of Technology, Cambridge, MA.

Wierzbicka, A. (1986). Human emotions: Universal or culture-specific? *American Anthropologist*, **88**, 584–594.

Wittgenstein, L. (2009). *Philosophical Investigations. The German Text, with an English translation by G.E.M. Anscombe and P.M.S. Hacker and Joachim Schulte*. Wiley-Blackwell, Chichester.

Wundt, W. (1896). *Grundriss der Psychologie*. Engelmann, Leipzig.

Xu, Y. (2010). In defense of lab speech. *Journal of Phonetics*, **38**, 329–336.

Zadeh, L. (1965). Fuzzy sets. *Information and Control*, **8**, 338–353.

3

Aspects of Modelling

My lord, facts are like cows. If you look them in the face hard enough, they generally run away.

(Dorothy L. Sayers, *Clouds of Witness*)

This chapter deals with the perspectives from which a phenomenon is viewed in models and theories; it does not deal with intrinsic characteristics of phenomena – these were addressed in Chapter 2. State-of-the-art and paralinguistic phenomena themselves are dealt with in Chapters 4 and 5.

This might seem to tear apart what belongs together. However, with this structure we want to disentangle aspects that are, as it were, external to paralinguistics proper, belonging to other fields of science such as psychology or medicine, or to general principles of methodology, for instance, whether data should be collected in the lab or in real-life settings.

3.1 Theories and Models of Personality

All the games people play now / Every night, every day now
Never meanin' what they say now / Never sayin' what they mean

(Joe South, 1968)

Personality is a term of uncertain origin derived from the Latin 'persona' ('mask', 'character' in a play, as in the phrase *dramatis personae*); thus, originally, the term did not denote the character of a person but the character that is played by her on stage – a nice metaphor for the dual character of paralinguistic phenomena, expressed also in the 'felt versus perceived' and 'intentional versus instinctual' taxonomies.

The following definition of the concept of 'personality' is given in Flett (2007, p. 4):

'Personality' refers to relatively stable individual differences that are believed to [be] present early in life and involves characteristics that generalize across time and across situations. We usually discuss personality in terms of the dispositioned factors and

Computational Paralinguistics: Emotion, Affect and Personality in Speech and Language Processing, First Edition.
Björn W. Schuller and Anton M. Batliner. © 2014 John Wiley & Sons, Ltd. Published 2014 by John Wiley & Sons, Ltd.

associated behaviours that distinguish us and make us different from other people, but
there are some personality characteristics and processes that may be at least somewhat
similar across individuals.'

The forebears of personality theories are all those eminent historical theories and figures in psychology – because psychology *per se* deals with the personality of humans: the psychoanalytic theories of Freud, Jung, Adler, Horney, Erikson and Fromm; the behaviouristic theories of Skinner and Eysenck; and the humanistic theories of Maslow, Rogers and Frankl. This list could be extended.

Allport was one of the early researchers to study personality traits (Allport and Allport 1921; Allport 1927) and to start with a lexical approach taking the words found in natural language as a basis for taxonomies (Allport and Odbert 1936). This work was continued by Norman (1967) who established specifications for seven categories: stable terms (biophysical traits); temporary states, moods and attitudes; social roles and relationships; evaluative terms and mere quantifiers; terms for anatomical medical, physical and grooming characteristics; ambiguous, vague and tenuously or obliquely metaphorical terms; and very difficult, obscure, and little-known terms. Such terms can be conceived of as mutually exclusive categories or, as Chaplin *et al.* (1988) suggest, as prototypes (see Section 2.10).

Following this early work, different granularities or levels of personality traits were proposed – cf. the title of Eysenck (1991): 'Dimensions of personality: 16, 5 or 3?' These descriptive, empirically oriented approaches towards personality eventually resulted in the now well-established so-called *five-factor model* of personality (the 'big five') which is based on a lexical tradition (clustering of lexical synonyms) and a questionnaire tradition (assessment of one's own or others' personality with the help of standardised questionnaires); see McCrae and John (1992). This five-factor model is elaborated on in Digman (1990) and John and Srivastava (1999) who favour a mnemonic convention besides the numbering convention which reflects the relative size of the factors in lexical studies; differently ordered, this results in an acronym that forms the *OCEAN* of personality dimensions:

E Extraversion (or 'extroversion'), energy, enthusiasm

A Agreeableness, altruism, affection

C Conscientiousness, control, constraint

N Neuroticism, negative affectivity, nervousness

O Openness, originality, open-mindedness

This five-factor model of personality, naturally enough, does not meet the high standards of biological taxonomies because it is not based on (fully) 'objective' measures, and its terminology might be culture- and language-dependent (John and Srivastava 1999). Thus the 'big five' are both established and still controversial; as Scherer (2013) puts it, 'the definition of the traits and their measurement remain highly debated. This is more of a problem for *symptom studies* in which the personality of the speakers used needs to be reliably measured than in *attribution studies* (appeal) in which there tends to be a certain level of agreement on a small set of popularly used personality descriptors'. Open questions, as already mentioned in Chapter 2, are for instance: universality, cross-linguistics, cross-cultural, and prototypical versus discrete.

The default methodology in paralinguistic approaches towards personality is to resort either to the 'big five' or to one or more subcategories where some intersubjectivity has been established, for instance, by agreed-upon scales. The 'Big Five Inventory' can be tested, for example, with a longer questionnaire consisting of 44 short phrases (John and Srivastava 1999), or a shorter one consisting of 10 short phrases (Rammstedt and John 2007) yielding lower effect sizes but still usable if time is limited. Studies addressing the automatic recognition of different types and different granularities of personality traits based on acoustic and linguistic information will be dealt with in Section 5.3.

3.2   Theories and Models of Emotion and Affect

Siobhan also says that if you close your mouth and breathe out loudly through your nose it can mean that you are relaxed, or that you are bored, or that you are angry and it all depends on how much air comes out of your nose and how fast and what shape your mouth is when you do it and how you are sitting and what you said just before and hundreds of other things which are too complicated to work out in a few seconds.

(Mark Haddon, *The Curious Incident of the Dog in the Night-Time*, 2003)

The autistic boy in Haddon's novel obviously has difficulties mapping the many formal elements that can be perceived onto the many different functions (emotions and *emotion-related states*) that can be indicated by them. As 'normal', 'typically developed' people we might not have the same problems in daily life but we definitely do have similar problems when trying to describe, annotate, and model something that is as pervasive and, at the same time, evasive, as emotion. It is pervasive because it is '. . . present in most of life, but absent when people are emotionless (which . . . happens rather rarely)' (Cowie *et al.* 2011, p. 14). It is evasive because there are too many and too different everyday conceptualisations of emotion. This two-headed state is mirrored in theoretical approaches as well. On the one hand, there are the well-known full-blown emotions – everybody agrees that they are emotions but only adherents to special theories claim that they are the only ones. On the other hand, there is something like *emotional intelligence* which is claimed to be as important as *cognitive intelligence* and to pervade the whole of life (Mayer *et al.* 2008; Salovey and Mayer 1990). This concept seems to be used by practitioners but is rather stigmatised by theoreticians because it is claimed to be too vague and too all-encompassing (Locke 2005).

A broad and a narrow definition of 'emotion' coexist not only in theoretical approaches but also in normal language use and thus in dictionaries. The following definition of a strong and a weak variant of emotion can be found in the Merriam-Webster online dictionary:

1 a obsolete : DISTURBANCE
   b : EXCITEMENT
2 a : the affective aspect of consciousness : FEELING
   b : a state of feeling
   c : a conscious mental reaction (as anger or fear) subjectively experienced as strong feeling usually directed toward a specific object and typically accompanied by physiological and behavioural changes in the body

The definition in the Oxford Dictionaries online reads differently but again, there is a broad and a narrow definition:

> noun: *a strong feeling deriving from one's circumstances, mood, or relationships with others:* she was attempting to control her emotions
> mass noun: his voice was shaky with emotion

- mass noun: *instinctive or intuitive feeling as distinguished from reasoning or knowledge:* responses have to be based on historical insight, not simply on emotion

Emotion can be viewed from different perspectives. Giving an overview of theories and models of emotion is not an easy task; statements like this often introduce introductory chapters in longer studies, or shorter introductions into the subject. This book is no exception. It is relatively straightforward to mention the 'forebears' of basic approaches – which we will do, following Cornelius (2000) – and to mention prominent representatives of present-day emotion research. This will be done in the following as well. As long as one is confined to the realm of one specific theory, matters can seem to be relatively straightforward. This holds for other fields as well, for instance, for different linguistic theories.

In the *Darwinian perspective* (Darwin 1872), emotions are evolved phenomena with important survival functions. This evolutionary perspective suggests conceiving emotions as rather universal; see Ekman *et al.* (1987) and the so-called 'big six' (happiness, sadness, fear, disgust, anger, and surprise). These are claimed to be basic and other emotions to be derived from them. In the *Jamesian perspective* (James 1884), emotions are considered to be experienced and constituted via bodily responses to the environment. In the *cognitive perspective* (Arnold 1960; Clore and Ortony 2000), emotions are constituted via appraisal by which events are judged as positive or negative. Finally, in the *social constructivist perspective* (Averill 1980), emotions are based on culture and learned social rules.

The mediating link between basic theories of emotion and computational approaches in paralinguistics is methodology. Implicitly, when we employ physiological signals as features or, even more importantly, as ground truth, we adhere to a Darwinian perspective. When we are interested in intentionally produced emotions, we adhere to a social constructivist perspective. The concept of *arousal* implies a sort of biologistic point of view, the concept of *valence* (to evaluate something as positive or negative) rather a cognitive one (appraisal).

It seems to us that the 'points of contact' between emotion theories and 'practical' computational paralinguistics do not necessarily concern pivotal, basic theoretical concepts but rather the edges of theories: lists of traits and states, types of annotation or modelling (categories versus dimensions), quantifications, and especially knowledge of the phenomena one wants to deal with. The ultimate criterion within computational paralinguistics is not adequacy for and within specific theoretical constructs but adequacy for specific applications and/or performance in classification and modelling. It is like being at the greengrocer: you do not care about the biological taxonomies of fruits or vegetables, you just pick the ones you want to eat. We will come back to this pivotal difference in perspective throughout the following chapters.

Cowie *et al.* (2011) try to disentangle the many (slightly) different conceptualisations of the terms 'feeling', 'affect', and 'emotion'. The prevailing language use is to reserve 'emotion' for some specific states; (Scherer (2005, 2013) presents such a prototypical conceptualisation of 'emotion'); and to use 'affect', most likely due to the impact of the seminal book by Picard (1997), in a broader sense, encompassing emotions and 'emotion-related states':

> *'Affect' is a word that deserves special attention, because it is much used in the area, and it has a very curious semantic profile. It is rarely used in everyday discourse. Insofar as it has a generally accepted meaning, it signifies something akin to emotion, but broader in some sense. Experts have taken it up and given it a great variety of more precise senses, often grounded in a theory which implies that emotional and emotion-related phenomena divide naturally in particular ways.*
>
> (Cowie *et al.* 2011, p. 12)

Note that in German, the meaning of 'affect' is extremely narrow: it means 'utmost emotionality' (you kill someone 'in affect'). A somehow fuzzy semantics also has the term 'emotion-related states' – often used in the same rather vague sense as 'affect' in English. In Scherer (2003) and Juslin and Scherer (2005), 'affective states' encompass personality traits and several different states such as emotion, mood, interpersonal stances, and attitudes (see Table 2.1).

Within such a broad conceptualisation of affect, traits and states have a complex relationship. Peculiarities of short(er) states are influenced by personality traits; this holds for the 'same' phenomena as well (Reisenzein and Weber 2009; Revelle and Scherer 2009); see also Endler and Kocovski (2001, p. 232) who refer to 'trait anxiety as an individual's predisposition to respond, and state anxiety as a transitory emotion characterized by physiological arousal and consciously perceived feelings of apprehension, dread, and tension'. The same holds for emotion-related states and traits (see our example of the 'interested person' in Section 2.1).

The big issues – still debated today – were introduced by Wundt (1896):

> *On the basis of* quality *we may distinguish certain fundamental emotional forms corresponding to the chief affective directions distinguished before. . . . This gives us pleasurable and unpleasurable, exciting and depressing, straining and relaxing emotions. It must be noted, however, that because of their more composite character the emotions, are always, even more than the feelings,* mixed *forms. Generally, only a* single *affective direction can be called the* primary *tendency for a particular emotion.*

Wundt's first two 'affective directions', 'pleasurable/unpleasurable' and 'exciting/depressing', are mirrored in the prevalent dimensional model with evaluation and arousal (cf. Section 2.5). Emotion categories are nowadays mostly processed as 'single', 'primary' emotion, most likely for practical reasons. More complex models are, for example, major versus minor emotions (Schuller *et al.* 2010a; Vidrascu and Devillers 2005), or composite models with graphical representations such as the 'emogram' (in analogy to 'sonagram') in Adelhardt *et al.* (2006) or the emotion profile in Mower *et al.* (2011).

Final Note on Theories and Models

We confine our short presentation of theories and models to personality and emotion/affect. As for the other paralinguistic phenomena, we do not necessarily need 'big' theories. Of course, there are psychological, sociological, cultural and literal theories and models, for instance, of age – to mention just a few fields that deal with age. As far as computational paralinguistics is concerned, age is mostly a straightforward phenomenon which can be modelled in years, or at the very beginning of life in months or other shorter time frames. Note that it is not straightforward to find out the exact age – it can deliberately be masked, and it can be indicated differently by females, males, or different cultures, and relevant parameters can be influenced by any other factors. A non-native accent or some variety of pathological speech can be complicated to describe and model computationally but the phenomenon behind it is often relatively straightforward. When it is not, this problem normally belongs to theories external to paralinguistics.

3.3 Type and Segmentation of Units

Being determines consciousness.

(Karl Marx)

The limits of my language means the limits of my world.

(Ludwig Wittgenstein)

Although the extraction and segmentation of units of analysis could simply be viewed as a technicality, we decided to treat it in this chapter on modelling. In so doing, we want to stress the impact of seemingly down-to-earth decisions on units of analysis. Figure 3.1 tries to give a 'practical' account of the alternatives, not a strict and complete theoretical taxonomy.

First of all, the scientific field – and thus the possibility of obtaining and processing data – predefines the *type* of data addressed; this is displayed on the vertical axis. Linguists and computer linguists normally deal with written *language*. Phoneticians – but often psychologists and clinicians as well – deal with spoken language, that is, *speech* which is often very restricted, such as prompted short utterances, words, or even sustained vowels. The same holds for non-verbal, vocal events such as laughter or coughing. Speech processing lies in some sense

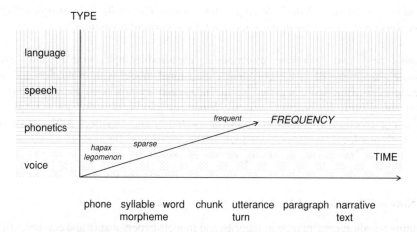

Figure 3.1 Type and time duration of units, and frequency

between linguistics and *phonetics*: sometimes it only deals with pure acoustic phenomena (phonetics), sometimes manually or automatically transcribed words or higher units are objects of investigation.

The impact of dealing with acoustic or linguistic information only is straightforward: in the former case, we cannot say anything on linguistic parameters that might be relevant; in the latter case, vice versa. In speech processing, when both acoustics and linguistics are being modelled, we face either sub-optimal parameter representation if we employ speech recognition, or a great effort of manual transcription and annotation.

On the horizontal axis, we display linguistically meaningful units, from shortest (left) to longest (right). A typical unit in clinical studies is sustained vowels (phones). Mostly, higher units are simply speech signals, either recorded and stored separately, or cut out from longer passages, using more or less 'intelligent' criteria. The strategies chosen of course depend on the phenomena one wants to model; it is less decisive for traits where we segment, than for states where the unit of investigation should be coextensive with the state one wants to model.

Pang and Lee (2008) demonstrate for text-based opinion mining and sentiment analysis the differences in analysis for different textual units. In Batliner *et al.* (2011) it is shown not only that both acoustic and linguistic phenomena contribute to the marking of emotion-related states but also that linguistic information – for instance, whether words are content words or function words – is implicitly modelled in acoustic parameters. In Batliner *et al.* (2010) units of different size such as words versus chunks are demonstrated to have different impact on the performance of emotion classification; Seppi *et al.* (2010) demonstrate that it is stressed syllables that carry emotion information, and not unstressed ones. All these results feature a stronger dovetailing between acoustics and linguistics than one might expect, having only studies in mind that address either/or.

Moreover, there are 'linguistically blind', technically oriented solutions towards smaller units. We can model single frames (mostly with a length of 10 ms), or fractions of longer time units (mostly of seconds), or we can subdivide the whole signal into equally spaced parts. This is a typical 'engineering solution'. Its advantages are that we do not need any other higher-level information. Its disadvantage is, of course, that we might lose information by cutting across meaningful units; this might level out with higher frequencies and longer durations of units.

The third dimension displayed in Figure 3.1 is frequency. High frequency is aimed at, but often we have to face the sparse-data problem. A single instance 'token' is a so-called *hapax legomenon*, a single instance 'type' is simply just the same unit (the vowel /a/, or the filled pause 'uhm', or the word 'stupid' occurring more than once).

Basically, each study within paralinguistics might be placed in the three-dimensional representation of Figure 3.1, according to the units that are modelled and addressed. This is a placement of the *formal* unit of analysis which can be related to the *function* we are interested in. So far, the combination of different (types of) technical segmentations has proved to be very promising, as was shown for emotion processing in Schuller and Rigoll (2006) and in Schuller *et al.* (2008) where absolute time intervals of equal length and relative time intervals were combined in a hierarchical approach. The same holds for higher-level information on syntactically or emotionally consistent chunks (Batliner *et al.* 2010) or for combining word- and utterance-level processing (Batliner *et al.* 2011); it remains to be seen whether this performance gain can be transferred onto fully automatic processing.

3.4   Typical versus Atypical Speech

I'm obsessively opposed to the typical.

(Lady Gaga)

In (computational) paralinguistics, we try to find, collect, annotate, extract features, describe, and contrast the phenomena processed with other ones. These can be regular, 'normal', frequent phenomena, or less normal and less frequent phenomena. Ben-Ze'ev (2000, p. 9) observes a frequent 'confusion of extreme, typical, and common cases of a mental category', as far as emotions are concerned. To speak about (proto)typical emotions can mean to speak about two different things: very extreme – and therefore infrequent – emotions such as utmost rage, or very frequent – and therefore less pronounced – emotions or emotion-related states such as interest or boredom. In automatic speech processing, the notion of 'typical versus atypical' speech seems to be simpler and mostly based on frequency, that is, on the availability of large quantities of data for training. Another use of 'typical' can be observed in speech pathology: a frequent combination is 'typical speech and language development' aiming to describe the normal development of speech and language in a child, and contrasting this with possible deviations thereof ('atypical speech/language'). This allows stigmatised and sometimes less politically correct terms such as 'pathological' to be avoided. For example, children and adults with autism spectrum condition are contrasted with those who are 'typically developed'. Note, however, that there are tendencies to replace 'atypical' with terms such as 'wider spectrum' in order to avoid any negative connotations. Irrespective of such considerations of political correctness, the term 'atypical' should not be taken as implying any assessment but rather as meaning simply 'less frequent'.

In all these constellations, the research interest is either to find out the differences between typical and atypical speech, or to classify cases as belonging to the one or the other class, in order to improve the processing of atypical speech, or to screen and monitor atypical speech. This is nicely illustrated in the editorial of a special issue on 'atypical' speech by Stemmer *et al.* (2010):

> *Unfortunately, the algorithms used in current systems for robust modeling, speaker normalization and adaptation have many limitations, in particular for speech that deviates significantly from the data in the training corpus. Atypical speakers like nonnative speakers, children, or members of the elderly population still lead to much higher error rates in state-of-the-art speech recognizers. . . .*

The phenomena dealt with in this special issue read like a catalogue of paralinguistic topics: speech recognition for non-native speakers (Doremalen *et al.* 2010); mismatches between adults' and children's speech (Ghai and Sinha 2010); breathy and whispery speech, both being special types of phonation, normally indicating speaker idiosyncrasies or specific functions (Ishi *et al.* 2010); speech recognition for the evaluation of voice and speech disorders in head and neck cancer (Maier *et al.* 2010); recognition of children's emotional speech, a combination of two atypical types, children's speech and emotional speech (Steidl *et al.* 2010); and speech recognition of ageing voices (Vipperla *et al.* 2010).

Thus, typicality lies in the eyes of the observer; 'typically developed' children are contrasted with handicapped children, but they are atypical if compared with adults, because they are less studied and thus there are less and smaller databases available with this type of speech.

Speech *per se* is not only typical or atypical, it can be either/or, seen from the point of view of dialogue modelling. In a dialogue – which can be taken as a simplification of a multi-party interaction with only two interaction partners – speech is normally directed at the dialogue partner. However, it can be 'private', 'egocentric', 'self-directed' speech, directed to the speaker herself, or it can be directed to someone other than the dialogue partner, also called 'off-talk' (Oppermann *et al.* 2001). This is a fully normal and adequate use of language; however, it has to be detected in an automatic dialogue system and thus modelled as 'atypical' – if not, misunderstandings might happen. We will come back to this type of speech in Section 5.8.

3.5   Context

Language is not merely a set of unrelated sounds, clauses, rules, and meanings; it is a total coherent system of these integrating with each other, and with behavior, context, universe of discourse, and observer perspective.

(Kenneth L. Pike)

There are different types of context: (1) phonetic/linguistic context, that is, what a speaker produced before or after the unit we want to analyse; (2) multimodal context, that is, which body posture, gestures, and facial gestures the speaker produces concomitantly, synchronously or before and after the unit we want to analyse; (3) immediate situational context, that is, the overall setting (communication partners, type of communication, room characteristics, etc.); and (4) general context in time and space, that is, generally speaking, in which historic/geographic situation the communication partners are – this can be narrow, concentrating on the speaker herself and her personal situation and what she has experienced in recent hours, it can be wide, including macro-sociological and political constellations, and it can simply be be narrowed down to membership of class, etc.

Paradigmatic and syntagmatic relations are basic to linguistic structure. A *paradigmatic* (vertical in Figure 3.2) relationship exists between elements that can substitute each other and result in another regular unit, for instance, the phonemes /g/ and /t/ in g*old* versus t*old*, or the words 'blue' and 'yellow' in *a* blue *shirt* versus *a* yellow *shirt*. A *syntagmatic* (horizontal in Figure 3.2) relationship exists between units that can be combined linearly (sequentially), for instance, *five + brown + bottles*. Moreover, there can be paradigmatic, concomitant relationships between paralinguistic and linguistic parameters, and within paralinguistic parameters, across paralinguistic functions. For instance, pitch excursion can indicate word stress, phrase accent, and/or higher emotional arousal. Breathy voice can indicate some pathology and/or intimacy and attractiveness.

Figure 3.2 illustrates some concordant or discrepant paradigmatic and syntagmatic relationships; for the sake of simplicity, we do not display immediate paradigmatic situational context, and display only linguistic context and not phonetic context such as high/low pitch maximum, modal/breathy voice and suchlike, which could modify the paralinguistic message. The reader can consider the different combinations of linguistic context and which of them are likely or not – given the linguistic message and given the different context in time and space.

An abundance of different types of context can be modelled – we should say 'could be modelled', because the context that has been modelled so far is mostly local or restricted to a few types that are relatively easy to deal with, such as immediate acoustic or linguistic

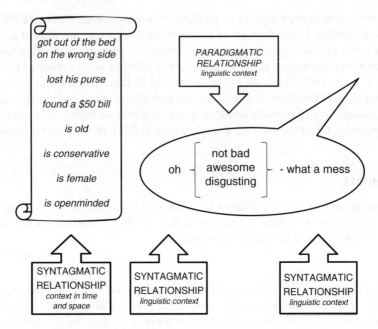

Figure 3.2   Paradigmatic and syntagmatic relationships

context to the left and right, or a separate processing of female and male subjects. Situational or 'historic' context is modelled rather seldom. This is simply due to the complexity of the task of extracting context features and combining this information in automatic processing. This holds especially for any type of wider syntagmatic modelling. We can easily imagine that somebody is more prone to anger when he got out of bed on the wrong side (which is just a metaphorical way of summarising syntagmatic contextual factors); it is almost impossible to model this in experiments or real-life studies, though. It is more promising in more structured settings such as interactions with an automatic call-centre system when the agent only has a limited range of actions available which can be traced back and modelled in relation to the user's reactions (Steidl *et al.* 2004; Walker *et al.* 2000). Moreover, there are quite a few studies on multimodal modelling of context which in itself is rather local (mostly paradigmatic, that is, concomitant); see, for example, the studies on focus of attention referred to in Section 5.8.

3.6   Lab versus Life, or Through the Looking Glass

The two requirements of naturalness and rigorousness appear to occupy the opposite ends of a continuous scale of gradual mutual exclusion. At one extreme we find complete naturalness coupled with complete lack of control; at the other end, complete control over the experimental design at the expense of naturalness and linguistic relevance.

(Lehiste 1963, quoted in Ladd 1996)

Laboratory (lab) speech was the main object of investigation for phonetics and paralinguistics until the 1980s. Branches of theoretical linguistics studied a few, often constructed, examples.

Even when doing field studies, aiming at dialects or less investigated languages, researchers did normally not record 'free' speech but mostly used a structured questionnaire. Moreover, analogue recording and processing devices (such as the oscillograph or the pitch-meter) and early computational processing of speech data normally implied tightly controlled – and not too many – lab data.

With the advent of digital recording and storing devices, and consequently the advent of specially tailored and eventually open-source software for processing large amounts of data, the situation changed; it now was possible to address large databases which did not necessarily have to be of excellent signal quality. However, there is still a divide between proponents of lab speech – who are normally oriented towards theoretical approaches – and proponents of real-life data – who are likely to be interested in applications. Of course, this is a rough picture and does not do justice to everything: for instance, there is research aimed at clinical applications, using highly controlled lab data. However, this is due to practical constraints and not necessarily to a strong inclination towards theory.

The difference between 'lab' and 'life' will accompany us throughout this book. At first sight, this is just a difference in data collection; in fact, it is much more. It pertains to conceptualisation, prototypicality of data, classes to be modelled, relevant features, etc. Often, this pairing goes along with acted versus realistic – but not necessarily: there are lab constellations which guarantee fairly realistic data, for instance in a so-called *Wizard of Oz* setting where the role of the system is played by a human operator (Fraser and Gilbert 1991); and conversely, there are recordings 'in the wild', for instance, from TV shows, where the expression of feelings and emotions is predefined by the stage director and thus acted and regulated by the participants; cf. the convenience databases mentioned by Scherer (2013).

As Xu (2010) puts it:

> *...because it allows systematic experimental control, lab speech is indispensable in our quest to understand the underlying mechanisms of human language. In contrast, although spontaneous speech is rich in various patterns, and so is useful for many purposes, the difficulty in recognizing and controlling the contributing factors makes it less likely than lab speech to lead to true insights about the nature of human speech.*

A discussion along similar lines can be found in Scherer (2013) for emotion research; but see Wilting *et al.* (2006) who use the the same elicitation procedure as Scherer (2013) and argue against using actors for the study of real emotions, based on their findings.

The problem is about experimental control and the observer's paradox (see Section 2.2), that we cannot observe spontaneous behaviour if we observe it – and the description of spontaneous behaviour 'in the wild' should, after all, be the ultimate goal. There is always a lacuna in the argumentation, whether we are describing the metabolism or the behaviour of mice in a cage, or the speech of humans in the lab, and want to extrapolate onto the metabolism or the behaviour of humans in real life. If we are interested in cerebral correlates of speech production, or in the speech of patients with severe disabilities, there is no way out of the lab. If acted (and thus controlled) data are a genuine object of investigation themselves (Section 2.2), there is no doubt at all that they should be investigated. Read data are easily obtainable and can be used, for instance, to augment training databases. Moreover, a sophisticated experimental setting might come close to modelling real-life data but there is no straightforward way to find out whether this is really the case. We do not agree with the claim of Xu (2010) that controlled lab data are necessary for really finding out something. Admittedly, methodologies, procedures, and the size of databases sometimes lag behind. Basically, however, there are means of dealing with noisy and less pronounced data.

Methodologies differ. Whereas the typical methodology for lab speech, especially in basic phonetic or psychological research, is a highly developed analysis of variance meeting the standards of high-impact psychological journals (results meeting alpha below 0.01, etc.), for less controlled real-life speech, highly developed statistical procedures for classification and regression, aiming at high(er) performance in recall, accuracy, unweighted average recall, etc., are employed. We will elaborate on these differences in Section 3.10.

3.7   Sheep and Goats, or Single Instance Decision versus Cumulative Evidence and Overall Performance

When the Son of Man comes in his glory, and all the angels with him, he will sit on his glorious throne. All the nations will be gathered before him, and he will separate the people one from another as a shepherd separates the sheep from the goats. He will put the sheep on his right and the goats on his left.

(Matthew 25: 31–33)

Doddington *et al.* (1998) extended the rudimentary menagerie of sheep and goats in automatic speech recognition (ASR) (Danis 1989) to two further types, for the speaker verification task:

Sheep comprise our default speaker type. . . . [They] dominate the population and systems perform nominally well for them.

Goats . . . are those speakers who are particularly difficult to recognize. . . .

Lambs . . . are those speakers who are particularly easy to imitate.

Wolves . . . are those speakers who are particularly successful at imitating other speakers.

Thus, the sheep are good – because they are easy to recognise; this can be extended to any paralinguistic problem. Goats are bad because they cannot be recognised. Lambs and wolves are surely interesting for specific paralinguistic tasks, not only for modelling speakers for entry control systems. For instance, comedians good at imitating celebrities have the capabilities of wolves, and the more particular the voices of celebrities are, the easier it might be to imitate them (lambs). However, normally we deal with groups of sheep or goats when trying to automatically classify paralinguistic phenomena; single sheep or goats are used for demonstration, or – albeit rather seldom – for detailed error analysis. The measure of success is overall accuracy or recognition, and an incorrect *single instance decision* is not detrimental; however, it can be, in special circumstances: remember Bill Gates' failing to demonstrate USB plug and play, crushing Microsoft Windows 98. Or demonstrating a system at a computer expo in the presence of a very important person, for instance, the prime minister tell him to use the system but his speech is not recognised. Telling a customer of an automated call centre that he is angry and trying to calm him down when he is not can be fatal. The failure of a lie detector in court can have tremendous consequences if its decision is taken as evidence.

Therefore, single instance decisions can be important – for better or for worse. The usual measure, however, is *overall performance*, and the criterion is something like a test of significance in basic psychological research, effect size, or percentage correct in automatic speech processing. All these criteria are based on some convention within the scientific community, and do have their shortcomings. If the number of instances is high, for a two-class problem, practically every second instance can be classified incorrectly, and still the difference can be 'significant'. Effect size should be a means of overcoming this deficit, but in practice it is not transformed into a percentage of instances, classified correctly or incorrectly. It might be a good strategy to conceptually convert any of these measures into a rough but intuitive figure: 'every xth case is classified incorrectly'. For instance, the recognition rate of automatic dictation systems was, at a certain time, above 95% correct, but acceptance was still rather low. So, one incorrect word out of 20 on average made such a bad impression on the customers. We will elaborate on the pros and cons of different measures of performance in Section 3.10.

Standard performance measures in computational paralinguistics are necessary for benchmarks and challenges, to enable strict comparability. The same holds for basic research. However, if we aim at specific applications, we should consider whether we should not, besides the standard measures, conceptually and algorithmically take into account whether we are aiming at good single instance decisions, or at overall assessment. Both points of view can be combined in repeated screening approaches because a sub-optimal one-time performance normally gives way to a more reliable estimation of tendencies if such measures are repeatedly applied.

To give one example, employing the detection of single instances of customers' anger in a call-centre application might be risky because false alarms are inevitable. Using such information in a cumulative way for checking call-centre agents' quality is promising; however, from an ethical point of view, such an application might be questionable.

3.8   The Few and the Many, or How to Analyse a Hamburger

Hamburgers! The cornerstone of any nutritious breakfast.

(Jules Winnfield in *Pulp Fiction*)

The approach of food critics towards the hamburger – if they care for such a thing at all – and the approach of food chemists is strikingly different. The food critic will comment on appearance and especially on taste, pointing out a few pivotal features. The food chemist uses a mass spectrometer and squishes everything, so the result will be some indeterminate porridge that will be analysed chemically. It is not easy to go back, from the mush analysed by the food chemists and their lists of hundreds of ingredients, to the more holistic description of the food critics.

In the historic development of computational paralinguistics, we took the road from the critic to the chemist, from the few to the many. In the 'pre-automatic' phase of emotion modelling (Frick 1985; Lieberman and Michaels 1962) the inventory of acoustic features was more or less predefined or at least inspired by basic (phonetic) research. Note that acoustic and linguistic features will be introduced and explained in detail in Chapters 8 and 9.

Hence, until the 1990s, features were 'hand-picked', expert-driven, and based on phonetic knowledge and models; this was especially true for pitch (contour) features which were often based on intonation models (see Mozziconacci 1998). To give some examples of developments in recent years: at the beginning of a 'real' automatic processing of emotion, Dellaert *et al.* (1996) used 17 pitch features; McGilloway *et al.* (2000) reduced 375 measures to 32 variables as robust markers of emotion; Batliner *et al.* (2000) used 27 prosodic features on the utterance level; Oudeyer (2003) 200 features and information gain for feature reduction; Schuller *et al.* (2005) 276 features and support vector machines with sequential floating forward search (SFFS); and Vogt and André (2005) 1280 features and correlation-based feature subset selection; classifiers and feature selection procedures will be explained in detail in chapter 11. The increase in the number of features employed in the Interspeech challenges 2009–2012 can be taken as paradigmatic example: 384 in the Interspeech 2009 Emotion Challenge (Schuller *et al.* 2009), 1582 in the Interspeech 2010 Paralinguistics Challenge (Schuller *et al.* 2010b), 4368 in the Interspeech 2011 Speaker State Challenge (Schuller *et al.* 2011), and 6125 in the Interspeech 2012 Speaker Trait Challenge (Schuller *et al.* 2012). Within a straightforward brute-force approach, 'more' seems to be better; this can change with a sophisticated feature selection; however, a reduced feature set, especially tailored for a specific data set, might go along with a loss of generalisation if applied for other (types of) data.

The traditional approach towards features was of course influenced by technical constraints and possibilities, and confined to processing only a few (Lieberman and Michaels 1962); later on, types of features were investigated, for instance, only fundamental frequency (F0) features, or prosodic features including intensity and duration modelling as well. Researchers from the ASR community naturally enough often start with the mel-frequency cepstral coefficient (MFCC), when they shift the focus onto paralinguistic phenomena. At the extremes nowadays, on the one hand, we can place very large feature sets within a brute-force approach – running the risk of not being able to say something about the most important features. On the other hand, we can employ just a few features, well known to be relevant to the actual task: for instance, speech tempo, which quite reliably can be modelled with syllables per second, has turned out to be very relevant to the assessment of non-native prosody in particular, and many types of deviant prosody in general. Pitch excursion or pitch mean and intensity are intuitively good predictors for arousal – what goes up must come down: the same way as a fast heart rate, indicating arousal, will slow down in the relaxation phase, a higher pitch mean value will normally drop in more relaxed, neutral speech. Employing just a few features of course does not tell us how far we could get with a more complete feature set. Thus, we badly need studies employing very large feature sets with many different types of features, including feature (type) selection and/or reduction, and especially an estimate of the impact of not only single features but also feature types. This can be done in a sort of univariate approach (see Schuller *et al.* 2007) where information gain ratio has been used for obtaining most important features within the same feature type; Batliner *et al.* (2011) used a sort of multivariate approach by doing SFFS for all feature types together and determining the 'share' of each feature type, that is, the percentage of features belonging to each feature type and surviving the feature selection procedure. In both cases, the FAU Aibo database was used. Classification performance for the different feature types was similar for both univariate and multivariate approaches. The multivariate approach also allows 'trading relations' (interdependencies) to be found, for instance, between acoustic and linguistic feature types. A straightforward example is unnormalised duration which is highly interrelated with word class, especially with content word (such as noun, adjective, or verb) versus function word (such as particle, pronoun, or interjection): content words are normally longer than function words and, at the same time, semantically more important. A similar approach is used in (Busso and Rahman 2012).

3.9 Reifications, and What You are Looking for is What You Get

There is an error; but it is merely the accidental error of mistaking the abstract for the concrete. It is an example of what I will call the 'Fallacy of Misplaced Concreteness'.

(Whitehead 1925, p. 52)

Life is largely a matter of expectation.

(Horace)

In this section, we comment on two fallacies that sometimes cannot be avoided; however, we should at least be aware of them.

The first is the reification (called the 'Fallacy of Misplaced Concreteness' by Whitehead) of terms and concepts. Let us discuss an example from prosody. ToBI ('TOnes and Break Indices') is a phonologically inspired intonation model (Silverman *et al.* 1992), characterised by a small set of conventions for describing prosodic grouping (breaks) and tonal events (tones). In this model, an H*L tone, for example, characterises an F0 (pitch) peak on a stressed syllable, followed by a falling pitch movement. However, we should always be aware that we do not produce a ToBI tone H*L – we produce something that can be described, within one specific model, with this terminology. The same holds for the much debated question of what an emotion is, after all: if we call almost everything 'emotional intelligence', then this use might be unfortunate but it cannot be wrong in itself. Terminology is a matter of convention, not of truth. (This recalls the famous distinction made by Householder (1952) between two extreme positions on the 'metaphysics of linguistics' – the *God's truth* position (a language *has* a structure) and the *hocus-pocus* position (the linguist *imposes* some sort of structure onto formless data). In their pure form, these positions are outdated; however, they are reflected in present-day theoretical debates.)

The second fallacy is to restrict the object of investigation to a specific phenomenon, a specific selection of subjects and tasks, and/or to restrict the features employed to only a few, and then to make claims going far beyond this specific constellation. It is a commonplace that very often, a specific type of (young, male, middle-class, Protestant, white/'Caucasian') American student is taken as a subject in an experiment, and, based on these results, statements about 'the human' are made.

The problem can be compared, technically speaking, with a situation where there is too much weight on testing the alternative hypothesis against a null hypothesis which simply states that there is not such a difference as formulated in the alternative hypothesis. Instead, we could consider more and different alternative hypotheses. Let us take the case of pitch/F0 against other prosodic parameters as an example. The term 'pitch accent' is rather unfortunate; as Boves *et al.* (1984, p. 20) claim:

There is, however, a real danger in a research strategy that concentrates fully on a single factor at the expense of all others . . . The fact that many authors confine the meaning of the term 'intonation' to pitch movement without any references whatsoever to other factors can be taken as a proof.

Nowadays, the term 'prosody' is normally used generically for pitch, energy, duration, and sometimes voice quality, whereas 'intonation' is confined to pitch phenomena. The problem is that even within intonation models, pitch cannot stand alone: it is naturally correlated with energy and duration, and it cannot be described at all without recourse to the time axis; in ToBI, this is basically a sequence of tones, that is, a coarse-grained before/after relationship (Batliner and Möbius 2005).

Thus, what you are looking for is what you get. In parallel to the acronym WYSIWYG (what you see is what you get), we can call this 'WYALFIWYG' (Batliner 1989). If you only look for the relevance of intonational (i.e., pitch) features, you will only get results for pitch features. This primacy of pitch resulted in a wealth of different intonation models, ToBI being arguably the most prominent, at the cost of the other prosodic factors. Falling for this WYALFIWYG phenomenon obscures the contributions of other prosodic parameters. In spite of several studies that employed more prosodic parameters apart from pitch, showing that pitch is not the most relevant parameter for phenomena such as boundaries or accents (see, for example, Batliner *et al.* 1999 2001; Kochanski *et al.* 2005), there are still (too) many studies conducted that only deal with pitch. We do not want to belittle the impact of pitch; we just want to point out that it is an empirical question which feature type(s) contribute(s) to what extent – which only can be answered reliably if we do not confine our approaches to just one out of several feature types.

3.10   Magical Numbers versus Sound Reasoning

All is number.

(Pythagoras)

A good decision is based on knowledge and not on numbers.

(Plato)

'The Answer to the Great Question... Of ... Life, the Universe and Everything... Is ... Forty-two.'

(Douglas Adams, *The Hitch-Hiker's Guide to the Galaxy*)

During corpus engineering (see Chapter 6) and the evaluation of results, based on the processing of our corpora, we are sort of 'accompanied by numbers' – considerations on the necessary number of speakers, utterances, labellers, necessary 'threshold values' such as levels of significance, and the semantics of these numbers. Here, we cannot give a full account of such measures and their pros and cons. However, we want to sketch the basic problems, give references, and suggest ways out.

All these numbers are not 'God's truth' measures (Section 3.9) but based on conventions that are more or less accepted in the respective fields of academia. As far as we can see, the stronger the measures (thresholds) are, the more debated they are. We want to deal with these measures under three headings: *rules of thumb* refer to (rather vague) requirements of the minimum/maximum number of speakers, utterances, labellers, features, and the like; *weak thresholds* refer to evaluation measures such as kappa, effect size, and classification performance; *strong thresholds* refer to tests of significance, normally the so-called null hypothesis decision or null hypothesis testing procedure.

Rules of Thumb

There is no agreed-upon catalogue of the minimum (or maximum) number of speakers, items (words, sentences, utterances, paragraphs on the one hand, and phenomena such as specific emotions and non-verbals on the other hand), annotators and annotations. The basic maxim is: the more, the better. A number of speakers equal to one would be rather exotic; however, there exist several well-known databases with a number of speakers well below 10 and there are some databases with only one or two annotators. Of course, all this depends on the availability of speakers and the difficulties of recording. To illustrate with two constellations: native or even non-native English recordings of adult speakers are relatively easy to obtain, especially if the data are scripted; on the other hand, non-scripted recordings of non-typically developed children are not that easy to obtain. Gibbon *et al.* (1997, pp. 107ff.) mention, as representative of automatic speech processing, one to five speakers for basic research and speech synthesis systems, five to 50 for experimental research, and more than 50 for speech recognition or speaker verification. These figures could constitute a rough guideline for computational paralinguistics as well. As for annotators, more than two will allow for majority voting, thus three could be taken as minimum requirement; again, it depends on the difficulty of the task – and perhaps simply on financial means – how many annotators we eventually can afford and do 'need'. For the perceptual evaluation of non-native prosody, Hönig *et al.* (2010 2012) showed that the loss of performance is low when we employ five annotators (compared to up to 60 annotators); this holds both for 'experts' (trained phoneticians) and 'naïve' labellers. For most tasks, employing some 10 annotators seems to be well above the minimum requirement and allows for outliers as well.

As far as number of cases/items versus number of features ('predictors') is concerned, we are often faced with a sparse-data problem on the one hand – especially when we aim for realistic data, some very interesting phenomena are simply rare – and with (too) many features, especially when we employ a brute-force approach. There exist some rules of thumb such as 'at least twice as many cases as there are features' – which are often violated. Such violations might not hurt for classifiers such as support vector machines (cf. Chapter 11) and simple linear classifiers; due to their nature, it certainly works for random forests because here, many decision trees are employed, each of them with only a few features out of the whole feature vector. Eventually, it is a matter of experience – and of scrutinising one's own results – whether such violations can be accepted or must be avoided.

Weak Thresholds

Arguably, of most interest for computational paralinguistics is finding an answer to the question 'How good are we?' because this tells us to what extent our models fit the data and whether we might be able to employ these models successfully in real-life applications. Typical evaluation measures are (different varieties of) kappa for inter-labeller correspondence, correlation measures (Pearson's r or Spearman's rho), and different measures of 'effect size', such as Cohen's d. Classification performance is mostly given as 'percentage correctly classified'. We now want to characterise these measures on an exemplary basis; note that several other measures are used besides, such as the so-called area under the 'receiver operating characteristic' (ROC) curve; see Section 11.3.3.

Kappa, alpha. Cohen's kappa (Cohen 1960, 1968) is a measure of inter-labeller agreement for categorical ratings with values between 1 (complete agreement) and 0 (no agreement) which was introduced into (computational) linguistics by Carletta (1996). An alternative for more than two raters is given in Fleiss (1971). Another alternative for different levels of measurement, again with values between 1 and 0, was introduced by Krippendorff (1970); see also Krippendorff (2004). An in-depth account of kappa is given in Gwet (2010). For Cronbach's alpha, see Cronbach (1951).

Correlation. Pearson's product-moment correlation coefficient r (Pearson 1901) measures the correlation, that is, the linear dependency, between two variables on an interval scale; a non-parametric alternative for ordinal data is Spearman's rank correlation coefficient rho (Spearman 1904). Both can have values between $+1$ and -1, ± 1 meaning a perfect correlation, 0 meaning no correlation at all.

Effect size. This gives a quantitative estimate of the difference between two groups and of the strength of a phenomenon (mirrored in this difference). Note that both correlation measures (r and rho) as well as the so-called 'explained variance' (r^2) (O'Grady 1982) and measures of classification performance estimate effect size as well. Here, we will only address one measure, namely the standardised mean difference between two groups, Cohen's d (Cohen 1992); see also Coe (2002), Ferguson (2009), Durlak (2009) and Ellis (2010). Fritz *et al.* (2012) point out that although estimates of effect size '... are useful for determining the practical or theoretical importance of an effect...' and are called for in the publication manuals of psychological journals, they are not used that often: 'The most often reported analysis was analysis of variance, and almost half of these reports were not accompanied by effect sizes.' Cohen's d can have values well above 1, although most often values below 1 are reported; we will come back to this below.

Classification performance. Weighted average recall (WAR), that is, the sum of the number of cases in the diagonal of the confusion matrix divided by the overall number of all cases, and *unweighted average recall* (UAR), that is, the unweighted (by number of instances in each class) mean of the percentage correctly classified in the diagonal of the confusion matrix, are two different measures for 'percentage correctly classified' (see Section 11.3.3). The chance baseline for two-class problems is either constituted by the larger class for WAR, or by 50% for UAR; we will concentrate on UAR because for this measure, chance level is only defined by the number of classes (50% for two classes, 33.3% for three classes, 25% for four classes, and so on) and not by the number of cases per class which varies across experiments. Comparable to correlations, kappa, and other measures, UAR has values up to 1 (or 100%). For the sake of the argument, in this section, we will assume a balanced distribution of the two classes, meaning that WAR = UAR.

Note that in different fields such as engineering, psychology, or clinical studies, there exist different terms for measures that combine and summarise the cell entries in a confusion matrix, especially for the standard setting of two classes, such as recall/hit, false alarms, true/false positives/negatives, and recall or precision or sensitivity or specifity (see Section 11.3.3). UAR was introduced for the processing of prosodic phenomena under the name of 'class-wise averaged recall' in Batliner *et al.* (1998). A full account of all these measures is, however, beyond the scope of this short section; some are dealt with in more detail below.

Strong Thresholds

Null hypothesis testing (NHT) or *null hypothesis decision* (NHD) is the preferred method in basic psychological research and related fields; p-values below a certain level (0.05, 0.01, or 0.001) are the preferred measure and function as strong thresholds. If p-values are above these levels, results are mostly deemed to be not significant, in both sense of the word: not statistically significant and not relevant. NHT has been critically evaluated for decades. According to Rozeboom (1960), the 'most basic error [of the NHT procedure] lies in mistaking the aim of a scientific investigation to be a *decision*, rather than a *cognitive* evaluation of propositions'. This criticism has been repeated by Cohen (1994), Gigerenzer (2004), and Hubbard and Lindsay (2008), amongst many others. These and other authors mention several deficiencies of NHT, amongst them that: (1) samples are almost never randomly drawn from the population; (2) the rejection of the null hypothesis does not affirm the alternate hypothesis; (3) p is not the probability that the null hypothesis is false; (4) levels of significance are pure conventions; (5) large sample sizes will almost always produce 'significant' results even if the effect is very low or non-existent. (Note that significance tests are dealt with in Section 11.3.4.) Adjustments of the significance level in the case of repeated measurements are often not done; however, these are problematic themselves (Pernegger 1998). Cumming *et al.* (2007) demonstrate that NHT still prevails in psychological journals – whereas they are not found that often in computational studies.

Instead, at least exact p-values should be reported, error bars and/or confidence intervals, and effect sizes given in combination with descriptive statistics. Confidence intervals 'can be...conceptually defined as a range of plausible values for the corresponding parameter (i.e., for the unstandardized size of the effect in the population' (Beaulieu-Prévost 2006, p. 12). They are advocated by Beaulieu-Prévost (2006), Brandstätter (1999) and Wilkinson (1999). A few simple significance tests are explained in more detail in Section 11.3.4.

The Semantics of Numbers

Different wordings are used but the semantics of the values for correlation, inter-labeller correspondence (kappa), and effect size (note that this can change for different computations of effect size) are similar; the following scale is often used for the assessment of correlations: 0.0–0.2, very low; 0.2–0.4, low; 0.4–0.6, medium; 0.6–0.8, good; and 0.8–1.0, excellent. Although d can have values far above 1.0, Cohen (1992) mentions the following ranking: 0.2, small; 0.5, medium; and 0.8, large. Note that such rankings were typically not intended to be taken literally – they more or less evolved from very weak into stronger thresholds; for instance, a value of 0.39 will be rightly interpreted differently from a value of 0.21 although both are in the same interval. Such evaluation metrics are often required by journal editors and reviewers, and they definitely should be computed in order to get a feeling for one's own data and processing steps. However, they have to be interpreted with care: a very low value, for instance for inter-labeller correspondence, will indicate some severe problems. Too high a value, on the other hand, can indicate a trivial problem, or simply a precisely defined task that may or may not be adequate for the phenomenon you want to model. A moderately low value can simply indicate a difficult labelling task. A trivial example is to assign the label 'angry' if you encounter the word *angry* or *anger*. This will yield a perfect inter-labeller correspondence but can simply be wrong if the preceding context is a negation (*not angry*), and one definitely will miss some instances of anger if it is expressed by different means. A less trivial example is the labelling of tones within the ToBI approach which is an intonation model where pitch values are mapped onto a small number of 'high' and 'low' tones; good inter-labeller correspondence might mean that the formal differences can be learnt quite well but we do not know whether functional differences correspond exactly to the same thresholds (Batliner and Möbius 2005).

A very interesting relationship of effect size measures with classification performance is discussed in Coe (2002):

> *Another way to conceptualise the overlap [of two experimental groups] is in terms of the probability that one could guess which group a person came from, based only on their test score – or whatever value was being compared. If the effect size were 0 (i.e., the two groups were the same) then the probability of a correct guess would be exactly a half – or 0.50. With a difference between the two groups equivalent to an effect size of 0.3, there is still plenty of overlap, and the probability of correctly identifying the groups rises only slightly to 0.56. With an effect size of 1, the probability is now 0.69, just over a two-thirds chance. . . . It is clear that the overlap between experimental and control groups is substantial (and therefore the probability is still close to 0.5), even when the [effect size] is quite large.*

The relationship of correlation r and effect size d with classification performance for a two-class problem is displayed in Table 3.1; for the assumptions needed for converting these values into each other, see Coe (2002). At first sight, this seems strange: an effect size greater than 0.8 is normally considered to be very good, yet it only corresponds to a classification performance of 0.66 in a balanced two-class problem. However, such an effect is definitely important if it pertains, for instance, to differences in the income of males and females, morbidity after cancer treatment, or – to illustrate with an example from computational paralinguistics – differences in the assessment of non-native traits between groups with or without specific pronunciation training in a given L2. In contrast, for instance, if a customer has to be convinced that automatic dictation systems are a good thing to use, 66% correct (i.e., every third word is incorrect) is a rather devastating outcome. Here again, we see the difference between single instance decisions and cumulative evidence (see Section 3.7).

A classification performance of 56% correct for a two-class-problem will quite often be statistically significantly different from chance level. This corresponds to an effect size of 0.3 – which is considered to be rather low. Even an effect size of 0.8, which is considered

Table 3.1 Relationship of correlation r and effect size d with classification performance (here, WAR = UAR) for two classes, according to Coe (2002)

Correlation r	Effect size d	Classification performance
0.0	0.0	0.50
0.10	0.2	0.54
0.15	0.3	0.56
0.29	0.6	0.62
0.37	0.8	0.66
0.45	1.0	0.69
0.57	1.4	0.76
0.71	2.0	0.84
0.83	3.0	0.93

to be good, corresponds only to 66% correct. Again, this does not mean that such results are useless. However, most of the time, either only significance is reported (in humanities), or percentage correct (in engineering approaches); neither tells the whole story. It depends on the (application) task whether such results are 'only for the records', or useful in the long run for science, or in the short run for some application task.

In computational approaches, it is nowadays quite common to evaluate the effect size used, that is, classification performance, not only in terms of absolute values (such as 57% for a two-class problem as 'poor', 73% as 'medium', or 95% as 'very good') but also relative to some baseline that has been defined by the authors themselves or by other authors. For instance, any performance obtained for the recognition of acted emotions cannot be realistically expected for realistic, spontaneous emotions. Often it is claimed that differences between new values obtained and such baselines should be proven to be 'statistically significant'. However, confidence intervals can be a reasonable alternative or addition (see Seppi *et al.* 2010).

Thus, we are faced with different traditional measures within, on the one hand, psychological, sociological, or medical studies, and on the other hand, studies within computer science or engineering, even if the topic can be very similar or identical: on the one hand, statistical significance from NHT, only sometimes with concomitant effect sizes, and on the other hand, classification performance (an effect size), only sometimes complemented with information on 'significant' differences between measures. Thus, NHT is primary in basic and to some extent in clinical studies, and secondary in computational approaches; for effect size measures (classification performance), it is the other way round. Note that this does not hold for correlations because these measure effect size by default.

A first step towards bridging the gap between all these disciplines would be to always report both types of measures, and especially relate the 'classic' effect size measures to classification performance as used within computer science (cf. Table 3.1). In classification, the *confusion matrix* is the 'mother of measures' and should be given if possible, ideally with number of test instances per class in a separate column/row, or directly in numbers of test instances; this provides the complete basis of calculation. All this will also help harmonise the different semantics of these different measures, including *p*-values, across fields, to put into perspective a 'statistically significant *p*-value' if the corresponding effect size is pretty low, and especially to level out the different semantics of 'good effect sizes' and 'low classification performance' if, in fact, they are equivalent (cf. Table 3.1).

Last but not least, we want to point out that decimal places have their own semantics in paralinguistic research: care should be taken not to report too many decimal places just because the software provides them. If, for instance, we report percentage correct for less than 100 experimental subjects (our 'cases'), then even one decimal place does not really make much sense – what is the semantics of 0.3 persons? In this case, two or three decimal places are simply nonsensical.

As a more precise rule of thumb, the number of decimal places N_{dec} in relation to the total number of test data instances N_{Test} can be determined as:

$$N_{dec} = \lfloor \log_{10} N_{Test} \rfloor - 1 \quad \text{and} \quad N_{Test} \geq 10.$$

Moreover, all these numbers should be given in a consistent way, that is, always with the same number of decimal places throughout the study, and always either rounded up or rounded down.

References

Adelhardt, J., Frank, C., Nöth, E., Shi, R. P., Zeissler, V., and Niemann, H. (2006). Multimodal emogram, data collection and presentation. In W. Wahlster (ed.), *SmartKom: Foundations of Multimodal Dialogue Systems*, volume 1, pp. 597–602. Springer, Berlin.

Allport, F. H. and Allport, G. W. (1921). Personality traits: Their classification and measurement. *Journal of Abnormal and Social Psychology*, **16**, 6–40.

Allport, G. W. (1927). Concepts of trait and personality. *Psychological Bulletin*, **24**, 284–293.

Allport, G. W. and Odbert, H. S. (1936). Trait-names: A psycho-lexical study. *Psychological Monographs*, **47**, i–171.

Arnold, M. B. (1960). *Emotion and Personality. Vol. 1, Psychological Aspects*. Columbia University Press, New York.

Averill, J. R. (1980). A constructivist view of emotion. In R. Plutchik and H. Kellerman (eds), *Emotion: Theory, Research and Experience*, pp. 305–339. Academic Press, New York.

Batliner, A. (1989). Eine Frage ist eine Frage ist keine Frage. Perzeptionsexperimente zum Fragemodus im Deutschen. In H. Altmann, A. Batliner, and W. Oppenrieder (eds), *Zur Intonation von Modus und Fokus im Deutschen*, pp. 87–109. Niemeyer, Tübingen.

Batliner, A. and Möbius, B. (2005). Prosodic models, automatic speech understanding, and speech synthesis: Towards the common ground? In W. Barry and W. Dommelen (eds), *The Integration of Phonetic Knowledge in Speech Technology*, pp. 21–44. Springer, Dordrecht.

Batliner, A., Kompe, R., Kießling, A., Mast, M., Niemann, H., and Nöth, E. (1998). M = Syntax + Prosody: A syntactic–prosodic labelling scheme for large spontaneous speech databases. *Speech Communication*, **25**(4), 193–222.

Batliner, A., Buckow, J., Huber, R., Warnke, V., Nöth, E., and Niemann, H. (1999). Prosodic feature evaluation: Brute force or well designed? In *Proc. of ICPhS*, volume 3, pp. 2315–2318, San Francisco.

Batliner, A., Fischer, K., Huber, R., Spilker, J., and Nöth, E. (2000). Desperately seeking emotions: Actors, wizards, and human beings. In *Proc. of the ISCA Workshop on Speech and Emotion*, pp. 195–200, Newcastle, Co. Down.

Batliner, A., Buckow, J., Huber, R., Warnke, V., Nöth, E., and Niemann, H. (2001). Boiling down prosody for the classification of boundaries and accents in German and English. In *Proc. of Eurospeech*, pp. 2781–2784, Aalborg.

Batliner, A., Seppi, D., Steidl, S., and Schuller, B. (2010). Segmenting into adequate units for automatic recognition of emotion-related episodes: a speech-based approach. *Advances in Human-Computer Interaction*. 15 pp.

Batliner, A., Steidl, S., Schuller, B., Seppi, D., Vogt, T., Wagner, J., Devillers, L., Vidrascu, L., Aharonson, V., and Amir, N. (2011). Whodunnit: Searching for the most important feature types signalling emotional user states in speech. *Computer Speech and Language*, **25**, 4–28.

Beaulieu-Prévost, D. (2006). Confidence intervals: From tests of statistical significance to confidence intervals, range hypotheses and substantial effects. *Tutorials in Quantitative Methods for Psychology*, **2**, 11–19.

Ben-Ze'ev, A. (2000). *The Subtlety of Emotions*. MIT Press, Cambridge, MA.

Boves, L., Have, B. L. t., and Vieregge, W. H. (1984). Automatic transcription of intonation in Dutch. In D. Gibbon and H. Richter (eds), *Intonation, Accent and Rhythm. Studies in Discourse Phonology*, pp. 20–45. de Gruyter, Berlin.

Brandstätter, E. (1999). Confidence intervals as an alternative to significance testing. *Methods of Psychological Research Online*, **4**, 33–46.

Busso, C. and Rahman, T. (2012). Unveiling the acoustic properties that describe the valence dimension. In *Proc. of Interspeech*, pp. 1179–1182, Portland, OR, USA.

Carletta, J. (1996). Assessing agreement on classification tasks: the kappa statistic. *Computational Linguistics*, **22**, 1–6.

Chaplin, W. F., John, O. P., and Goldberg, L. R. (1988). Conceptions of states and traits: Dimensional attributes with ideals as prototypes. *Journal of Personality and Social Psychology*, **54**, 541–557.

Clore, G. L. and Ortony, A. (2000). Cognition in emotion: Always, sometimes, or never? In R. D. Lane and L. Nadel (eds), *Cognitive Neuroscience of Emotion*, pp. 24–61. Oxford University Press, Oxford.

Coe, R. (2002). It's the effect size, stupid: What effect size is and why it is important. Paper presented at the Annual Conference of the British Educational Research Association, University of Exeter, England, 12–14 September, accessed from www.leeds.ac.uk/educol/documents/00002182.htm on 5 October 2012.

Cohen, J. (1960). A coefficient of agreement for nominal scales. *Educational and Psychological Measurement*, **20**, 37–46.

Cohen, J. (1968). Weighted kappa: Nominal scale agreement with provision for scaled disagreement or partial credit. *Psychological Bulletin*, **70**(4), 213–220.

Cohen, J. (1992). A power primer. *Psychological Bulletin*, **112**, 155–159.

Cohen, J. (1994). The earth is round ($p < .05$). *American Psychologist*, **49**, 997–1003.

Cornelius, R. R. (2000). Theoretical approaches to emotion. In *Proc. of the ISCA Workshop on Speech and Emotion*, pp. 3–10, Newcastle, Co. Down. ISCA.

Cowie, R., Sussman, N., and Ben-Ze'ev, A. (2011). Emotion: Concepts and definitions. In P. Petta, C. Pelachaud, and R. Cowie (eds), *Emotion-Oriented Systems: The Humaine Handbook*, Cognitive Technologies, pp. 9–30. Springer, Berlin.

Cronbach, L. J. (1951). Coefficient alpha and the internal structure of tests. *Psychometrika*, **16**(3), 297–334.

Cumming, G., Fidler, F., Leonard, M., Kalinowski, P., Christiansen, A., Kleinig, A., Lo, J., McMenamin, N., and Wilson, S. (2007). Statistical reform in psychology: Is anything changing? *Psychological Science*, **18**, 230–232.

Danis, C. M. (1989). Goats to sheep: can recognition rate be improved for poor Tangora speaker? In *Proc. of the Workshop on Speech and Natural Language Association for Computational Linguistics HLT '89*, pp. 145–150, Stroudsburg, PA, USA.

Darwin, C. (1872). *The Expression of the Emotions in Man and Animals*. John Murray, London.

Dellaert, F., Polzin, T., and Waibel, A. (1996). Recognizing emotion in speech. In *Proc. of ICSLP*, pp. 1970–1973, Philadelphia.

Digman, J. M. (1990). Personality structure: Emergence of the five-factor model. *Annual Review of Psychology*, **41**, 417–440.

Doddington, G., Liggett, W., Martin, A., Przybocki, M., and Reynolds, D. (1998). SHEEP, GOATS, LAMBS and WOLVES. A statistical analysis of speaker performance in the NIST 1998 Speaker Recognition Evaluation. In *Proc. of ICSLP*, Sydney.

Doremalen, J. v., Cucchiarini, C., and Strik, H. (2010). Optimizing automatic speech recognition for low-proficient non-native speakers. *EURASIP Journal on Audio, Speech, and Music Processing*, **2010**.

Durlak, J. A. (2009). How to select, calculate, and interpret effect sizes. *Journal of Pediatric Psychology*, **34**(9), 917–928.

Ekman, P., Friesen, W. V., O'Sullivan, M., Chan, A., Diacoyanni-Tarlatzis, I., Heider, K., Krause, R., LeCompte, W. A., Pitcairn, T., Ricci-Bitti, P. E., Scherer, K. R., Tomita, M., and Tzavaras, A. (1987). Universals and cultural differences in the judgments of facial expressions of emotion. *Journal of Personality and Social Psychology*, **53**, 712–717.

Ellis, P. D. (2010). *The Essential Guide to Effect Sizes*. Cambridge University Press, Cambridge.

Endler, N. S. and Kocovski, N. L. (2001). State and trait anxiety revisited. *Journal of Anxiety Disorders*, **15**, 231–245.

Eysenck, H. (1991). Dimensions of personality: 16, 5 or 3? – Criteria for a taxonomic paradigm. *Personality and Individual Differences*, **12**, 773–790.

Ferguson, C. J. (2009). An effect size primer: A guide for clinicians and researchers. *Professional Psychology: Research and Practice*, **40**, 532–538.

Fleiss, J. L. (1971). Measuring nominal scale agreement among many raters. *Psychological Bulletin*, **76**, 378–382.

Flett, G. L. (2007). *Personality Theory & Research*. Wiley, Mississauga, ON.

Fraser, N. and Gilbert, G. (1991). Simulating speech systems. *Computer Speech and Language*, **5**(1), 81–99.

Frick, R. (1985). Communicating emotion: the role of prosodic features. *Psychological Bulletin*, **97**, 412–429.

Fritz, C. O., Morris, P. E., and Richler, J. J. (2012). Effect size estimates: Current use, calculations, and interpretation. *Journal of Experimental Psychology: General*, **141**, 2–18.

Ghai, S. and Sinha, R. (2010). Exploring the effect of differences in the acoustic correlates of adults' and children's speech in the context of automatic speech recognition. *EURASIP Journal on Audio, Speech, and Music Processing*, **2010**.

Gibbon, D., Moore, R., and Winski, R. (eds) (1997). *Handbook of Standards and Resources for Spoken Language Systems*. Mouton de Gruyter, Berlin.

Gigerenzer, G. (2004). Mindless statistics. *Journal of Socio-Economics*, **33**, 587–606.

Gwet, K. L. (2010). *Handbook of Inter-Rater Reliability*. Advanced Analytics, Gaithersburg, MD.

Hönig, F., Batliner, A., Weilhammer, K., and Nöth, E. (2010). How many labellers? Modelling inter-labeller agreement and system performance for the automatic assessment of non-native prosody. In *Proc. of SLATE*, Tokyo.

Hönig, F., Batliner, A., and Nöth, E. (2012). Automatic assessment of non-native prosody annotation, modelling and evaluation. In *Proc. of the International Symposium on Automatic Detection of Errors in Pronunciation Training (ISADEPT)*, pp. 21–30, Stockholm.

Householder, F. W. (1952). Review of: Methods in Structural Linguistics. Zellig S. Harris. University of Chicago Press, 1951. *International Journal of American Linguistics*, **18**, 260–268.

Hubbard, R. and Lindsay, R. M. (2008). Why P values are not a useful measure of evidence in statistical significance testing. *Theory and Psychology*, **18**, 68–88.

Ishi, C. T., Ishiguro, H., and Hagita, N. (2010). Analysis of the roles and the dynamics of breathy and whispery voice qualities in dialogue speech. *Journal on Audio, Speech, and Music Processing*, **2010**.

James, W. (1884). What is an emotion? *Mind*, **9**, 188–205.

John, O. P. and Srivastava, S. (1999). The big five trait taxonomy: History, measurement and theoretical perspectives. In L. A. Pervin and O. P. John (eds), *Handbook of Personality Theroy and Research*, pp. 102–138. Guilford Press, New York.

Juslin, P. N. and Scherer, K. R. (2005). Vocal expression of affect. In J. A. Harrigan, R. Rosenthal, and K. R. Scherer (eds), *New Handbook of Methods in Nonverbal Behavior Research*, pp. 65–135. Oxford University Press, Oxford.

Kochanski, G., Grabe, E., Coleman, J., and Rosner, B. (2005). Loudness predicts prominence; fundamental frequency lends little. *Journal of the Acoustical Society of America*, **11**, 1038–1054.

Krippendorff, K. (1970). Estimating the reliability, systematic error and random error of interval data. *Educational and Psychological Measurement*, **30**(1), 61–70.

Krippendorff, K. (2004). Reliability in content analysis. Some common misconceptions and recommendations. *Human Communication Research*, **30**(3), 411–433.

Ladd, D. R. (1996). Introduction to Part I. 1.1. Naturalness and spontaneous speech. In Y. Sagisaka, N. Campell, and N. Higuchi (eds), *Computing Prosody. Approaches to a Computational Analysis and Modelling of the Prosody of Spontaneous Speech*, pp. 3–6. Springer, New York.

Lehiste, I. (1963). Review of K. Hadding-Koch, Acoustic-phonetic studies in the intonation of Southern Swedish. Lund: Gleerup. *Language*, **39**, 352–360.

Lieberman, P. and Michaels, S. B. (1962). Some aspects of fundamental frequency and envelope amplitude as related to the emotional content of speech. *Journal of the Acoustical Society of America*, **34**(7), 922–927.

Locke, E. A. (2005). Why emotional intelligence is an invalid concept. *Journal of Organizational Behavior*, **26**, 425–431.

Maier, A., Haderlein, T., Stelzle, F., Nöth, E., Nkenke, E., Rosanowski, F., Schützenberger, A., and Schuster, M. (2010). Automatic speech recognition systems for the evaluation of voice and speech disorders in head and neck cancer. *EURASIP Journal on Audio, Speech, and Music Processing*, **2010**. 7 pp.

Mayer, J. D., Salovey, P., and Caruso, D. R. (2008). Emotional intelligence – new ability or eclectic traits? *American Psychologist*, **63**, 503–517.

McCrae, R. R. and John, O. P. (1992). An introduction to the five-factor model and its applications. *Journal of Personality*, **60**, 175–215.

McGilloway, S., Cowie, R., Doulas-Cowie, E., Gielen, S., Westerdijk, M., and Stroeve, S. (2000). Approaching Automatic recognition of emotion from voice: A rough benchmark. In *Proc. of the ISCA Workshop on Speech and Emotion*, pp. 207–212, Newcastle.

Mower, E., Mataric, M. J., and Narayanan, S. S. (2011). A framework for automatic human emotion classification using emotion profiles. *IEEE Transactions on Audio, Speech and Language Processing*, **19**, 1057–1070.

Mozziconacci, S. (1998). *Speech variability and emotion: production and perception*. Ph.D. thesis, Technical University Eindhoven.

Norman, W. T. (1967). 2800 personality trait descriptors: Normative operating characteristics for a university population. The University of Michigan. College of Literature, Science, and the Arts, Department of Psychology.

O'Grady, K. E. (1982). Measures of explained variance: Cautions and limitations. *Psychological Bulletin*, **92**, 766–777.

Oppermann, D., Schiel, F., Steininger, S., and Beringer, N. (2001). Off-talk – a problem for human-machine-interaction. In *Proc. of Eurospeech*, pp. 2197–2200, Aalborg.

Oudeyer, P.-Y. (2003). The production and recognition of emotions in speech: Features and algorithms. *International Journal of Human-Computer Studies*, **59**(1-2), 157–183.

Pang, B. and Lee, L. (2008). Opinion mining and sentiment analysis. *Foundations and Trends in Information Retrieval*, **2**(1–2), 1–135.

Pearson, K. (1901). On lines and planes of closest fit to systems of points in space. *Philosophical Magazine*, **2**, 559–572.

Pernegger, T. V. (1998). What's wrong with Bonferroni adjustment. *British Medical Journal*, **316**, 1236–1238.

Picard, R. (1997). *Affective Computing*. MIT Press, Cambridge, MA.

Rammstedt, B. and John, O. P. (2007). Measuring personality in one minute or less: A 10-item short version of the Big Five Inventory in English and German. *Journal of Research in Personality*, **41**, 203–212.

Reisenzein, R. and Weber, H. (2009). Personality and emotion. In P. J. Corr and G. Matthews (eds), *The Cambridge Handbook of Personality Psychology*, pp. 54–71. Cambridge University Press, Cambridge.

Revelle, W. and Scherer, K. (2009). Personality and emotion. In *Oxford Companion to the Affective Sciences*, pp. 1–4. Oxford University Press, Oxford.

Rozeboom, W. (1960). The fallacy of the null-hypothesis significance test. *Psychological Bulletin*, **57**, 416–428.

Salovey, P. and Mayer, J. D. (1990). Emotional Intelligence. *Imagination, Cognition and Personality*, **9**, 185–211.

Scherer, K. R. (2003). Vocal communication of emotion: A review of research paradigms. *Speech Communication*, **40**, 227–256.

Scherer, K. R. (2005). What are emotions? And how can they be measured? *Social Science Information*, **44**(4), 695–729.

Scherer, K. R. (2013). Vocal markers of emotion: Comparing induction and acting elicitation. *Computer Speech and Language*, **27**, 40–58.

Schuller, B. and Rigoll, G. (2006). Timing levels in segment-based speech emotion recognition. In *Proc. of Interspeech*, pp. 1818–1821, Pittsburgh.

Schuller, B., Müller, R., Lang, M., and Rigoll, G. (2005). Speaker independent emotion recognition by early fusion of acoustic and linguistic features within ensembles. In *Proc. of Interspeech*, pp. 805–809, Lisbon.

Schuller, B., Batliner, A., Seppi, D., Steidl, S., Vogt, T., Wagner, J., Devillers, L., Vidrascu, L., Amir, N., Kessous, L., and Aharonson, V. (2007). The relevance of feature type for the automatic classification of emotional user states: Low level descriptors and functionals. In *Proc. of Interspeech*, pp. 2253–2256, Antwerp, Belgium.

Schuller, B., Wimmer, M., Mösenlechner, L., Kern, C., Arsić, D., and Rigoll, G. (2008). Brute-forcing hierarchical functionals for paralinguistics: A waste of feature space? In *Proc. of ICASSP*, pp. 4501–4504, Las Vegas, NV.

Schuller, B., Steidl, S., and Batliner, A. (2009). The Interspeech 2009 Emotion Challenge. In *Proc. of Interspeech*, pp. 312–315, Brighton, UK.

Schuller, B., Zaccarelli, R., Rollet, N., and Devillers, L. (2010a). CINEMO – A French spoken language resource for complex emotions: Facts and baselines. In *Proc. of LREC*, pp. 1643–1647, Valletta, Malta.

Schuller, B., Steidl, S., Batliner, A., Burkhardt, F., Devillers, L., Müller, C., and Narayanan, S. (2010b). The Interspeech 2010 Paralinguistic Challenge. In *Proc. of Interspeech*, pp. 2794–2797, Makuhari, Japan.

Schuller, B., Batliner, A., Steidl, S., Schiel, F., and Krajewski, J. (2011). The Interspeech 2011 Speaker State Challenge. In *Proc. of Interspeech*, pp. 3201–3204, Florence.

Schuller, B., Steidl, S., Batliner, A., Nöth, E., Vinciarelli, A., Burkhardt, F., Son, R. v., Weninger, F., Eyben, F., Bocklet, T., Mohammadi, G., and Weiss, B. (2012). The Interspeech 2012 Speaker Trait Challenge. In *Proc. of Interspeech*, Portland, OR.

Seppi, D., Batliner, A., Steidl, S., Schuller, B., and Nöth, E. (2010). Word accent and emotion. In *Proc. of Speech Prosody*, Chicago.

Silverman, K., Beckman, M., Pitrelli, J., Ostendorf, M., Wightman, C., Price, P., Pierrehumbert, J., and Hirschberg, J. (1992). ToBI: A standard for labeling English prosody. In *Proc. of ICSLP*, pp. 867–870, Banff, Alberta.

Spearman, C. (1904). The proof and measurement of association between two things. *American Journal of Psychology*, **15**, 72–101.

Steidl, S., Ruff, C., Batliner, A., Nöth, E., and Haas, J. (2004). Looking at the last two turns, I'd say this dialogue is doomed — measuring dialogue success. In P. Sojka, I. Kopeček, and K. Pala (eds), *Proc. of Text, Speech and Dialogue (TSD)*, pp. 629–636, Berlin.

Steidl, S., Batliner, A., Seppi, D., and Schuller, B. (2010). On the impact of children's emotional speech on acoustic and language models. *EURASIP Journal on Audio, Speech, and Music Processing, Special Issue on Atypical Speech*, **2010**.

Stemmer, G., Nöth, E., and Parsa, V. (2010). Editorial: Atypical speech. *EURASIP Journal on Audio, Speech, and Music Processing*, **2010**.

Vidrascu, L. and Devillers, L. (2005). Annotation and detection of blended emotions in real human-human dialogs recorded in a call-center. In *Proc. of the IEEE International Conference on Multimedia and Expo, ICME, Amsterdam*, pp. 944–947.

Vipperla, R., Renals, S., and Frankel, J. (2010). Ageing voices: The effect of changes in voice parameters on ASR performance. *EURASIP Journal on Audio, Speech, and Music Processing*, **2010**.

Vogt, T. and André, E. (2005). Comparing feature sets for acted and spontaneous speech in view of automatic emotion recognition. In *Proc. of Multimedia and Expo (ICME05)*, pp. 474–477, Amsterdam.

Walker, M. A., Langkilde, I., Wright, J., Gorin, A., and Litman, D. (2000). Learning to predict problematic situations in a spoken dialogue system: Experiments with how may I help you? In *Proc. of NAACL-00*, pp. 210–217, Seattle.

Whitehead, A. N. (1925). *Science and the Modern World*. Macmillan, New York.

Wilkinson, L. (1999). Statistical methods in psychology journals: guidelines and explanations. *American Psychologist*, **54**, 594–604.

Wilting, J., Krahmer, E., and Swerts, M. (2006). Real vs. acted emotional speech. In *Proc. of Interspeech*, pp. 805–808, Pittsburgh.

Wundt, W. (1896). *Grundriss der Psychologie*. Engelmann, Leipzig.

Xu, Y. (2010). In defense of lab speech. *Journal of Phonetics*, **38**, 329–336.

4

Formal Aspects

4.1 The Linguistic Code and Beyond

It is one of the aims of linguistics to define itself, to recognise what belongs within its domain. In those cases where it relies upon psychology, it will do so indirectly, remaining independent.

(Ferdinand de Saussure)

Here, we want to give a short and necessarily rough account of some basic principles of linguistics, especially of linguistic structure, contrasting them with the basic principles of paralinguistics; note that in this usage, linguistics encompasses both phonetics and linguistics proper – the same way as paralinguistics as we understand it encompasses aspects encoded in voice and speech, and additionally in (written) language.

The classic sub-systems of linguistic structure are phonetics/phonology, morphology, syntax, and semantics; moreover, there is pragmatics as a borderline system in transition to all the other cultural systems. The building blocks of linguistic structure are *distinctive* units; phonemes are the smallest distinctive segmental units in the sound system (phonology), morphemes are the smallest distinctive, semantically meaningful units in the language system; they are distinctive because if we exchange them with some other distinctive unit, the meaning changes. Note that in this usage, 'distinctive' is used in its linguistic meaning, 'non-distinctive' as well; thus, linguistically non-distinctive elements and features can very well distinguish paralinguistic functions. Typically, this distinction is, however, not all-or-nothing but a matter of degree.

Distinctive feature theory is based in the structuralism 'founded' by de Saussure, and in the Prague School of phonology, and was formalised for the first time in 1941 by Jakobson (1969). It was later employed in (lexical) semantics. Examples are: [+syllabic, +high, +front, −round] denoting the phoneme /i/ represented by the phone [i], and [+human, +female, +adult] denoting a woman. The distinction between the words 'rabid' and 'rapid' can be claimed to be based on the distinction between [+voiced] and [−voiced]. However, Lisker and Abramson (1964) and Lisker (1986) showed that quite a lot of features change when going from one to another category; thus, it is much more than only the distinction in voicing.

Computational Paralinguistics: Emotion, Affect and Personality in Speech and Language Processing, First Edition. Björn W. Schuller and Anton M. Batliner. © 2014 John Wiley & Sons, Ltd. Published 2014 by John Wiley & Sons, Ltd.

The term *denotation* refers to the 'referential meaning' of a linguistic unit; the unit is normally a word or any higher constituent/expression. The referent can be concrete and unique (e.g., a person, a thing), or an (abstract) concept (e.g., the horizon). The term *connotation* refers to the more or less vague emotive and affective meanings associated with a linguistic unit/expression. Linguistics deals with denotation, and paralinguistics with connotations. These terms are described in Bussmann (1996):

> ... *denotation refers to the constant, abstract, and basic meaning of a linguistic expression independent of context and situation, as opposed to the connotative, i.e. subjectively variable, emotive components of meaning. Thus, the denotation of night can be described as the 'period of time from sunset to the following sunrise,' while the connotation may include such components as 'scary', 'lonely', or 'romantic.'*

A reference book on all aspects of phonetics is Hardcastle *et al.* (2010); sounds and sound systems of languages are extensively dealt with in Ladefoged and Maddieson (1996). A reader covering all aspects of linguistics is Aronoff and Rees-Miller (2001). Some of the classic books on linguistics are: Saussure (1916), Sapir (1921), and Bloomfield (1933) on American and classic structuralist linguistics. Standard introductions to modern linguistics are Lyons (1968) and Fromkin *et al.* (2002). For an overview, we can refer to Crystal (1997) which is suitable for the non-technical reader, and Newmeyer (1988) for an in-depth account. A survey of the world's major languages is given in Comrie (2009).

Good guides through the – sometimes not fully consistent – linguistic terminology are the following dictionaries of linguistics and phonetics: Trask (1996), Bussmann (1996), and Crystal (2008). The terminology of communication disorders in the fields of speech, language and hearing is covered in Nicolosi *et al.* (2004).

Especially relevant for paralinguistics are the fields of linguistics that are marginal to theoretical core linguistics, but close to or identical with paralinguistics, namely pragmatics and another 'hyphenated' linguistics, namely sociolinguistics. *Pragmatics* mainly deals with the use of linguistic means in conversation, thus the topics are close to or sometimes identical with social signal processing; Morris (1938) introduced the term 'pragmatics' into a general theory of semiotics, consisting of syntax, semantics and pragmatics. Standard introductions to the field are Levinson (1983), Mey (2001) and Horn and Ward (2005). *Sociolinguistics* mainly deals with linguistic and phonetic traits characterising different social variables such as class, regional variants, gender, and the like; the early use of the term is ascribed to Currie (1952) although in Hymes (1979) it is attributed to T.C. Hudson in 1939. Standard handbooks and readers are: Coulmas (1997), Paulston and Tucker (2003), Ammon *et al.* (2004) and Wodak *et al.* (2011). Trudgill (2004) mentions different sub-fields of sociolinguistics: *macro-sociolinguistics* encompassing linguistic varieties used by large groups of speakers such as dialects or regional variants; *micro-sociolinguistics* dealing with face-to-face-interactions (discourse and conversation analysis); *ethnomethodology* dealing with 'talk' but not language; and much more. Sociophonetics can be seen as a parallel field to sociolinguistics, focusing on spoken language, or as part of sociolinguistics (Foulkes *et al.* 2010); as often, there is a narrow meaning of 'linguistics' (dealing with written/'natural' language) and a broad meaning ('linguistics' dealing with all aspects of spoken or written language). All this nicely illustrates that sociolinguistics could be conceived of as part of paralinguistics, with a special focus on the non-private aspects, that is, on interaction within a societal setting. Note that we do not want to usurp this or other fields and incorporate them into paralinguistics; we just want to point out that there is a considerable overlap, and that it definitely is worthwhile to look for studies conducted within these fields, when we are interested in specific paralinguistic phenomena.

Within a variety – be this a language, a dialect, or a sociolect – there are necessary, regular elements that have to be there; and there are free variants for these elements that are not linguistically distinctive. The more intrinsic variety in a phoneme of a language, or in any other phonetic parameter, without 'crossing the border' towards another category, the more likely it might be that variation within the categories can be employed for indicating paralinguistic functions and differences. Varieties within phonemic categories are called *free variants* or *allophones*. Allophones of /r/ are a typical example and will be dealt with in Section 4.2.1. Suprasegmental parameters such as voice quality and pitch can often be employed for indicating paralinguistic functions because their functional load within the linguistic system is low; however, tone languages, for example, restrict the degrees of freedom.

Moreover, as far as the lexicon is concerned, the language user is free to choose amongst a multitude of (types of) words; thus, the choice of words and word classes can indicate paralinguistic functions as well.

In the next two sections, we will deal with such elements and parameters, thereby covering distinctions 'beyond the linguistic code'; we first address exemplary phonetic phenomena in Section 4.2, and then linguistic phenomena in Section 4.3. The next two sections deal with deviations from the 'correct' linguistic code (disfluencies in Section 4.4), and with phenomena external to the linguistic code, namely non-verbal, vocal events (Section 4.5). The form of all these phenomena can – but need not – indicate some paralinguistic function.

4.2   The Non-Distinctive Use of Phonetic Elements

In this section, we want to illustrate the rather formal approaches towards phonetic phenomena in paralinguistics, presenting one exemplar for the segmental level (/r/ variants), one for the supra-segmental, prosodic level (pitch), and one for both (voice quality, especially laryngealisations, which are both segmental and supra-segmental phenomena).

4.2.1   Segmental Level: The Case of /r/ Variants

The quantity of consonants in the English language is constant. If omitted in one place, they turn up in another. When a Bostonian 'pahks' his 'cah', the lost r's migrate southwest, causing a Texan to 'warsh' his car and invest in 'erl wells'.

(Author unknown)

/r/ sounds (rhotics) form a class of sounds that can only be characterised by their being '. . . written with a particular character in orthographic systems derived from the Greco-Roman tradition, namely the letter "r" or its Greek counterpart *rho*' (Ladefoged and Maddieson 1996, p. 215). As for manner of articulation, the class consists of trills, flaps, fricatives, and approximants; place of articulation can be alveolar, retroflex, and uvular – there are even labialised variants. Thus, manner and place of articulation vary widely. Rhotics can be deleted, especially after low, back vowels such as /a/. The bilabial trill is not part of rhotics – it is practically never a phoneme in any language, either. The most frequent type is the voiced alveolar trill, and it is most typical for languages to have only one /r/ sound (Maddieson 1984, pp. 78ff.). Lindau

(1985) concludes that 'there is no physical property that constitutes the essence of all rhotics. Instead, the relations between members of the class are more of a family resemblance.'

Perceptually, different types of /r/ sounds are clearly distinct, but normally, they still belong to the same phoneme. These different types are therefore free variants – free to indicate paralinguistic functions (Scobbie 2006). Diachronically, the dorsal variants of /r/, such as [R], seem to have originated in Parisian French in the seventeenth century and have spread throughout western Europe, replacing the alveolar variants such as [r] (Chambers and Trudgill 1980). Thus, a prestigious form of speech that manifested itself in the speech of outstanding figures such as kings and politicians was taken over, spreading to other countries/languages, and social classes, the same way as table manners spread from the higher classes to the lower classes (Elias 1939). Van Hout and Van de Velde (2001) give a concise introduction to the manifold aspects of /r/ sounds in western Europe, their origin and possible reasons for sound changes – a development that, for instance, still can be observed in southern parts of Germany, where dialectal forms give way to standard forms. Note that it can also be the other way round: The vowels of Queen Elizabeth II have not influenced the standard southern British accent of the 1980s, but the reverse; we can speculate that this development can be taken as an implicit social signal, not within one communication but 'long term', of democratisation by diminishing the distance between the Queen's speech and the speech of speakers who are younger and lower in the social hierarchy (Harrington *et al.* 2000).

In a seminal study on sociolinguistic variables in the city of New York, Labov (2006) shows that the preservation or deletion of /r/ in final or post-vocalic, pre-consonantal position (car, cart) is a strong indicator of social class membership. In southern American English, /r/ deletion in such positions characterises dialects.

Now what does all this mean for computational paralinguistics? As far as we can see, specific phenomena such as /r/ variants are normally not yet explicitly employed in computational paralinguistic approaches. To distinguish language varieties such as regional accents or sociolects, methods from automatic speech recognition (ASR) are applied. Modelling separate phenomena such as /r/ variants might simply be too complex and manual modelling is required. However, we should keep in mind that computational 'macro-approaches' towards modelling and discriminating varieties of speech such as regional accents always are based on a multitude of such 'micro-distinctions'.

Explicit modelling of specific segmental distinctions is done, however, in approaches towards computer-aided/assisted pronunciation training in foreign language teaching, in order to obtain highly reliable classifications of non-native pronunciation, and to give corrective feedback to the learner (see Cucchiarini *et al.* 2009), because such specific mispronunciations can be marked non-native traits, sometimes fossilised, and do have a great impact on the impression of non-nativeness.

4.2.2 Supra-segmental Level: The Case of Pitch and Fundamental Frequency – and of Other Prosodic Parameters

'You crazy', said Max. It was either a statement or question.

(John le Carré, *Tinker Tailor Soldier Spy*)

'So you're our man, then' he said. It was half statement, half question.

(Josef Skvorecky, *The Engineer of Human Souls*)

In this section, we want to discuss the interplay of linguistic and paralinguistic functions of pitch; sometimes, we have to refer to the other main supra-segmental parameters of duration, intensity, rhythm, and voice quality because all these usually function as a bundle of parameters, more or less highly correlated with each other.

The *lexical* function of pitch is to distinguish between segmentally identical words, for instance, in 'real' tone language such as Chinese where tones are distinguished by their shape (contour) and their pitch range (register), and in a 'simpler' variety, to distinguish between accent I and accent II words in Swedish and Norwegian. Another lexical function is to signal *stress* (*word accent position*), in conjunction with the other prosodic parameters – sometimes called pitch accent, a term that should, in our opinion, be avoided if we are talking, for instance, about English or German. This serves to tell words apart as well, for example, *OBject* versus *obJECT* in English, or *TENor* versus *TenOR* in German. Languages have typical positions of word accents. This is sometimes a rule: first syllable in Czech, penultimate syllable in Polish, last in French. Sometimes there are preferred positions which can be more or less formulated as rules, for instance, in English or German. The signalling of the *phrase accent*, that is, of the focal unit, serves to indicate the most important (often, the 'new') semantic message. *Sentence mood*, especially question versus non-question, is, across languages, very often indicated by high versus low final pitch movements; note that this is a (strong) tendency, not a rule without exception, as Studdert-Kennedy and Hadding (1973) and Hadding-Koch and Studdert-Kennedy (1964) have pointed out; moreover, questions can be signalled by specific particles, for instance, in Finnish. We will address a similar formal means, namely the so-called 'uptalk', below.

The indication of all these functions is layered, following more or less strict rules, from word level to phrase and utterance level. Together, they form a language-specific rhythm that helps the child acquire his/her first language, and that can be one of the big obstacles to acquiring a native pronunciation in a foreign language. Note that even on this language-specific layer, there can be a considerable degree of freedom, with respect to not only the pitch register but also positions of focal accents and the like. In an intonation model, there might be only one possibility for producing a sentence 'out of the blue' or within a specific context; however, in reality, there are often several alternatives. The cross-language and, by tendency, universal utilisation of pitch from an ethological perspective is addressed in Ohala (1984).

The marking of paralinguistic functions is, so to speak, *modulated onto* the linguistic layer; this will now be discussed, more or less following the structure of Chapter 5.

The most important biological trait primitive is the difference between adults and children, and between females and males, in pitch register, that is, average pitch height: some 100–150 Hz in male speech, 150–250 Hz in female speech, and 300–500 Hz in children's speech. A full and very detailed account of the 'Physiological, aerodynamic, and acoustic differences between male and female voices' is given in Kreiman and Sidtis (2011, pp. 124ff.). For instance, a man's larynx is about 20% larger than a woman's. However, this does not fully explain the differences between pitch height in males and females; there are additional cultural factors as well. Van Bezooijen (1995) has found that Japanese women have higher pitches than Dutch women; this has been traced back to the assumption that 'Japanese women raise their pitch in order to project a vocal image associated with feminine attributes of powerlessness' (Van Bezooijen 1995, p. 253). Similar stereotypes exist in the United States as well, but with opposed tendencies. Imagine the high-pitched attractive voice of Marilyn Monroe versus the equally attractive but lower-pitched voice of Lauren Bacall. It seems that, on average, female pitch has been lowered in the US and other Western countries since the 1950s due to the changing role of women in society.

Pitch is a good use case to demonstrate that what we produce is not the same as what we can measure (fundamental frequency, F0) or what we perceive (pitch), although there is a regular and highly positive correlation between these parameters. To take such correspondences into account, Scherer (2003) advocates the Brunswikian lens model with distal indicators and proximal percepts. There are other models for different sender–receiver characteristics; it will do as well to be familiar with the differences between articulatory/phonatory mechanisms, the acoustics of the phonetic signal, and perceptive phonetics. There are different perceptually adequate pitch scales: the *mel scale*, based on a subjective magnitude estimation of pitch; the *Bark scale*, based on measurements of the so-called critical bands; and the *equivalent rectangular bandwidth* (ERB) *scale*, based on measurements of the bandwidth of the auditory filters. All of them are logarithmic above 500 Hz; below 500 Hz, the mel and Bark scales are linear, and the ERB scale between linear and logarithmic; details can be found in Zwicker and Fastl (1999) and Hermes and Gestel (1991).

To take into account the different pitch registers of males and women – whether caused only by physiological factors or also by social stereotypes – we have to either model females and males separately, or normalise F0 with respect to different reference values such as speaker-specific baselines, that is, the lowest possible F0 value for a speaker, or to the F0 mean value of the unit of analysis, for instance, the utterance; moreover, we can use the transformations mentioned above. Most important is the normalisation to some gender-, speaker-, or utterance-specific reference value; however, a perceptually adequate modelling can have some added value as well. In automatic speech processing, speaker normalisation is often done for all acoustic-prosodic features.

Irrespective of the normalisation chosen, we have to distinguish whether differences in the expression of emotion or any other paralinguistic trait are 'simply' due to these bodily differences (biological trait primitive), whether they are due to a straightforward stereotype that women generally speak with higher pitch (cultural trait primitive), or whether they are really due to the employment of different features by male and female speakers (culturally or personally influenced differences in the use of pitch for indicating affect and emotions). It is thus not really of interest just to report that when using non-normalised pitch values, a separate modelling of females and males yields better performance. However, to disentangle different conditioning factors and features and to attribute to them relative importance, would be of great interest. We will return in Section 4.2.3 to a tendency in female speech to lower pitch in such a way as to result in irregular phonation (laryngealisations).

As for personality traits, we can expect that, for instance, less lively characters are mirrored in lower pitch register and especially lower pitch range; recall the extraversion, energy, enthusiasm factor in the OCEAN dimensions mentioned in Section 3.1. As for emotions, raised pitch is a relatively stable indicator of arousal, be this negative (anger) or positive (joy). In parallel, lowered pitch indicates sadness. Such results are consistent for acted data and have been repeatedly reported by Scherer and colleagues; they can more or less be expected for non-acted data as well. Note that the marking of emotion seems to comply with the linguistic structure; Seppi *et al.* (2010) have shown that it is stronger for content words than for function words, and, within content words, stronger for the word accent syllable. This corresponds to the assumption that paralinguistic structure is modulated onto linguistic structure, and not distributed randomly or independently. Of course, this correspondence can be out of tune in non-typical (deviant, non-native) speech; in turn, this can be taken as an indicator of non-typical speech.

So far we have dealt with pitch in the speech of individuals. Of course, it serves, together with other prosodic features, specific functions in dyadic and multi-party interactions; there, it is a complex interplay of *entrainment* – the tendency of dialogue partners to adapt to the other's speaking style – and signalling different roles. Entrainment can be observed for all acoustic, prosodic and linguistic parameters – pausing, back-channelling, adaptation of tempo and pitch, and choice of words. This is a dynamic process that can – but need not – change the initial attribution of social status which can be either symmetric or asymmetric (higher/lower status). Lower pitch (going over to laryngealised speech; see Section 4.2.3) can signal higher status, both in humans (Ohala 1984) and in other mammals (Wilden *et al.* 1998).

Returning to a linguistic function of pitch, namely the indication of *questions*, in Batliner (1989), two pragmatic (i.e., paralinguistic) dimensions of meaning for questions were established: first, *expectancy of answer* with the extremes of fully open questions at one end, and fully rhetorical questions at the other end; and second, *indicated interest in an answer*, with no interest at all in rhetorical questions at one end, and very strong interest at the other end. The higher the expectancy or the interest is, the higher the (final) pitch rise may be. In German, modal particles can help to disambiguate such nuances; in English, it is mostly the job of intonation. Kreiman and Sidtis (2011, p. 301) reflect on the theoretical consequences. We can consider linguistic categories such as question versus non-question as categorical, and paralinguistic meanings as graded. This means that questions are questions but with a more or less pronounced additional meaning, for instance, additional expression of surprise or bewilderment which are indicated by higher than usual final pitch values. In fact, we can analytically distinguish the linguistic from the paralinguistic constellation; however, we doubt that speakers do the same. In experiments, it depends on the interest of the researcher and, thus, on the experimental design; recall WYALFIWYG in Section 3.9. So it boils down to a matter of methodological considerations what to attribute to which component. For description, the suggestion of Kreiman and Sidtis (2011, p. 301) to distinguish categorical linguistics from graded paralinguistics might be a good choice. In practice, we most likely will do linguistic and paralinguistic analyses and processing alongside each other, and will model pitch values and final pitch curves separately in both components.

Final rises for indicating questions should be distinguished from *uptalk* characterising statements, a more or less regular final pitch rise in regional or social varieties. It is well known to exist in varieties of Irish English, for instance, in the Belfast dialect, and in some varieties of Swiss German (in connection with a final question tag *oder?*, functioning in the same way as *isn't it?*); further references can be found in Foulkes and Docherty (2006) who summarise this and other similar phenomena under the heading 'sociophonetics'. It seems to be spreading among both the male and female population in southern California and does not function as continuation or question marker but rather as a symptom of group membership (Kreiman and Sidtis 2011, pp. 129, 272). It might be difficult to distinguish these two functions – indicating questions or indicating group membership – just by analysing the pitch contour. However, in specific cases, a pragmatic test might do: unless we can assume a rhetorical question, such an intonation contour in a statement about oneself (*I am hungry*) suggests uptalk and not questioning.

In computational paralinguistics, we can model pitch implicitly, together with a multitude of other features, with or without trying to assess its impact for the function we are interested in. Nowadays, this is the usual brute-force approach. In basic research, we often address single formal parameters such as pitch. In so doing, we can find out more about the behaviour of this single feature/parameter. Yet, this comes at the cost of disregarding other parameters; for instance, in order to be able to distinguish between uptalk and question intonation, we might need additional information on the variety spoken by the speaker.

4.2.3   In Between: The Case of Other Voice Qualities, Especially Laryngealisation

Is Chantal the brunette that speaks in the slow, deep, creaky voice? She pronounces every word like its torture.

(Pomme de Divan, Forum, Candid Reality Shows)

According to Abercrombie (1967, p. 91), voice quality '. . . refers to those characteristics which are present more or less all the time that a person is talking; it is a quasi-permanent quality running through all the sound that issues from his mouth'. Laver (1980), Crystal (2008), and Kreiman and Sidtis (2011) all subscribe to this definition which encompasses pitch, loudness, tempo, and timbre. Gordon and Ladefoged (2001) deal with cross-linguistic aspects of voice quality. Gerratt and Kreiman (2001) try to establish a taxonomy of non-modal phonation, and Kreiman *et al.* (1993, 2007) address perceptual assessment of voice quality.

It is difficult, if not impossible, to extract surgically, as it were, pitch from other voice qualities; for instance, jitter (micro-variations of F0) is sometimes attributed to pitch, sometimes to voice quality. Pitch is often modelled under the assumption that we are dealing with the typical, 'modal' variety of phonation; if not, and if extraction 'fails', we have to reconstruct pitch by extrapolating pitch across unvoiced passages, and/or by getting rid of octave errors that can often be observed in irregular phonation; examples of such irregular passages are given in Figure 4.1.

In Section 4.2.2 we dealt with pitch and mostly with upward deviations indicating gender-specificity, cultural stereotypes, and markers of social group membership. We now address the opposite: downward deviation, resulting in *irregular phonation*. Other non-typical voice qualities will also be mentioned but not dealt with in depth. This means that we do not want to get rid of such irregularities but to investigate them as is and to find out which paralinguistic function they indicate.

'Phonation concerns the generation of acoustic energy . . . at the larynx, by the action of the vocal folds' (Laver 1994, p. 132). The three main sub-types of *phonation* are: 'normal' *modal* voice, *laryngealisation* (frequency range below the modal voice), and *falsetto* (frequency range above the modal voice). Laryngealisation shows up as irregular voiced stretches of speech. Mostly, it does not disturb pitch perception but is perceived as supra-segmental, differently shaped irritation modulated onto the pitch curve which can be found both in typical and atypical speech. This irregular phonation is still less well known and understood than typical, modal phonation. Its occurrences and distributions nicely demonstrate the interplay of local (partly segmental in the phonological sense) and global phenomena. Such an interplay also exists for pitch; however, it is practically impossible to disentangle both factors on the time axis because pitch is ubiquitous – if there is voicing, then there is pitch. Irregular phonation, however, can be restricted to specific segments or to specific local contexts; examples are given below.

Laver (1968) gives a good overview of classic approaches to voice quality as an index of biological, psychological and social characteristics of a speaker. Painter (1991) describes the different types in terms of laryngeal configurations, based on video recordings.

There are various terms for this phenomenon used more or less synonymously: irregular phonation, laryngealisation, *glottalisation, creak, vocal fry, creaky voice, pulse register*, etc. We use 'laryngealisation' as a catch-all term for all these phenomena originating in the larynx that show up as irregular voiced stretches of speech. Irrespective of terminology, the definition of sub-types is mostly holistic, taking into account physiological, acoustic, and perceptual aspects. Batliner *et al.* (1994) developed a taxonomy strictly based on *formal* characteristics of the time signal along six dimensions: number of glottal pulses; degree of damping; amplitude and F0 variations, both intrinsic (paradigmatic aspect) and with respect to left and right context (syntagmatic aspect). These dimensions allow us to establish five distinct types: *glottalisation, diplophonia, damping, sub-harmonic*, and *aperiodicity*, in addition to a less distinct waste-paper-basket category; they are displayed in Figure 4.1. Phonation outside the boxes is regular (modal), inside the boxes irregular (laryngealised).

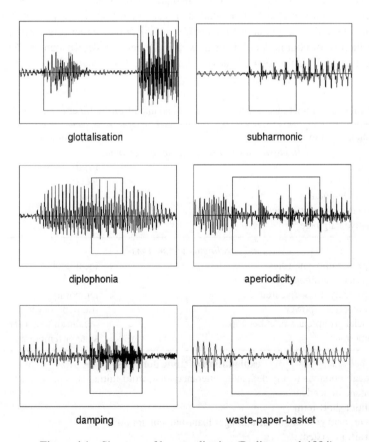

Figure 4.1 Six types of laryngealisation (Batliner *et al.* 1994)

There have been a few attempts to classify laryngealisations automatically (Ishi *et al.* 2005; Kießling *et al.* 1995). Normally, there is no explicit modelling and classification of laryngealisations in automatic approaches to paralinguistics.

As for the *functional* aspect, one can try and find different functions for this formal phenomenon. In spite of the fact that it is largely unnoticed by speakers even if they employ it themselves, there are a plethora of such functions; the following presentation is adapted from Batliner *et al.* (2007) and extended.

Table 4.1 displays different functions of laryngealisations which can be linguistic or paralinguistic. They can be caused either by greater effort or by relaxation; in the former case, they go together with *accentuation* (prominence) which is, of course, a *local* phenomenon. (Actually, it might be that laryngealisation does not denote accentuation but can be accompanied by it, in the case of low and/or back vowels such as [a] in a stressed syllable (cf. below); note that in stressed position, laryngealisation cannot be caused by relaxation.) A typical place for relaxation is *the end of an utterance* (Böhm and Shattuck-Hufnagel 2007); *turn-taking* can thus be signalled to the dialogue partner; this is again a *local* phenomenon. Local and Kelly (1986) report that different types of laryngealisations are used in (British and American) English conversations for holding the floor (filled pauses with glottal closure, no evidence of creaky phonation) and for yielding the floor (filled pauses with lax creaky phonation, no glottal closure). *Word boundaries* in the hiatus,that is, a word final vowel followed by a word initial vowel, can be marked by laryngealisations in German. Boundary marking with such irregular phonation which is, of course, *local*, is dealt with in Huber (1988), Kushan and Slifka (2006), and Ní Chasaide and Gobl (2004). It is well known that back *vowels* such as [a] tend to be

Table 4.1 Functions of laryngealisations, adapted from Batliner *et al.* (2007)

phenomenon	time domain
linguistic functions: phonotactics, grammar, . . .	
vowels	local
accentuation	local
word boundaries	local
the end of an utterance	local
native language	local
paralinguistic functions	
speaker idiosyncrasies	local–global
speaker pathology	trait
too many drinks/cigarettes	temporary
competence/power	trait/temporary
social group/class membership	local/trait/temporary
emotions	state or temporary

explanation of 'time domains'
local: phonotactically definable (utterance-final, word-initial, etc.) or phone-dependent
state: short-term
temporary: medium-term, longer than state but not trait
trait: persistent

more laryngealised than front vowels such as [i] (*local* phenomenon). A language-specific use of laryngealisations either can be due to phonotactics, as in German, where every vowel in word-initial position is 'glottalised', or phonemes can be laryngealised, as in the case of the vowels in Jalapa Mazatec (Ladefoged and Maddieson 1996, p. 317) which can be creaky (laryngealised), breathy (murmured), or modal (plain); this is a *local* phenomenon, denoting the *native language*. Normally, specific segments which are laryngealised characterise languages (see, for vowels, Gerfen and Baker 2005); in Danish, the glottal catch or 'stød' (Fischer-Jørgensen 1989) can be found in vowels and consonants. Paralinguistic functions are displayed in the middle part of Table 4.1. Laver (1994, pp. 194ff.) lists different uses and functions of 'creak' phonation, amongst them the paralinguistic function 'bored resignation' in English Received Pronunciation, 'commiseration and complaint' in Tzeltal, and 'apology or supplication' in an Otomanguean language of Central America. Laryngealisations can be a marker of personal identity and/or social class; often, they indicate higher-class speech. However, in the same way as table manners spread from higher to lower classes, specific phonation types can be employed by other social groups, such as young females (see below). Wilden *et al.* (1998) refer to evidence, not just for human voices but for mammals in general, that 'non-linear phenomena' (i.e., irregular phonation/laryngealisation) can denote individuality and status, that is, pitch is used as an indicator of a large body size and/or social dominance: '... subharmonic components might be used to mimic a low-sounding voice'. *Emotional states* such as *despair*, *boredom* and *sadness*, are *short-term* or *temporary*. Bad news is communicated with a breathy and creaky voice (Freese and Maynard 1998), boredom with a lax creaky voice, and, to a smaller extent, sadness with a creaky voice (Gobl and Ní Chasaide 2003). Drioli *et al.* (2003) report for perception experiments with synthesised stimuli that disgust is conveyed with a creaky voice. Erickson *et al.* (2004) found, for one female Japanese speaker, creaky voice in imitated sadness but not in spontaneous sadness; thus they assume a social connotation of creaky voice. Display of boredom and upper-class behaviour might coincide; the same can happen if someone who permanently uses laryngealisations as a speaker-specific trait tells a sad story. On the other hand, at first sight, speakers who exhibit laryngealisations as an idiosyncratic trait can make a sad impression without actually being sad. A common denominator for some of the paralinguistic functions of laryngealisations might be inactivity/ passivity in mood (boredom, sadness, etc.) corresponding to relaxation — which is one of the possible physiological sources of laryngealisations. However, this is not a must: if laryngealisations are used to signal competence/power, then the basic attitude can be composure and need not be passivity. However, if speakers habitually produce laryngealised speech, then it is at least very difficult to distinguish other functions.

Note that all these characteristics which *per se* are *not* characteristics of single speakers can – apart perhaps from the language-specific phonemes – be used more or less distinctly by different speakers (Batliner *et al.* 2007). As for the para- and extralinguistic function of laryngealisations, speakers can simply use them throughout to a higher extent; such *speaker idiosyncrasies* are *local/traits*. Some children display severe phonation irregularities such as harsh voice and/or laryngealisations, due to lack of control of laryngeal settings. 'Creaky superstars' like Tom Waits or Leonard Cohen, who claimed that his voice got even lower after having stopped smoking, are well known. Laryngealisations can be a consistent trait throughout phonation, or may only be employed in utterance-final position; note that this is not a new observation – it was also observed in the actress Mae West who impersonated a both attractive and dominant person. The same can be observed in female professionals in the US – a famous example being the editor of the *New York Times*. Yuasa (2010, p. 315) found that female speakers of American English in California used a creaky voice to a higher extent than comparable American male or Japanese female speakers. A new stereotype seems to be emerging with the 'female creaky voice as hesitant, nonaggressive, and informal but also educated, urban-oriented, and upwardly mobile'. In a study by Wolk *et al.* (2012), two-thirds of a sample of young adult Standard American English speakers used vocal fry, especially at the end of sentences.

Anecdotal evidence shows that *nasalisation* in young German women might also be an adapted trait to signal group membership, initiated by the advent of TV and especially early evening soaps. The use of laryngealisations by younger American woman might serve the same purpose. We can speculate that there are two tendencies. First, in the US, pitch in women's voices has lowered in the last 50 years. Second, as an alternative to higher pitch as an indication of femininity, there is a trend towards laryngealisation, that is, lowering pitch, to signal competence.

Both trends – using uptalk and producing laryngealisations – help in stabilising group membership. There might be an additional ethological factor: low voice means big and powerful. This can be adapted by groups that are not big and powerful, such as female teenagers. A parallel can be found in diachronic linguistic change, which is always characterised by opposite tendencies, for instance, towards simplification (and thus deletion of phonemes or morphemes) on the one hand, and by evolving complexity, on the other hand. Laryngealised speech might indicate a trend towards complexity. Thus, again, we are faced with both multi-functionality of this phenomenon, and with changes that only can be explained if we consider the cultural and historical context.

Other types of phonation serve the same or similar purposes. *Breathy voice* (Campbell and Mokhtari 2003; Laver 1991) is well known, with a higher proportion of noise (harmonics-to-noise ratio) than usual; famous speakers (and singers) are Harry Belafonte, Brigitte Bardot, and Carla Bruni. Breathiness conveys both attractiveness and intimacy (Batliner *et al.* 2006). We can speculate about the origin of this personality trait. In intimate situations (mother–child or male–female interaction), the partners are close together and subtle, less prominent voice characteristics can be employed. In contrast (see Section 5.3), leadership and charisma are correlated with voice characteristics such as greater loudness, higher and more variable pitch (Weninger *et al.* 2012) – these being features that can be heard across distances greater than those found in intimate scenarios. Less attractive and not yet fully understood as an indicator of personality seems to be another phonation type, namely *hyperfunctional dysphonia*, where the voice is hoarse and too forceful – famous speakers being Paris Hilton and Heidi Klum – with women considerably overrepresented by a factor of up to 8 (Wilson *et al.* 1995); at least, such voice qualities are 'ear-catching', simply because they are atypical.

We have chosen to detail the different functions of laryngealisations in order to demonstrate the advantage and disadvantage of a formal approach: only by starting with one specific formal phenomenon and trying to find out where and when it is employed and by which groups of speakers do we get a complete account of this formal characteristic. Laryngealisations differ from rhotics, which are a segmental phenomenon, and from pitch, which is a supra-segmental parameter: they can be confined to segments or to specific local contexts, they can be global, or they can combine segmental and global characteristics. However, we do not yet know the specific relevance of this formal characteristic for indicating specific functions that is, its contribution to automatic modelling; this can only be achieved when we model all features that are possibly relevant for a specific function.

Within automatic speech processing and computational paralinguistics we can (1) try to get rid of laryngealisations, as indicated above; (2) try to model them implicitly, simply by using brute force and a very large feature set, for classifying any function; and (3) try to model them more or less explicitly, aiming at specific functions such as those mentioned in this section.

4.3   The Non-Distinctive Use of Linguistics Elements

A shprakh iz a dialekt mit an armey un flot. – A language is a dialect with an army and navy.

(Max Weinreich, who attributed this statement to a participant in one of his seminars)

In traditional linguistics, the system of a language is conceived of as being rather monolithic, based on paradigmatic and syntagmatic distinctive traits and elements. There are many more degrees of freedom in the use of a language. Each variety thereof could be seen as a language system in itself, although there have been many attempts to define the relationship of language with dialect or regional accents or varieties. Varieties can be regional (horizontal) or social (vertical), or both; additionally, and within these horizontal and vertical varieties, there are individual varieties characterising a single speaker. For each of these varieties, there exist linguistically distinctive and non-distinctive traits. Again, it is a matter of which language system or variety we take as ground, that is, as typical, when we model atypical variations within this system (in phonological terminology, allophones or free variants) as figure standing out from the ground.[1]

In this section, we deal with units of mostly but not exclusively written ('natural') language that can indicate more or less specific paralinguistic functions, again on an exemplary basis, that is, at the word level, and thus with the lexicon, with word classes, and at the phrase level. We will concentrate on object language, that is, on the use of linguistic elements by individuals or groups, and not on metalanguage, which is employed for describing paralinguistic phenomena and dealt with in Section 5.3.

4.3.1   Words and Word Classes

Rose is a rose is a rose is a rose.

(Gertrude Stein)

The lexicon can change much more easily than the phonology or the syntax of a language; new words emerge, and other words vanish. Moreover, the word is the basic unit in automatic speech recognition (ASR) – no wonder that this, especially in earlier times, has also been called 'automatic word recognition' (Clapper 1971) – and this knowledge can be exploited in computational paralinguistics. For the lexicon, we can do *tokenisation*, that is, map the text onto word classes (e.g., to *part-of-speech* (POS) classes), and *stemming*, that is, cluster

---

[1] On figure–ground organisation in gestalt psychology, see Koffka (1935).

morphological variants of a word by its stem into a *lexeme*. With both strategies, we reduce the number of entries in the vocabulary and at the same time provide more training instances per class. Rough estimates of the size of the lexicon are 500 000 items for English and German, and 1000 items for the basic vocabulary needed to understand some 85% of a text. The most frequent words are function words (articles, personal pronouns, etc.). Although the lexicon of a speech corpus or of a specific application such as call-centre interaction is much smaller than the lexicon of a language, this illustrates the problem: the lexicon is an open class, and we will often face many unknown, so-called *out-of-vocabulary (OOV)* words. The lexicon of a language or of a language variety is composed of many entries that differ in connotation, not in denotation. The users of this language (variety) can choose amongst them in order to express their likes and dislikes. In our context, the semantic connotations of content words are more interesting than those of function words; as for (relative or absolute) frequency, both content and function words are relevant. Thus, there is a nested type–token relationship. By way of illustration, let us return to the *woman* versus *slut* example from Section 1.2: *woman* is the denotation type, and one token denoting this type, but with rather specific connotations, is *slut*. Looking at such type–token relationships, we might be able to find out something about the 'social psychology behind languages' (Batliner 1981). As for language usage, we can have a look at how often a specific word (type), such as *slut*, has been used by a specific speaker (token) or by groups of speakers. This is a genuine topic in paralinguistic research which, however, normally addresses not single words but bunches of words with either negative or positive connotations (valence), or indicating specific traits and states or types of speech/language.

Single words or just a few words can serve to indicate the language or regional dialect and thereby the nationality of speakers; so far, this information has been used more by humans than within computational paralinguistic approaches. For instance, during the few decades when there were two German republics, the GDR and the FRG, it was not morphology or syntax that drifted apart, but the lexicon. Thus, the use of *Broiler* instead of *Brathähnchen* (roast chicken), or of *Plaste* instead of *Plastik* (plastic) revealed that the speaker came from the GDR.

The first applied sociologists (known to us) were those Israelites who used the word *shibbóleth* to distinguish friends from enemies, within a rather straightforward application:

> *The Gileadites captured the fords of the Jordan leading to Ephraim, and whenever a survivor of Ephraim said, 'Let me cross over', the men of Gilead asked him, 'Are you an Ephraimite?' If he replied, 'No,' they said, 'All right, say "Shibboleth."' If he said, 'Sibboleth,' because he could not pronounce the word correctly, they seized him and killed him at the fords of the Jordan. Forty-two thousand Ephraimites were killed at that time.*
>
> (Judges 12:5)

This can be compared to today's (biometric) access control by checking a (phonetic) password, or to a rudimentary classification into accents/dialects/languages.

The Ritchie Boys (after Camp Ritchie, Maryland), a special military intelligence unit in the US army composed of Jewish immigrants, were in Europe after the invasion of Normandy, close to the front lines, interrogating prisoners and collecting information. They normally preserved a strong German, non-native accent. When returning to their unit – sometimes from behind enemy lines – and upon being asked for the password, in a few cases, they were mistakenly taken for the enemy and shot dead because of this strong German, that is, non-native, accent. This can be seen as another, tragic application of the shibboleth principle. We can imagine a similar application which can also have a severe impact on people. It is well known that banks have algorithms for distinguishing high-risk and low-risk credit users. Residential area is one important criterion – and it is, in specific constellations, highly correlated with regional accents. When automatic procedures are employed for recognising a stigmatised regional variant, for instance within a call-centre application, the only difference might be that they are not based on a single word but on a bunch of features. The similarity will be a considerable percentage of false alarms.

The Merriam-Webster online defines *shibboleth* as 'a use of language regarded as distinctive of a particular group' – thus we could say that what we want to do in paralinguistics is to find shibboleths that distinguish different groups, whether it be different social classes, learners of a foreign language with different proficiency levels, or different emotions. In the same way, we might speak of 'bags of shibboleths', echoing *bags-of-words*. Note that in dialectology, the term *isogloss* is used for the regional, geographic distributions of specific words, and *isophone* for specific sounds. Both can be considered as shibboleths distinguishing dialects. Thus, 'language' as used in the definition of Merriam-Webster online given above is used in a broad sense: the use of specific words (lexemes) or phrases, or the use of specific phonetic varieties of such linguistic units.

It is intuitively plausible that such shibboleths, especially those with strong connotations, can be very useful in modelling paralinguistics. The problem is, however, that they might be sparse – even *hapax legomena*, that is, found only once in the database; thus chances are high that they have to be modelled as OOV words. If we wanted to avoid OOV words, we would have to include all words in the lexicon that possibly might occur. However, we would then expand the lexicon enormously, and this is not a good idea either because it means that the word error rate goes up: there are too many words that can be confounded with each other. On the positive side, there is even quite a lot of information in the distinction between content word and function word, and in a rough assignment of words to a few cover classes. This will now be exemplified with two studies.

Chung and Pennebaker (2007) used a text analysis program called 'Linguistic Inquiry and Word Count' (LIWC) for counting both content and style words within any given text, and for mapping about 2000 words or word stems onto (negative and positive) emotion and function word classes. The most commonly used words are function words (I, the, and, to, a, of, that, in, it, my, . . .), that is, (personal) pronouns, articles, conjunctions, or prepositions. They found across multiple studies and experiments that the use of first person singular is associated with negative affective states; that the combined use of first person singular pronouns and exclusive words predicts honesty; that, within dyads, the person who uses 'I' words to a lesser extent tends to be the higher-status participant; and that females tend to use first person singular pronouns at a consistently higher rate than do males.

In Batliner *et al.* (2011) and Schuller *et al.* (2007) experiments are reported using a database with German children giving commands to Sony's pet robot Aibo; this is the same database as used in the Interspeech 2009 Emotion Challenge (Schuller *et al.* 2009b); further details can be found in Batliner *et al.* (2008), Steidl (2009) and in Section 2.13.1 above. Note that before this challenge, only a subset of the whole database was used for processing. A large feature vector with more than 4000 features modelling different types of acoustic and linguistic parameters was employed for classifying automatically three emotion (affect) classes, the social emotion *motherese, emphatic* as a pre-stage to negative, and the negative emotion *angry*, together with the default case *neutral*. Six POS classes were annotated and modelled: nouns, inflected adjectives and participles, not-inflected adjectives and participles, verbs, auxiliaries, and particles/interjections. A cross-tabulation of the four emotion categories with the six POS classes illustrates the high impact of POS due to the unbalanced distribution: more adjectives for *motherese* (e.g., *good boy*), more verbs for *emphatic* (e.g., *stop*), and more nouns and fewer particles for *angry* (e.g., vocative *Aibo*). In Schuller *et al.* (2007) performance is reported for employing the feature sets separately, in Batliner *et al.* (2011) the features were employed in combination and different measures for the contribution of the single types were reported. POS features obviously model (positive or negative) valence and syntactic/semantic salience to a high extent: only employing these six POS features to distinguish the four classes yielded a competitive classification performance, compared to only employing the prosodic features of duration, energy, or spectrum or cepstrum (e.g., MFCC) separately. This shows the high impact of such a rough measure which seems to be quite robust against ASR errors. This might be due to a high contingency of duration characteristics with POS main classes: content words are, on average, longer than function words. Thus, even misrecognised words provide salient information if confusion occurs only within cover classes.

4.3.2   Phrase Level: The Case of Filler Phrases and Hedges

No, I say, dash it!
                                (Berti Wooster in P.G. Woodhouse, *Right Ho, Jeeves*)

Filler phrases are words or combinations of words (phrases) that are *in some respect sort of* not very important for interpreting the referential meaning of an utterance, *I'd say*. Lakoff (1973a) introduced the term *hedges* for deintensifying words such as *sort of, technically, strictly speaking, in some respects, in a sense*, and tried to establish semantic criteria. In German, well-known fillers are *irgendwie* 'somehow' or 'sort of' (some time ago rather fancy in the post-flower-power sub-culture), and *sozusagen* 'in a way' (idiosyncratic habit, very conspicuous in TV interviews and talk shows). There is no clear dividing line between pure fillers and modal particles (Fischer 2007) or sentence adverbs. The term 'hedges' seems to be used – somewhat vaguely – for 'a class of devices that supposedly soften utterances by signalling imprecision and noncommitment' (Dixon and Foster 1997, p. 90), 'expressing the speaker's lack of commitment to an entire proposition' (Ranganath *et al.* 2013, p. 97), and *sort of* used synonymously for 'fillers'.

The very fact that fillers/hedges do not have a specific syntactic function or semantic meaning, apart from this vague desintensifying function which, however, obscures this very function if employed too often, makes them well suited for indicating paralinguistic functions. Thus while, for a strictly semantic analysis, we might be tempted to throw them away, for a paralinguistic analysis, it is worth making them the object of investigation. This has been done by three studies to which we now turn.

The observation made by Lakoff (1973b) that women use hedges more often than men in order to sound feminine, based rather on anecdotal evidence, has been examined by (Dixon and Foster 1997, p. 89), employing a sample of South African students:

> *The results showed that contextual influences eclipsed the effects of gender; in fact, no main effects were found for speaker gender. . . . perhaps reflecting differences in social status, both sexes used* sort of *to express tentativeness more frequently when talking to male addressees. When speaking to female addressees, on the other hand, men deployed facilitative* you know *hedges more readily than women.*

This is in line with studies that scrutinised the prejudice that women are more talkative than men. When we try to summarise the outcomes, it has to be 'it always depends', namely on the specific circumstances. Thus, phenomena such as hedges are definitely paralinguistic markers, but we have to take a close look at the sample, the context, and the scenario.

Salager-Meyer (2011) investigated the use of hedges in research papers by academics belonging to different nationalities. She concludes that

> *research papers in French use much more prescriptive, authoritarian, and categorical language than those written by English-speaking colleagues. . . . Arrogance and over self-confidence (that is, a lack of hedging devices) have also been noted in Finnish academic writing and in research papers written in Bulgarian and English by Bulgarian-speaking scientists when compared with research papers written in English by native-English-speaking scientists, thus suggesting that Finnish and Bulgarian academic writers show a higher degree of commitment and, consequently, a lower degree of deference, toward the discourse community than their English counterparts.*
>
> (Salager-Meyer 2011, p. 36)

The author herself adds the caveat that several factors might additionally be taken into account, such as the writer's status, age, and sex.

Ranganath *et al.* (2013) studied, among other phenomena, the use of hedges in a speed dating scenario:

> *Across the studies and labelers, friendly men tend to use less hedges, less* uh, *less* you know, *and have more varied intensity . . . by far the strongest linguistic association with awkward men and women is the use of hedges. . . . results suggest that hedges are also used metalinguistically to indicate the speaker's psychological distancing from or discomfort with the situation; words which are distancing at the semantic or pragmatic level acquire the metapragmatic connotations of distancing.*
>
> (Ranganath *et al.* 2013, pp. 103, 111)

Again, what we see is that the use of hedges is not gender-specific but rather characterises personality.

Hedges seem to have similar distributions and functions as some disfluencies (see Section 4.4); words such as *well, I say,* filled pauses such as *uh, uhm,* and hesitations/lengthening often serve the same purpose.

4.4   Disfluencies

Well, why is life worth living? That's a very good question. Uhm, well, there are certain things I–I guess that make it worthwhile. Uh, like what? Okay. Uhm, for me, ah, ooh, I would say – what, Groucho Marx, to name one thing.

(Woody Allen, *Manhattan*)

As always, there are different terms, partly synonyms, to describe the phenomenon we want to deal with in this section. We decided in favour of 'disfluencies' denoting all those phenomena that can be observed in spontaneous, not preplanned, not acted, not prompted speech, and in 'spontaneous writing' which has not been corrected in a second pass. Disfluencies deviate from 'correct' language use, although they are very common and only 'atypical' in the sense that they have not been that often object of investigation in automatic speech processing. The quotation by Woody Allen represents speech, translated into orthography. Similar things can happen in careless, 'spontaneous' typing, so one of these sentences, if written carelessly, could look like this: *'well there ar ecetiran things that mammke it worhtwlhie.'* Although we want to concentrate on speech in this section, we point out that there are definitely parallels in writing that can be employed, for instance, in forensics; think of blackmailers making a telephone call or writing a letter on a typewriter: here, speaker and writer identification or verification monitor both the manifestations of dynamic processes, be it prosodic peculiarities or the timing of the keystrokes manifested in density and blackening.

Fluency denotes the undisturbed flow in producing correct speech, *disfluency* denotes problems pertaining to segments, prosody, and grammar. Spontaneous speech differs from preplanned (read) speech in several respects. Overall, articulation and pronunciation can be less pronounced and slurred, with more deletions and contractions. Moreover, it can display so-called *agrammatical* phenomena such as (more) *unfilled pauses* (silence) and *filled pauses* (*uhm, well*), and *hesitations* (lengthening of syllables) to provide additional time for planning; *repetitions, slips of the tongue, false starts* with an *interruption point* and *repair* (fresh start), because of planning difficulties; at the syntactic level, non-standard grammar – either agrammatical or following the rules of some sub-standard variety, and more frequent use of *ellipsis* (deletion of pronouns, etc.); and more colloquial choices from the lexicon. Similar phenomena can be observed in less formal, more colloquial written language, such as emails and blogs; here, typing errors are the equivalent of agrammatical phenomena in spontaneous speech.

Two early studies dealt with hesitations: Goldman-Eisler (1961) with the distribution of filled and unfilled pauses, and Martin and Strange (1968) with the varying frequency of hesitations in spontaneous speech, due to different experimental conditions. Wingate (1984) pointed out that the term 'dysfluency' should be used only for pathological, that is, deviant speech (see Section 5.6), because the prefix 'dys' denotes abnormality, for instance, in stuttering, whereas 'dis' denotes 'apart' or 'not', that is, simply 'non-fluency'. Levelt (1983) developed a three-stage model of self-repairs in speech: monitoring of own's own speech and the interruption of the speech flow as first phase; in the second phase, hesitations, pauses and the so-called *edit terms* (*uhm* or *no, I mean*) can be observed; then, the repair follows in the third phase. This model prepared the ground for the subsequent, elaborated and refined, models. Self-corrections concern not only the chain of words but also prosody as the carrier of words in speech; this is the topic of Levelt and Cutler (1983). The phenomena are embedded in a theory of 'Speaking: From Intention to Articulation' in Levelt (1989).

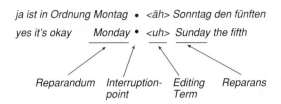

Figure 4.2   A speech repair

Commonly, a repair is segmented into four parts that have to be detected and processed differently:

- **reparandum**, the "wrong" part of the utterance;
- **interruption point (IP)**, the boundary marker at the end of the reparandum;
- **edit(ing) term**: special phrases which indicate a repair, such as *well, I mean* or filled pauses such as *uhm, uh* (optional, most of the time missing);
- **reparans**, the correction of the reparandum.

Such repairs can be very complex; a simple example is given in Figure 4.2 from the Verbmobil scenario (Spilker *et al.* 2001).

In the 1990s, ASR researchers began to be interested in spontaneous, that is, less regular speech data. This interest initiated studies on prosodic differences between spontaneous and read speech (Batliner *et al.* 1995a, 1997; Blaauw 1995; Daly and Zue 1992; Silverman *et al.* 1992), and on the modification of syntactic phenomena such as sentence mood within and by spontaneous speech (Batliner *et al.* 1993). For dealing with the syntax and pragmatics of spontaneous speech, other and more shallow approaches have been developed (Batliner *et al.* 1998). As spontaneous speech turned out to be 'syntactically deficient', another emerging topic was the problem of how to deal with such agrammatical phenomena in order to obtain grammatically 'clear' speech which could be passed on to higher modules for further processing in end-to-end systems such as Verbmobil (Wahlster 2000).

Spontaneous speech phenomena in general are dealt with in Llisterri (1992) and Shriberg (2005). Hindle (1983) developed a first system of rules for resolving non-fluencies in speech. Further approaches towards reaching this goal are reported in Bear *et al.* (1992), Heeman and Allen (1999), Nakatani and Hirschberg (1993), Shriberg (1994), and Batliner *et al.* (1995b). Spilker *et al.* (2001) describe the processing of speech repairs in the Verbmobil system and point out and evaluate the problems one has to face when modelling speech repairs in a real and full end-to-end system.

In 'normal' automatic speech processing, disfluencies are not an object of investigation but something one wants to get rid of. However, in paralinguistics, their (speaker-specific) frequencies, and their distributions can of course indicate specific paralinguistic functions such as emotions, or especially experiencing flow (versus becoming stuck) in learning environments. They can as well be taken as characterising speaker idiosyncrasies, that is, be employed for speaker verification or identification, and for personality trait investigations. Cucchiarini *et al.* (2000) found out that *rate of speech* correlates highly with perceived fluency. This has been corroborated by Hönig *et al.* (2012) with a very large feature vector with and without duration/tempo features. Devillers *et al.* (2005) use manually annotated disfluencies for classifying automatically emotions within an emergency call-centre scenario. Naturally enough, these features contribute less to classification performance in another scenario with another type of interaction (giving commands to a robot), and thus a lower overall frequency of disfluencies (Batliner *et al.* 2011).

It seems to us that the focus of interest has changed in automatic speech processing away from a rather detailed modelling and processing of dialogues and especially disfluencies, to a coarser-grained processing where phenomena are modelled more implicitly. In the same vein, disfluencies still seem yet to be the Cinderella of computational paralinguistics; this might be due to the complexity of the task of modelling them explicitly. However, they definitely are modelled implicitly as well within approaches towards automatic assessment of non-native prosody, for instance, with features representing speech tempo, rhythm, and duration (Hönig *et al.* 2012).

4.5   Non-Verbal, Vocal Events

Mr. Swinhoe informs me that he has often seen the Chinese, when suffering from deep grief, burst out into hysterical fits of laughter.

(Darwin 1872)

As pointed out earlier, 'non-verbal' is often used in a very broad sense, meaning 'everything that happens within human–human communication and which is not strictly verbal, that is, belonging to linguistics'; we want to confine the realm of 'non-verbal' here to 'vocal' events. So far we have addressed 'verbal, vocal events'; in this section we turn to 'non-verbal, vocal events'. In their pure form, non-verbals can be segmented on the time axis; in this case, verbals, that is, words, and non-verbals are mutually exclusive. Thus it is possible to model and process non-verbal events along the lines of words in automatic speech processing. Sometimes non-verbals such as filled pauses or some types of *affect bursts* or 'vocal outbursts' (Scherer 1994; Schröder 2000) such as *oh, wow* cannot only be segmented out from the word chain but comply with the phonotactics of a language as well; sometimes they do not, as is the case for coughing or sobbing.

Quite a lot of non-verbals exist in a less pure form as well, not separable from but synchronous with speech; these include laughing, sobbing, crying, coughing, which all can be separate events or *modulated onto the word chain*. However, this phenomenon differs from other paralinguistic parameters that are modulated onto the word chain as well (e.g., pitch) or voice quality (e.g., harshness, breathiness, or laryngealisations). The latter only co-occur together with a verbal event – there is no stand-alone pitch or laryngealisation, whereas non-verbal events can be stand-alone as well. We can peel these non-verbals away from their carrier, that is, from words, and produce them alone, with a sort of neutral phonation as carrier, and vice versa. However, this is only possible if there is no articulation plus phonation yet, as is the case for the so-called affect bursts which have to be conceived as holistic events.

Isolated non-verbals that are not embedded in an utterance can have *functions* similar to (short) utterances and can be replaced by them; imagine *wow* instead of *that's really awesome*, sighing instead of *I'm really sorry for you*, *uhm* instead of *well*, or laughing instead of saying *I don't think so* or *that's funny*. Of course, connotations can differ: while *well* instead of *uhm* will not make much difference, breaking into laughter instead of saying *that's funny* does.

A frequent function of non-verbals might be the same as that of disfluencies: to indicate planning, or to hold the floor. This replacement test holds of course for non-vocal/non-verbal phenomena as well (shrugging, rising the eyebrows, head-shaking); however, these are outside the vocal code.

Examples of some recent approaches that deal with non-verbals include sighs and yawns (Russell *et al.* 2003), cries (Pal *et al.* 2006), hesitations and consent (Schuller *et al.* 2009a), so-called 'grunts' (Campbell 2007a; Ward 1998, 2000a,b), and coughs (Matos *et al.* 2006). Provine (2012) lists laughter amongst other 'curious behaviour' such as yawning and hiccupping. Even silence can be conceived of as a non-verbal, paralinguistic phenomenon (Ephratt 2011) – after all, in a strict sense, it is not possible not to communicate in a dyadic or multi-party scenario (Watzlawick *et al.* 1967). Arguably, *laughter* is the prototypical non-verbal event, and therefore the exemplar we want to illustrate in the following.

The layman's idea of the default phonetic form of laughter is not wrong; it is often ortho-graphically described as 'hahaha' ([h@|h@|h@] in SAMPA (Speech Assessment Methods Phonetic Alphabet) notation). This form follows a syllable structure frequently occurring in many languages. Of course, there are many more phonetic forms of laughter – voiced, unvoiced or mixed – which do not necessarily obey phonotactic rules. The acoustics of laughter are described in Bacharowski and Smoski (2001) and Trouvain (2001, 2003) and in further studies referred to in these articles. Nwokah *et al.* (1999) studied mother–infant interactions, coining the term *speech-laughs* for laughter modulated onto speech, and described the phonetic forms of this type of laughter; before, laughter had been studied rather as an isolated event, not modulated onto speech (Trouvain 2001).

Normally, laughter is conceived of as a non-linguistic or paralinguistic event; it has been studied extensively by non-linguists, such as biologists and psychologists, because it can be found in other primates as well. As one way to express emotions (especially joy), it was dealt with by Darwin (1872); studies on its acoustics, however, as well as on its position in linguistic context – in the literal meaning of the word (where it can be found in the word chain; see Provine (1993) and Batliner *et al.* (2013)), and in the figurative sense (status and function) – started more or less at the same time as automatic speech processing started to deal with paralinguistic phenomena (Provine 1996). Rees and Monrouxe (2010) deal with the 'construction of power, identity and gender through laughter within medical workplace learning encounters', while Smoski and Bacharowski (2003) address the function of laughter as reflecting a mutually positive stance between social partners. Holt (2010) covers reciprocal and non-reciprocal laughters. Vettin and Todt (2004, p. 93) suggest that '... laughter in conversation may primarily serve to regulate the flow of interaction and to mitigate the meaning of the preceding utterance'. Szameitat *et al.* (2009) establish different acoustic profiles of distinct emotional expressions in laughter, and Kuiper and Martin (1998) and Bacharowski and Owren (2001) relate laughter to positive and negative affect.

The context of laughter is addressed in Campbell *et al.* (2005) and Campbell (2007b), namely different types of laughter and their function or different addressees in communication, and in Laskowski and Burger (2007) who deal with the distribution of laughter within multi-party conversations. The automatic classification of laughter is dealt with in Batliner *et al.* (2013), Kennedy and Ellis (2004), Laskowski (2009), Laskowski and Schultz (2008), Petridis and Pantic (2008, 2011), Truong and Leeuwen (2005, 2007), and further studies referred to in these articles. Detecting, that is, locating in the word chain, laughter and especially speech-laughs might not be easy; it is easier to decide whether laughter or speech-laughs occurred somewhere in an utterance (Batliner *et al.* 2013), and this might suffice in most applications.

Naturally enough, different frequencies of laughter are reported in the literature, depending on scenario and context: Petridis and Pantic (2008) use seven sessions from the Augmented Multi-party Interaction Meeting corpus, where subjects were recruited for the task, and pre-select those 40 laughter segments that do not co-occur with speech and are 'clearly audible' (total duration 58.4 seconds). Kennedy and Ellis (2004) report '1926 ground truth laughter events' found in 29 meetings (about 25 hours), the so-called 'Bmr subset' of the ICSI Meeting Recorder Corpus, divided into 26 train and 3 test meetings. Laskowski and Schultz (2008) report for the same partition 14.94% 'of vocalisation time spent in laughter' for train, and 10.91% for test; another subset of the ICSI meeting data (the so-called 'Bro subset') contains only 5.94% laughter. This is due to different types of interaction and participants who were more or less familiar with each other. On the other hand, only few laughter instances were found in 'transcript data of jury deliberations from both the guilt-or-innocence and penalty phases of [a] trial' (Keyton and Beck 2010, 386): '51 laughter sequences across 414 transcript pages''. In Batliner *et al.* (2013), 0.6% of all tokens that are either words or laughter instances are either speech-laughs or laughter; this low frequency is due to the task of the children in this database who had to give commands to a pet robot. In contrast to Nwokah *et al.* (1999), almost no combination of *motherese* with laughter, that is, speech-laughs indicating intimacy, could be found. This illustrates that generic statements about frequencies, occurrences and combinations cannot be made; it always depends on the type of interaction and scenario.

4.6   Common Traits of Formal Aspects

No man is an island.

(John Donne)

Each man is a whole Universe.

(Yuri Borev)

There are different systems of description for formal aspects which more or less overlap. Examples can be found in Sections 4.4 and 4.5. It is not possible to find a system that is consistent in itself and fully defined, because of the fringe phenomena that can be attributed to one system or another; a description as family resemblances seems to be a better choice. Moreover, we can describe classes based on their form or on their function: 'grunts' is a rather formal term, whereas 'hedges' denotes functions. 'Disfluencies' is formal and functional at the same time.

Kreiman and Sidtis (2011, p. 181) write with respect to recognising speaker identity from voice: 'Studies examining such small sets of features (whether in animal or human vocalization) miss the point that perceptually important features emerge idiosyncratically in different patterns, and that it is the unique relationship of a constellation of parameters taken from a very large set that signals a given unique pattern.' More successful than trying to recognise an individual speaker might be to try and characterise social group traits from voice, or linguistic functions such as focal parts or the end of the turn (yielding the floor) all of which are not idiosyncratic but 'cross-individual'; therefore, we simply have more instances for training. However, there will be more variety. Thus, it is always an empirical question which can only be answered reliably if we employ a very large feature vector, and not only just a few features, or only one specific type of features such as pitch (see Section 3.9).

Very often we eventually have to disentangle personal traits (speaker idiosyncrasies) from supra-individual traits. Speakers can develop personal traits – employ specific kinds of rhotics, or display above average percentages of segmental laryngealisations or specific pitch contours. However, all this is of course influenced by supra-individual tendencies, such as group behaviour, regional variants, or pathologies – similarly to the way personality influences and shapes the individual expression of emotions. When we look closely at such an individual speech behaviour, we could call this *micro-paralinguistics*. In contrast, the usual approach is *macro-paralinguistics*, levelling out individual traits and aiming at tendencies across single speakers. In practice, individual traits are – naturally enough – important if it is about one suspect (forensics), or one patient (diagnostics and speech therapy); see Section 3.8.

There are single Lego bricks that can be assembled to construct an excavator that can be used to transport sand from A to B. The same Lego brick can be used for constructing different devices which in turn can be used for the same or for different purposes. In the same way, there are single phonetic and linguistic parameters that can be used to build different linguistic constructs that in turn can serve different linguistic and/or paralinguistic functions. Thus, we have to keep in mind this multi-functionality of single (formal) parameters and of bundles of parameters, when doing computational paralinguistics.

References

Abercrombie, D. (1967). *Elements of General Phonetics*. Edinburgh University Press, Edinburgh.

Ammon, U., Dittmar, N., Mattheier, K. J., and Trudgill, P. (eds) (2004). *Sociolinguistics / Soziolinguistik. An International Handbook of the Science of Language and Society / Ein internationales Handbuch zur Wissenschaft von Sprache und Gesellschaft*. Walter de Gruyter, Berlin.

Aronoff, M. and Rees-Miller, J. (eds) (2001). *The Handbook of Linguistics*. Blackwell, Malden, MA.

Bacharowski, J.-A. and Owren, M. J. (2001). Not all laughs are alike: Voiced but not unvoiced laughter readily elicits positive affect. *Psychological Science*, **12**, 252–257.

Bacharowski, J.-A. and Smoski, M. J. (2001). The acoustic features of human laughter. *Journal of the Acoustical Society of America*, **110**(3), 1581–1597.

Batliner, A. (1981). Sexismus und Häufigkeit. *Deutsche Sprache*, **9**, 312–328.

Batliner, A. (1989). Wieviel Halbtöne braucht die Frage? Merkmale, Dimensionen, Kategorien. In H. Altmann, A. Batliner, and W. Oppenrieder (eds), *Zur Intonation von Modus und Fokus im Deutschen*, pp. 111–162. Niemeyer, Tübingen.

Batliner, A., Weiand, C., Kießling, A., and Nöth, E. (1993). Why sentence modality in spontaneous speech is more difficult to classify and why this fact is not too bad for prosody. In *Proc. of the ESCA Workshop on Prosody*, pp. 112–115, Lund.

Batliner, A., Burger, S., Johne, B., and Kießling, A. (1994). MÜSLI: A classification scheme for laryngealizations. In *Proc. of the ESCA Workshop on Prosody*, pp. 176–179, Lund.

Batliner, A., Kompe, R., Kießling, A., Nöth, E., and Niemann, H. (1995a). Can you tell apart spontaneous and read speech if you just look at prosody? In A. J. R. Ayuso and J. M. L. Soler (eds), *Speech Recognition and Coding: New Advances and Trends*, NATO ASI Series F, pp. 321–328. Springer, Berlin.

Batliner, A., Kießling, A., Burger, S., and Nöth, E. (1995b). Filled pauses in spontaneous speech. In *Proc. of ICPhS*, volume 3, pp. 472–475.

Batliner, A., Kießling, A., Kompe, R., Niemann, H., and Nöth, E. (1997). Tempo and its change in spontaneous speech. In *Proc. of Eurospeech*, pp. 763–766, Rhodes.

Batliner, A., Kompe, R., Kießling, A., Mast, M., Niemann, H., and Nöth, E. (1998). M = Syntax + Prosody: A syntactic–prosodic labelling scheme for large spontaneous speech databases. *Speech Communication*, **25**(4), 193–222.

Batliner, A., Biersack, S., and Steidl, S. (2006). The prosody of pet robot directed speech: Evidence from children. In *Proc. of Speech Prosody*, pp. 1–4, Dresden.

Batliner, A., Steidl, S., and Nöth, E. (2007). Laryngealizations and emotions: How many babushkas? In *Proc. of the International Workshop on Paralinguistic Speech – between Models and Data (ParaLing'07)*, pp. 17–22, Saarbrücken.

Batliner, A., Steidl, S., Hacker, C., and Nöth, E. (2008). Private emotions vs. social interaction — a data-driven approach towards analysing emotions in speech. *User Modeling and User-Adapted Interaction*, **18**, 175–206.

Batliner, A., Steidl, S., Schuller, B., Seppi, D., Vogt, T., Wagner, J., Devillers, L., Vidrascu, L., Aharonson, V., and Amir, N. (2011). Whodunnit: Searching for the most important feature types signalling emotional user states in speech. *Computer Speech and Language*, **25**, 4–28.

Batliner, A., Steidl, S., Eyben, F., and Schuller, B. (2013). On laughter and speech laugh, based on observations of child-robot interaction. In J. Trouvain and N. Campbell (eds), *The Phonetics of Laughing*. Saarland University Press, Saarbrücken. To appear; see http://www5.informatik.uni-erlangen.de/Forschung/Publikationen/2011/Batliner11-OLA.pdf.

Bear, J., Dowding, J., and Shriberg, E. (1992). Integrating multiple knowledge sources for detection and correction of repairs in human computer dialogs. In *Proc. of ACL*, pp. 56–63, Newark, Delaware.

Blaauw, E. (1995). *On the Perceptual Classification of Spontaneous and Read Speech*. OTS dissertation series, Utrecht.

Bloomfield, L. (1933). *Language*. Holt, Rinehart and Winston, New York.

Böhm, T. and Shattuck-Hufnagel, S. (2007). Listeners recognize speakers' habitual utterance final voice quality. In *Proc. of the International Workshop on Paralinguistic Speech – between Models and Data (ParaLing'07)*, pp. 29–34, Saarbrücken.

Bussmann, H. (ed.) (1996). *Routledge Dictionary of Language and Linguistics*. Routledge, New York.

Campbell, N. (2007a). On the use of nonverbal speech sounds in human communication. In *Proc. of the International Workshop on Paralinguistic Speech – between Models and Data (ParaLing'07)*, pp. 23–28, Saarbrücken.

Campbell, N. (2007b). Whom we laugh with affects how we laugh. In J. Trouvain and N. Campbell (eds), *Proc. of the Interdisciplinary Workshop on The Phonetics of Laughter*, pp. 61–65, Saarbrücken.

Campbell, N. and Mokhtari, P. (2003). Voice quality: The 4th prosodic dimension. In *Proc. of ICPhS*, pp. 2417–2420.

Campbell, N., Kashioka, H., and Ohara, R. (2005). No laughing matter. In *Proc. of Interspeech*, pp. 465–468, Lisbon.

Chambers, J. and Trudgill, P. (1980). *Dialectology*. Cambridge University Press, Cambridge.

Chung, C. and Pennebaker, J. (2007). The psychological functions of function words. In K. Fiedler (ed.), *Social Communication*, pp. 343–359. Psychology Press, New York.

Clapper, G. (1971). Automatic word recognition. *IEEE Spectrum*, **8**, 57–69.

Comrie, B. (ed.) (2009). *The World's Major Languages*. Routledge, London. 2nd edition.

Coulmas, F. (ed.) (1997). *The Handbook of Sociolinguistics*. Blackwell, Oxford.

Crystal, D. (1997). *The Cambridge Encyclopedia of Language*. Cambridge University Press, Cambridge.

Crystal, D. (2008). *A Dictionary of Linguistics and Phonetics*. Blackwell, Oxford, 6th edition.

Cucchiarini, C., Strik, H., and Boves, L. (2000). Quantitative assessment of second language learners' fluency by means of automatic speech recognition technology. *Journal of the Acoustical Society of America*, **107**(2), 989–999.

Cucchiarini, C., Neri, A., and Strik, H. (2009). Oral proficiency training in Dutch L2: The contribution of ASR-based corrective feedback. *Speech Communication*, **51**, 853–863.

Currie, H. (1952). A projection of sociolinguistics. The relationship of speech to social status. *Southern Speech Journal*, **18**, 28–37.

Daly, N. and Zue, V. (1992). Statistical and linguistic analyses of F0 in read and spontaneous speech. In *Proc. of ICSLP*, volume 1, pp. 763–766, Banff, Alberta.

Darwin, C. (1872). *The Expression of the Emotions in Man and Animals*. John Murray, London.

Devillers, L., Vidrascu, L., and Lamel, L. (2005). Challenges in real-life emotion annotation and machine learning based detection. *Neural Networks*, **18**, 407–422.

Dixon, J. A. and Foster, D. H. (1997). Gender and hedging: From sex differences to situated practice. *Journal of Psycholinguistic Research*, **26**, 89–107.

Drioli, C., Tisato, G., Cosi, P., and Tesser, F. (2003). Emotions and voice quality: Experiments with sinusoidal modeling. In *Proc. of VOQUAL'03*, pp. 127–132, Geneva.

Elias, N. (1939). *Über den Prozeß der Zivilisation. Soziogenetische und psychogenetische Untersuchungen. Band 1: Wandlungen des Verhaltens in den weltlichen Oberschichten des Abendlandes / Band 2: Wandlungen der Gesellschaft: Entwurf zu einer Theorie der Zivilisation*. Verlag Haus zum Falken, Basel.

Ephratt, M. (2011). Linguistic, paralinguistic and extralinguistic speech and silence. *Journal of Pragmatics*, **43**, 2286–2307.

Erickson, D., Yoshida, K., Menezes, C., Fujino, A., Mochida, T., and Shibuya, Y. (2004). Exploratory study of some acoustic and articulatory characteristics of *sad* speech. *Phonetica*, **63**, 1–25.

Fischer, K. (2007). Grounding and common ground: Modal particles and their translation equivalents. In A. Fetzer and K. Fischer (eds), *Lexical Markers of Common Grounds, number 3 in Studies in Pragmatics*, pp. 47–66. Elsevier, Amsterdam.

Fischer-Jørgensen, E. (1989). Phonetic analysis of the stød in standard Danish. *Phonetica*, **46**, 1–59.

Foulkes, P. and Docherty, G. (2006). The social life of phonetics and phonology. *Journal of Phonetics*, **34**, 409–438.

Foulkes, P., Scobbie, J., and Watt, D. (2010). Sociophonetics. In W. Hardcastle, J. Lavel, and F. Gibon (eds), *Handbook of Phonetic Sciences, 2nd edition*, pp. 703–754. Blackwell, Oxford.

Freese, J. and Maynard, D. W. (1998). Prosodic features of bad news and good news in conversation. *Language in Society*, **27**, 195–219.

Fromkin, V., Rodman, R., and Hyams, N. (2002). *An Introduction to Language*. Thomson/Heinle, Boston. 7th edition.

Gerfen, C. and Baker, K. (2005). The production and perception of laryngealized vowels in Coatzospan Mixtec. *Journal of Phonetics*, pp. 311–334.

Gerratt, B. R. and Kreiman, J. (2001). Toward a taxonomy of nonmodal phonation. *Journal of Phonetics*, **29**, 365–381.

Gobl, C. and Ní Chasaide, A. (2003). The role of voice quality in communicating emotion, mood and attitude. *Speech Communication*, **40**, 189–212.

Goldman-Eisler, F. (1961). A comparative study of two hesitation phenomena. *Language and Speech*, **4**, 18–26.

Gordon, M. and Ladefoged, P. (2001). Phonation types: a cross-linguistic overview. *Journal of Phonetics*, **29**, 383–406.

Hadding-Koch, K. and Studdert-Kennedy, M. (1964). An experimental study of some intonation contours. *Phonetica*, **11**, 175–185.

Hardcastle, W. J., Laver, J., and Gibbon, F. E. (eds) (2010). *The Handbook of Phonetic Sciences, 2nd Edition*. Wiley-Blackwell, Oxford.

Harrington, J., Palethorpe, S., and Watson, C. I. (2000). Does the Queen speak the Queen's English? *Nature*, **408**, 927–928.

Heeman, P. A. and Allen, J. F. (1999). Speech repairs, intonational phrases, and discourse markers: Modelling speakers' utterances in spoken dialogue. *Computational Linguistics*, **25**(4), 527–571.

Hermes, D. J. and Gestel, J. C. v. (1991). The frequency scale of speech intonation. *Journal of the Acoustical Society of America*, **90**, 97–102.

Hindle, D. (1983). Deterministic parsing of syntactic nonfluencies. In *Proc. of ACL*, pp. 123–128, MIT, Cambridge, Massachusetts.

Holt, E. (2010). The last laugh: Shared laughter and topic termination. *Journal of Pragmatics*, **42**, 1513–1525.

Hönig, F., Batliner, A., and Nöth, E. (2012). Automatic assessment of non-native prosody annotation, modelling and evaluation. In *Proc. of the International Symposium on Automatic Detection of Errors in Pronunciation Training (ISADEPT)*, pp. 21–30, Stockholm.

Horn, L. R. and Ward, G. (eds) (2005). *The Handbook of Pragmatics*. Blackwell, Oxford.

Huber, D. (1988). *Aspects of the Communicative Function of Voice in Text Intonation*. Ph.D. thesis, Chalmers University, Göteborg/Lund.

Hymes, D. (1979). The origin of 'sociolinguistics'. *Language in Society*, **8**, 141–141.

Ishi, C., Ishiguro, H., and Hagita, N. (2005). Proposal of acoustic measures for automatic detection of vocal fry. In *Proc. of Interspeech*, pp. 481–484, Lisbon.

Jakobson, R. (1969). *Kindersprache, Aphasie und allgemeine Lautgesetze*. Suhrkamp, Frankfurt am Main.

Kennedy, L. and Ellis, D. (2004). Laughter detection in meetings. In *Proc. of the ICASSP Meeting Recognition Workshop*, pp. 118–121, Montreal.

Keyton, J. and Beck, S. J. (2010). Examining laughter functionality in jury deliberations. *Small Group Research*, **41**, 386–407.

Kießling, A., Kompe, R., Niemann, H., Nöth, E., and Batliner, A. (1995). Voice source state as a source of information in speech recognition: Detection of laryngealizations. In A. Rubio Ayuso and J. López Soler (eds), *Speech Recognition and Coding: New Advances and Trends*, volume 147 of *NATO ASI Series F*, pp. 329–332. Springer, Berlin.

Koffka, K. (1935). *Principles of Gestalt Psychology*. Harcourt Brace, New York.

Kreiman, J. and Sidtis, D. (2011). *Foundations of Voice Studies: An Interdisciplinary Approach to Voice Production and Perception*. Wiley-Blackwell, Malden, MA.

Kreiman, J., Gerratt, B., Kempster, G., Erman, A., and Berke, G. (1993). Perceptual evaluation of voice quality: Review, tutorial, and a framework for future research. *Journal of Speech and Hearing Research*, **36**, 21–40.

Kreiman, J., Gerratt, B., and Ito, M. (2007). When and why listeners disagree in voice quality assessment tasks. *Journal of the Acoustical Society of America*, **122**, 2354–2364.

Kuiper, N. A. and Martin, R. A. (1998). Laughter and stress in daily life: Relation to positive and negative affect. *Motivation*, **22**, 133–153.

Kushan, S. and Slifka, J. (2006). Is irregular phonation a reliable cue towards the segmentation of continuous speech in American English? In *Proc. of Speech Prosody*, pp. 795–798, Dresden.

Labov, W. (2006). *The Social Stratification of English in New York City*. Cambridge University Press, Cambridge.

Ladefoged, P. and Maddieson, I. (1996). *The Sounds of the World's Languages*. Blackwell, Oxford.

Lakoff, G. (1973a). Hedges: A study in meaning criteria and the logic of fuzzy concepts. *Journal of Philosophical Logic*, **2**, 458–508.

Lakoff, R. (1973b). Language and woman's place. *Language in Society*, **2**, 45–80.

Laskowski, K. (2009). Contrasting emotion-bearing laughter types in multiparticipant vocal activity detection for meetings. In *Proc. of ICASSP*, pp. 4765–4768, Taipei, Taiwan.

Laskowski, K. and Burger, S. (2007). Analysis of the occurence of laughter in meetings. In *Proc. of Interspeech*, pp. 1258–1261, Antwerp.

Laskowski, K. and Schultz, T. (2008). Detection of laughter-in-interaction in multichannel close-talk microphone recordings of meetings. In A. Popescu-Belis and R. Stiefelhagen (eds), *Machine Learning for Multimodal Interaction*, Lecture Notes in Computer Science 5237, pp. 149–160. Springer, Berlin.

Laver, J. (1968). Voice quality and indexical information. *British Journal of Disorders in Communication*, **3**, 43–54.

Laver, J. (1980). *The Phonetic Description of Voice Quality*. Cambridge University Press, Cambridge.

Laver, J. (1991). *The Gift of Speech*. Edinburgh University Press, Edinburgh.

Laver, J. (1994). *Principles of Phonetics*. Cambridge University Press, Cambridge.

Levelt, W. (1983). Monitoring and self-repair in speech. *Cognition*, **14**, 41–104.

Levelt, W. (1989). *Speaking: From Intention to Articulation*. MIT Press, Cambridge, MA.

Levelt, W. J. and Cutler, A. (1983). Prosodic marking in speech repair. *Journal of Semantics*, **2**, 205–218.

Levinson, S. C. (1983). *Pragmatics*. Cambridge University Press, Cambridge.

Lindau, M. (1985). The story of r. In V. A. Fromkin (ed.), *Phonetic Linguistics. Essays in Honour of Peter Ladefoged*, pp. 157–168. Academic Press, Orlando, FL.

Lisker, L. (1986). "Voicing" in English: A catalogue of acoustic features signalling /b/ versus /p/ in trochees. *Language and Speech*, **29**, 3–11.

Lisker, L. and Abramson, A. S. (1964). A cross-language study of voicing in initial stops: Acoustical measurements. *Word*, **20**, 384–422.

Llisterri, J. (1992). Speaking styles in speech research. In *ELSNET/ESCA/SALT Workshop on Integrating Speech and Natural Language*, pp. 1–28, Dublin.

Local, J. and Kelly, J. (1986). Projection and 'silences': notes on phonetic and conversational structure. *Human Studies*, **9**, 185–204.

Lyons, J. (1968). *Introduction to Theoretical Linguistics*. Cambridge University Press, Cambridge.

Maddieson, I. (1984). *Patterns of Sounds*. Cambridge University Press, Cambridge.

Martin, J. G. and Strange, W. (1968). Determinants of hesitations in spontaneous speech. *Journal of Experimental Psychology*, **76**, 474–479.

Matos, S., Birring, S., Pavord, I., and Evans, D. (2006). Detection of cough signals in continuous audio recordings using hidden Markov models. *IEEE Transactions on Biomedical Engineering*, **53**, 1078–1083.

Mey, J. L. (2001). *Pragmatics: An Introduction*. Blackwell, Oxford.

Morris, C. W. (1938). *Foundations of the Theory of Signs*. University of Chicago Press, Chicago.

Nakatani, C. and Hirschberg, J. (1993). A speech-first model for repair detection and correction. In *Proc. of ACL*, pp. 46–53, Columbus, OH.

Newmeyer, F. J. (ed.) (1988). *Linguistics: The Cambridge Survey. Vols I–IV*. Cambridge University Press, Cambridge.

Ní Chasaide, A. and Gobl, C. (2004). Voice quality and f_0 in prosody: Towards a holistic account. In *Proc. of Speech Prosody*, Nara, Japan.

Nicolosi, L., Harryman, E., and Kresheck, J. (2004). *Terminology of Communication Disorders: Speech-Language-Hearing*. Lippincott Williams & Wilkins, Philadelphia.

Nwokah, E. E., Hsu, H.-C., and Davies, P. (1999). The integration of laughter and speech in vocal communication. *Journal of Speech, Language, and Hearing Research*, **42**, 880–894.

Ohala, J. J. (1984). An ethological perspective on common cross-language utilization of F_0 of voice. *Phonetica*, **41**, 1–16.

Painter, C. (1991). The laryngeal vestibule, voice quality and paralinguistic markers. *European Archives of Oto-Rhino-Laryngology*, **248**, 452–458.

Pal, P., Iyer, A., and Yantorno, R. (2006). Emotion detection from infant facial expressions and cries. In *Proc. of ICASSP*, pp. 809–812, Toulouse.

Paulston, C. B. and Tucker, G. R. (eds) (2003). *Sociolinguistics: The Essential Readings*. Blackwell, Malden, MA.

Petridis, S. and Pantic, M. (2008). Audiovisual laughter detection based on temporal features. In *Proc. of ICMI'08*, pp. 37–44, Chania, Greece.

Petridis, S. and Pantic, M. (2011). Audiovisual discrimination between speech and laughter: Why and when visual information might help. *IEEE Transactions on Multimedia*, **13**, 216–234.

Provine, R. (1993). Laughter punctuates speech: linguistic, social and gender contexts of laughter. *Ethology*, **15**, 291–298.

Provine, R. (2012). *Curious Behavior: Yawning, Laughing, Hiccupping, and Beyond*. Harvard University Press, Cambridge, MA.

Provine, R. R. (1996). Laughter. *American Scientist*, **84**, 38–47.

Ranganath, R., Jurafsky, D., and McFarland, D. A. (2013). Detecting friendly, flirtatious, awkward, and assertive speech in speed-dates. *Computer Speech and Language*, **27**, 89–115.

Rees, C. E. and Monrouxe, L. V. (2010). "I should be lucky ha ha ha ha": The construction of power, identity and gender through laughter within medical workplace learning encounters. *Journal of Pragmatics*, **42**, 3384–3399.

Russell, J., Bachorowski, J., and Fernandez-Dols, J. (2003). Facial and vocal expressions of emotion. *Annual Review of Psychology*, **54**, 329–349.

Salager-Meyer, F. (2011). Scientific discourse and contrastive linguistics: hedging. *European Science Editing*, **37**, 35–37.

Sapir, E. (1921). *Language: An Introduction to the Study of Speech*. Harcourt Brace, New York.

Saussure, F. de (1916). *Cours de linguistique générale*. Payot, Paris.

Scherer, K. R. (1994). Affect Bursts. In M. v. Goozen, N. E. v. d. Poll, and J. A. Sergeant (eds), *Emotions*, pp. 161–193. Lawrence Erlbaum, Hillsdale, NJ.

Scherer, K. R. (2003). Vocal communication of emotion: A review of research paradigms. *Speech Communication*, **40**, 227–256.

Schröder, M. (2000). Experimental study of affect bursts. In *Proc. of the ISCA Workshop on Speech and Emotion*, pp. 132–137, Newcastle, Co. Down.

Schuller, B., Batliner, A., Seppi, D., Steidl, S., Vogt, T., Wagner, J., Devillers, L., Vidrascu, L., Amir, N., Kessous, L., and Aharonson, V. (2007). The relevance of feature type for the automatic classification of emotional user states: Low level descriptors and functionals. In *Proc. of Interspeech*, pp. 2253–2256, Antwerp, Belgium.

Schuller, B., Müller, R., Eyben, F., Gast, J., Hörnler, B., Wöllmer, M., Rigoll, G., Höthker, A., and Konosu, H. (2009a). Being bored? Recognising natural interest by extensive audiovisual integration for real-life application. *Image and Vision Computing*, **27**(12), 1760–1774.

Schuller, B., Steidl, S., and Batliner, A. (2009b). The Interspeech 2009 Emotion Challenge. In *Proc. of Interspeech*, pp. 312–315, Brighton, UK.

Scobbie, J. (2006). (R) as a variable. In K. Brown (ed.), *Encyclopedia of Language and Linguistics, Second Edition*, volume 10, pp. 337–344. Elsevier, Oxford.

Seppi, D., Batliner, A., Steidl, S., Schuller, B., and Nöth, E. (2010). Word accent and emotion. In *Proc. of Speech Prosody*, Chicago.

Shriberg, E. (1994). *Preliminaries to a Theory of Speech Disfluencies*. Ph.D. thesis, University of California at Berkeley.

Shriberg, E. (2005). Spontaneous speech: How peoply really talk and why engineers should care. In *Proc. of Interspeech*, pp. 1781–1784, Lisbon, Portugal.

Silverman, K., Blaauw, E., Spitz, J., and Pitrelli, J. (1992). A prosodic comparison of spontaneous speech and read speech. In *Proc. of ICSLP*, pp. 1299–1302, Banff, Alberta.

Smoski, M. J. and Bacharowski, J.-A. (2003). Antiphonal laughter between friends and strangers. *Cognition and Emotion*, **17**, 327–340.

Spilker, J., Batliner, A., and Nöth, E. (2001). How to repair speech repairs in an end-to-end system. In *Proc. of the ISCA Workshop on Disfluency in Spontaneous Speech*, pp. 73–76, Edinburgh.

Steidl, S. (2009). *Automatic Classification of Emotion-Related User States in Spontaneous Children's Speech*. Logos Verlag, Berlin.

Studdert-Kennedy, M. and Hadding, K. (1973). Auditory and linguistic processes in the perception of intonation contours. *Language and Speech*, **16**, 293–313.

Szameitat, D., Alter, K., Szameitat, A., Wildgruber, D., Sterr, A., and Darwin, C. (2009). Acoustic profiles of distinct emotional expressions in laughter. *Journal of the Acoustical Society of America*, **126**, 354–366.

Trask, R. (1996). *A Dictionary of Phonetics and Phonology*. Routledge, London.

Trouvain, J. (2001). Phonetic aspects of 'speech laughs'. In *Proc. of the Conference on Orality and Gestuality Orage 2001*, pp. 634–639, Aix-en-Provence.

Trouvain, J. (2003). Segmenting phonetic units in laughter. In *Proc. of ICPhS*, pp. 2793–2796, Barcelona.

Trudgill, P. (2004). Sociolinguistics: An overview. In *Sociolinguistics/Soziolinguistik. An International Handbook of the Science of Language and Society/Ein internationales Handbuch zur Wissenschaft von Sprache und Gesellschaft*, pp. 1–5. Walter de Gruyter, Berlin.

Truong, K. and Leeuwen, D. v. (2005). Automatic detection of laughter. In *Proc. of Interspeech*, pp. 485–488, Lisbon.

Truong, K. P. and Leeuwen, D. A. v. (2007). Automatic discrimination between laughter and speech. *Speech Communication*, **49**, 144–158.

Van Bezooijen, R. (1995). Sociocultural aspects of pitch differences between Japanese and Dutch women. *Language and Speech*, **38**, 253–265.

Van Hout, R. and Van de Velde, H. (2001). Patterns of /r/ variation. In R. Van Hout and H. Van de Velde (eds), *'r-atics. Sociolinguistic, phonetic and phonological characteristics of /r/*, in *Etudes & Travaux*, **4**, 1–9.

Vettin, J. and Todt, D. (2004). Laughter in conversation: Features of occurrence and acoustic structure. *Journal of Nonverbal Behavior*, **28**, 93–115.

Wahlster, W. (ed.) (2000). *Verbmobil: Foundations of Speech-to-Speech Translations*. Springer, Berlin.

Ward, N. (1998). The relationship between sound and meaning in Japanese back-channel grunts. In *Proc. of the 4th Annual Meeting of the (Japanese) Association for Natural Language Processing*, pp. 464–467, Fukuoka.

Ward, N. (2000a). Issues in the transcription of English conversational grunts. In *Proc. of the 1st SIGdial workshop on Discourse and dialogue (SIGDIAL)*, pp. 29–35, Stroudsburgh, PA.

Ward, N. (2000b). The challenge of non-lexical speech sounds. In *Proc. of ICSLP*, pp. 571–574, Beijing.

Watzlawick, P., Beavin, J., and Jackson, D. D. (1967). *Pragmatics of Human Communications*. W.W. Norton, New York.

Weninger, F., Krajewski, J., Batliner, A., and Schuller, B. (2012). The voice of leadership: Models and performances of automatic analysis in online speeches. *IEEE Transactions on Affective Computing*, **3**, 496–508.

Wilden, I., Herzel, H., Peters, G., and Tembrock, G. (1998). Subharmonics, biphonation, and deterministic chaos in mammal vocalization. *Bioacoustics*, **9**, 171–196.

Wilson, J. A., Deary, I. J., Scott, S., and MacKenzie, K. (1995). Functional dysphonia. *British Medical Journal*, **311**, 1039–1040.

Wingate, M. E. (1984). Fluency, disfluency, dysfluency, and stuttering. *Journal of Fluency Disorders*, **9**, 163–168.

Wodak, R., Johnstone, B., and Kerswill, P. (eds) (2011). *The Sage Handbook of Sociolinguistics*. Sage, London.

Wolk, L., Abdelli-Beruh, N., and Slavin, D. (2012). Habitual use of vocal fry in young adult female speakers. *Journal of Voice*, **26**, 111–116.

Yuasa, I. P. (2010). Creaky voice: A new feminine voice quality for young urban-oriented upwardly mobile American women? *American Speech*, **85**, 315–337.

Zwicker, E. and Fastl, H. (1999). *Psychoacoustics – Facts and Models*. Springer, Berlin.

5

Functional Aspects

*The most intelligible factor in language is not the word itself, but the tone, strength,
modulation, tempo with which a sequence of words is spoken – in brief, the music behind
the words, the passions behind the music, the person behind these passions: everything,
in other words, that cannot be written.*

(Friedrich Nietzsche)

The 'big' topics, when it comes to functions of paralinguistics, are arguably personality and
emotion. Besides these, there are more basic phenomena such as age and gender, and there
are more specific phenomena, such as deviant speech and discrepant communication. In this
chapter we start with the basic phenomena, which we subdivide into *biological trait primitives*
and *cultural trait primitives*. This should be taken as a convenient distinction, not as some
theoretical foundation; as always, there are fringe phenomena at the border between these
two types, and biological traits are modified by culture. Then, we will address *personality*
and *emotion*. Sections 5.5–5.8 deal with specific aspects. *Sentiment analysis* is similar to the
modelling of emotion and affect in speech and language; until recently (see Wöllmer *et al.*
2013), it has been confined to written language, and has evolved rather independently of
approaches used for speech. So far, the phenomena addressed are communicative in a broad
sense: they characterise human beings, constituting the edge conditions for communication,
but they are not only constituted within communicative settings. For all traits and states dealt
with so far, there are 'normal' ways of expressing them. *Deviant* speech/language is different,
due to factors such as idiosyncrasies or pathology, or to the use of a language variety that
is not one's native language; another umbrella term is *atypical* speech. *Social signals* are
constituted and used within communicative settings; in the default case, they are used in an
honest way, that is, as agreed upon in society. However, there are different ways of *discrepant
communication* (see Section 5.8), such as irony, sarcasm, and deceptive speech (lying).

The different functions dealt with in this chapter can be imagined as having a layered figure–
ground relationship (Figure 5.1). First come biological trait primitives that are preshaped by
nature – of course, each individual has degrees of freedom but a child, for instance, cannot
employ an adult voice. Then come cultural trait primitives such as gender-specificity or native
language(s) – preshaped as well, within a specific societal setting. Personality develops and

Computational Paralinguistics: Emotion, Affect and Personality in Speech and Language Processing, First Edition.
Björn W. Schuller and Anton M. Batliner. © 2014 John Wiley & Sons, Ltd. Published 2014 by John Wiley & Sons, Ltd.

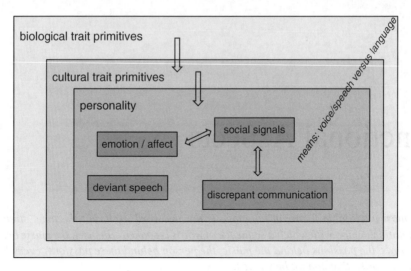

Figure 5.1   Layered figure–ground relationship of paralinguistic functions

evolves based on this ground, and (the expression of) emotions are strongly influenced by personality. All this can be studied by looking at speech, or by looking at language (sentiment analysis); of course, written language as a means of communication is less immediate than speech. There is a socio-cultural understanding of typical and atypical (deviant) speech. Speech and language in social interaction can again be looked at as figure against the ground that is built from all these components, evolving and instantiated within social interaction.

Of course, it is impossible to explicitly model this complex and layered figure–ground relationship, which is responsible for the difficulties in finding out the relevant parameters. Normally, we will just stay within one layer and try out to find one (complex) form, that is, a bundle of formal elements, indicating one specific function. In some studies, two layers are combined, for instance, when addressing the ability of autistic children to express emotions. The arrows in Figure 5.1 only denote some of the most important relationships.

Adjacent layers are not easily distinguished from each other. Intuitively, we know what belongs to biology and what to culture: which sex humans belongs to and their body height are determined by biology and genes; the same holds for age. These traits can either be dichotomised (sex) or measured (e.g., in centimetres or kilograms). Native language and cultural background, however, are determined by region, upbringing, and culture. Such traits cannot be measured exactly. Both biology and culture heavily shape and influence speech and language, and speech and language characteristics can be used to find out about biological and cultural traits.

At second glance, it is almost impossible to really disentangle biological and cultural traits. Whereas 'sex' refers to biological differences, 'gender' – albeit often used as a politically more correct term instead of 'sex' – refers to the cultural and social roles of men and women. Speech and language are both not independent of biological and especially cultural constraints, and effort is needed to disentangle the two factors. Moreover, terms such as 'transgender' and 'transsexual' indicate that even gender and sex are not binary. A straightforward distinction can, however, be made between the research interests. We will speak about 'biological trait primitives' if we are (only) interested in the sex of a speaker, or his/her physical weight, size, or height. We will speak about 'cultural trait primitives' if we are interested in the cross-individual characteristics of *groups of speakers* – such groups can be constituted by one sex/gender, for instance, if we are interested in average pitch height of women in one society and its change across generations; and we can try to find out how each gender employs some of the formal features exemplified in Chapter 4 differently, when indicating one and the same function.

Of course, the constituting traits of personality have much to do with cultural stereotypes; the fuzzy criterion distinguishing the two is whether we are aiming at the individual (personality) or at the group (cultural trait). We want to illustrate the difference with typical exemplars. On the one hand, there are cultural traits signalling group affiliation. The group can be very large, such as 'young female teenagers in some parts of the United States' (see Section 4.2.3). On the other hand, the group can be rather small – even twins can develop traits that distinguish them from others. Specific personality traits that are normally labelled based on others' assessments are likeability, charisma, or attractiveness. These traits can also be characterised by specific voice characteristics. They are normally conceived of as characterising individuals.

In this chapter, we will try to present the most important phenomena (functions) and illustrate them with studies. This cannot be done in an exhaustive way; yet, the chances are high that the reader, interested in some specific function, will get a fairly complete picture when starting with the literature referred to, ramifying the search into other studies. Of course, this holds especially for the references that present a more complete overview and literature survey on their topic than space allows us to present ourselves. For instance, Kreiman and Sidtis (2011) provide excellent overviews and literature surveys on many of the topics dealt with in this chapter. Whereas in Chapter 4, which focused on form, references were mostly made to studies from within phonetics and linguistics (and in some cases, engineering), naturally enough, in this chapter, which focuses on function, more references are given from psychology, sociology, medicine, and engineering.

5.1 Biological Trait Primitives

> *The first rule of thumb for all radio personalities is to look absolutely nothing like how they sound.*
>
> (Strong Bad, Homestar Runner)

The term *race* – a candidate for a biological trait primitive – is much debated and controversial (James 2011). Hao (2002) found significant differences for gender and race in certain vocal tract dimensions and formant frequencies, comparing controlled samples of white American, African American, and Chinese male and female speakers. However, paralinguistic studies on race seem to be sparse; this might mainly be due to the term being – rightly – stigmatised, but also to the difficulties of disentangling race from *ethnicity* or social group. It seems to be rather a matter of specifities of ethnic and/or social groups and of pertinent stereotypes – see Popp *et al.* (2003, p. 317) who found that there are consistent beliefs about speech style for gender and especially for race: 'Black speakers, both women and men, were rated as more direct and emotional, and less socially appropriate and playful, than White speakers.' Kreiman and Sidtis (2011, p. 149) summarise several studies on race and voice and conclude that ' . . . cues to a speaker's race are articulatory and learned, rather than anatomic and innate'. The speech of ethnic groups is often largely coextensive with regional and/or social accents which will be dealt with in Section 5.2.

In the following, we will address *sex* (often called *gender*), *age*, and physical traits such as *height, size*, and *weight*, and combinations thereof. Speaker recognition/detection/verification can be conceived of as a special case of looking at combined traits, not across speakers but within one speaker; here, the interest is not in finding out about characteristics of groups but in employing such characteristics to identify an individual. Müller (2007, p. V) uses the term *speaker classification* as '... assigning a given speech sample to a particular class of speakers...' and *speaker recognition* as a sub-field of speaker classification where the class 'speaker' has only one element, and the class 'non-speaker' more than one. In this very broad sense, all functional aspects dealt with in this chapter are aspects of speaker classification.

Thus, we want to employ speech parameters to find biological traits – or, to use a somewhat awkward term, 'biologically defined groups'; conceptually, this can be distinguished from using speech characteristics to find common traits in 'culturally defined groups'. These groups can be more or less coextensive.

Male and female voices differ due to physiological differences. The pharynx of males is longer on average than the pharynx of women, resulting in an average fundamental frequency (F0) of 110 Hz for males and 220 Hz for women. The vocal tract of females is shorter on average than the vocal tract of males, resulting in about 20% higher formants for females than for males (Kreiman and Sidtis 2011, pp. 124ff.). Note, however, that even this seemingly straightforward physiological fact can be modified by cultural factors: the male–female formant differences are much smaller amongst Danish than amongst Russian speakers (Johnson 2005). Apart from physiological variations, there are some intervening factors that contribute to deviations and especially to systematic differences: age, generation, and culture – age being a biological trait as well, whereas generation should be conceived as a cultural trait (see Sections 4.2.2 and 5.2). Age differences are summarised in Kreiman and Sidtis (2011, pp. 110ff.): F0 is high for children and decreases for adults aged under 60 years; after age 60, F0 decreases for females but increases for males. For adults, control of pitch, loudness, formants, voice quality, and tempo are quite stable, in contrast to the greater variability in children and seniors. Formant frequencies of children are considerably higher than those of adults (Li and Russell 2001). A comprehensive overview of the age-related acoustic variation in temporal as well as in laryngeally and supralaryngeally conditioned aspects of speech is given in Schötz (2007); see also Schötz (2006).

There is a long tradition of studies dealing with characteristics of vocal parameters with respect to age and sex/gender. In a study of two cohorts younger and older than 80, Mysak (1959) found a progressive upward trend in pitch as a function of age. In an early attempt towards the automatic recognition of sex from speech, Wu and Childers (1991) and Childers and Wu (1991) obtained up to 100% correct classification of speaker gender from vowels. Torre and Barlow (2009) address effects of age and sex on acoustic properties of speech, finding significant sex-by-age group interactions for F0, F1, and voice onset time.

The variable sex is often controlled for in studies within computational paralinguistics by separately modelling females and males. With the increasing interest in 'atypical' speech and the availability of databases spanning across age groups including children (Burkhardt *et al.* 2010; Schuller *et al.* 2010b), and with the additional interest in cross-linguistic/cross-cultural issues, it might be both necessary and interesting to examine pertinent differences. Several studies have addressed age and sex/gender, modelling them separately or together. Perceived age was classified automatically in Minematsu *et al.* (2002a,b, 2003). Metze *et al.* (2007) report on a first comparative study of speaker age recognition, using German data annotated with age and gender labels. In the Interspeech 2010 Paralinguistic Challenge (Schuller *et al.* 2010b), the 'aGender' corpus was used (Burkhardt *et al.* 2010) for a combined age and gender classification task. Various features and classification techniques have been used by the participants of this challenge (Bocklet *et al.* 2010; Kockmann *et al.* 2010; Li *et al.* 2010; Lingenfelser *et al.* 2010; Meinedo and Trancoso 2010; Nguyen *et al.* 2010; Porat *et al.* 2010). The database, baselines and results of the challenge are covered in Schuller *et al.* (2013b).

Other topics are the mismatches between adults' and children's speech and the use of normalisation techniques to overcome them (Ghai and Sinha 2010), or the mismatch between adult and older voices and its influence on automatic speech recognition (Vipperla *et al.* 2010).

As for physical size, formant frequencies seem to be more highly correlated with body size than pitch, both in animals – see the results reported in Fitch (1997) for rhesus macaques – and in humans (Evans *et al.* 2006; Gonzalez 2004; Krauss *et al.* 2002; Rendall *et al.* 2007). Possible reasons are discussed in Kreiman and Sidtis (2011). Speakers can vary pitch to a considerable extent, indicating different linguistic (stress, focus, sentence mood) or paralinguistic (emotion, affect) functions; thus, a possible indication of size can be distorted. Another reason might be that the larynx can grow independently of the body. This does not hold for the vocal tract, though. Bruckert *et al.* (2006) show that women use voice parameters, such as formant frequencies and dispersion, to assess male speakers' characteristics, such as age and weight. Judgements of pleasantness were based mainly on intonation. Mporas and Ganchev (2009) propose a regression-based scheme for the direct estimation of the height of unknown speakers from their speech, achieving an averaged relative error of approximately 3%.

5.1.1   Speaker Characteristics

We treat speaker characteristics in this section as well although it is shaped by all possible factors. Dellwo *et al.* (2007) sketch both the foundations of speech production and the factors responsible for individual speaker traits – basically the same factors that we try to structure in this chapter. First come the – biologically conditioned – primitives (especially sex and age), then the cultural primitives such as native language, regional dialect, and social variety. These traits are modified by personality and short-term affective states, which in turn are interdependent with deviant states (e.g., health conditions), the actual communicative situation, and specific intentions sometimes creating discrepant speech. Schultz (2007) establishes a taxonomy for speaker characteristics along similar lines, taking into account physiological, psychological, individual and collective factors, and discusses pertinent applications and approaches within automatic speech processing. The other big topic, speaker classification in forensic phonetics and acoustics, is covered by Jessen (2007) in the same book; in practical forensic work, this is still done by phonetics experts and not by machines. The typical forensic application is *speaker identification*, a well-known case being the Lindbergh–Hauptmann trial; see Solan and Tiersma (2003) and Kreiman and Sidtis (2011, pp. 237ff.). The son of the national hero Charles Lindbergh was kidnapped and killed, and later at trial, Lindbergh identified the voice of the man to whom ransom was paid, saying *Hey, doctor*, as belonging to the main suspect, Bruno Hauptmann, who was eventually executed. The quality of speaker identification, both by experts and by machines, is much debated; so far, reliability seems to be higher if speakers are excluded as suspects than the other way round (Clifford 1980; McGehee 1937).

The other field of speaker recognition is *speaker verification*, used in access control systems. The system has to verify that the speaker belongs to a specific, normally limited group of speakers who are allowed access.

An extensive survey of automatic speaker recognition and its fundamentals is given in Beigi (2011). Taking a broader view, we can refer to Perrot *et al.* (2007) who address vocal disguise and automatic detection, that is, how a voice can be transformed by electric scrambling or intentional modification in order to falsify or conceal speaker identity, and how such a disguised voice can be distinguished from the original voice. Majekodunmi and Idachaba (2011) discuss speaker recognition in the context of other approaches towards revealing the identity of individuals, such as fingerprint, face, and iris recognition.

5.2   Cultural Trait Primitives

At a VIP dinner last night an American woman asked me 'where are you from?' I said Australia, she said 'wow your English is amazing'.

(Mark Webber)

The prototypical cultural trait primitives are the first language of a speaker and its varieties. Often, the classic 'standard' language is, however, a construct, codified and taught especially in its written form and some standard pronunciation, and used as a sort of *lingua franca* (common language, interlingua) within and across language communities. A straightforward application for *language identification* (Muthusamy *et al.* 1994; Timoshenko 2012) within automatic speech processing is the monitoring of radio or telephone communication by governmental agencies. Admittedly, there is a weak borderline between a language as a genuine object of investigation within linguistics and pure automatic speech recognition (ASR), and language varieties as genuine objects of sociolinguistics and paralinguistics. Originally, both structural linguistics and ASR made the (contrafactual) assumption of a monolithic system that only allows specific realisations. Paralinguistics comes in when we deal with all the different varieties of a language which are normally conceived of as fully fledged linguistic systems – thus, the difference between languages and their varieties is due to historical and especially political factors rather than to linguistic characteristics. The varieties can be seen – again – as figure before the ground. The traditional sociolinguistic *deficit hypothesis* that these varieties are deficient in some respect has been replaced by the *difference hypothesis* that they all are of equal value – notwithstanding the fact that they can be more or less stigmatised.

A *dialect* is a regional variety of a language that differs with respect to lexicon, grammar, and pronunciation from other dialects. As already mentioned in Section 4.3, these differences can be pronounced or weak. Sometimes, the term *vernacular* is used for native, regional varieties such as 'African American Vernacular English'. *Regional accents* differ only with respect to pronunciation; the term is widely used for English (British, North American) regional varieties (Wells 1982), to distinguish local varieties or larger varieties such as Northern versus Southern English, or American versus Australian English; as a linguistic term, it is less common for referring to other language communities, such as German, although its use by Rakić *et al.*

(2011) is perhaps justified by the decline of dialects which have given way to less distinct regional varieties. For French, Woehrling and Mareüil (2006) distinguished three accents – Northern French, Southern French and Romande Swiss – based on a perceptual evaluation and using clustering and multidimensional techniques.

A *sociolect* is the variety of a social group with a local or a wider regional distribution, for example, an ethnic (minority) group, the working class, or any specific sub-culture.

Apart from denoting the variety of a specific group and from its use within grammar and prosody where it is used for different types of accentuation (word accent, phrase accent, focal accent), the term 'accent' can also mean *non-native or foreign accent*, denoting segmental or prosodic traits that do 'not belong to' the language that is spoken, indicating that it is not the speaker's first language. Non-native traits are taken as deficient at least as long as the speaker wants to improve his or her pronunciation, lexicon, and grammar. Immigrants displaying more or less strong non-native traits might be seen as speaking a somehow deficient variety; if these individuals group together, forming a community, then their speech can evolve into a vernacular. In the end, it might be an individual decision – and based on that, a decision of group identification, self-affirmation, and eventually political correctness – whether the speakers themselves and others regard a specific variety as something that should be taken care of, or as something that expresses their identity.

Thus, the difference between non-native = deviant and non-native = vernacular can only be defined 'in the eyes of the speakers themselves', not by using any linguistic criterion. In practice, approaches towards vernaculars describe differences, approaches towards non-native (second) languages (L2) produced by language learners assess differences between the native (first) language (L1) and L2. As always, this is not a clear demarcation – the difference can, however, be illustrated with prototypical examples, and it is visible when looking at the different approaches towards non-native or regional accents and the like. We want to cover non-native speech in Section 5.6 because it is normally seen – even by the speakers themselves – as something they rather would like to improve, that is, as deviant/deficient.

A straightforward approach towards recognising local varieties is using methods developed within automatic language identification, as done by Hanani *et al.* (2013) who apply a state-of-the-art language identification system for recognising 14 regional accents of British English, and for distinguishing the two largest ethnic groups in the city of Birmingham, the 'Asian' and 'white' communities. They obtain good performance for both tasks, compared with other approaches and human recognition accuracy.

Gillick (2010) showed significant differences in word usage patterns within conversational American English, derived from the output of a speech recognition system and yielding a classification accuracy between 60% and 82% for predicting demographic traits of individual speakers (gender, age, education level, ethnicity, and geographic region).

Prejudgements about regional accents exist and can lead to more or less unfavourable biases in job interviews, as shown by Markley (2000) for US regional accents, and by Rakić *et al.* (2011) for German regional accents (dialect regions). In a meta-analysis, Fuertes *et al.* (2012) found 20 studies comparing and establishing the effects of standard accents versus non-standard (foreign or minority) accents on interpersonal evaluation, and classified these characteristics as belonging '... to one of three domains that have been traditionally discussed in this area, namely status (e.g., intelligence, social class), solidarity (trustworthiness, in-group–out-group member), and dynamism (level of activity and liveliness)'.

Thus, a promising avenue for research within computational paralinguistics and regional accents/dialects can be seen within social-psychological basic research; further possible applications are of course within speaker identification (forensics), and within human–machine communication: call-centre monitoring and other types of monitoring, and generally, adaptation (entrainment) within human–computer interaction.

5.2.1   Speech Characteristics

So far we have discussed regional and social varieties of languages, not telling apart subgroups of speakers such as young versus old, or male versus female that can be characterised by the use of specific speech markers. Group characteristics are based on and evolve into cultural and personality stereotypes, but personality is dependent on biological traits (sex, age, etc.) as well. In practice, it might not really be possible to disentangle all these factors because for characterising and classifying or recognising an individual personality we have to model and train with groups of speakers. Moreover, speakers can permanently display specific speech characteristics such as breathy voice (in which case it is a personality trait such as attractiveness), or they can employ them within specific interaction scenarios (in which case we speak of a social signal such as intimacy).

In parallel to Section 5.1.1 where we addressed speaker characteristics, here we want to illustrate speech traits characterising groups of speakers – or prejudices about such traits – with a few exemplars; other phenomena and aspects will be covered especially in Sections 5.3 and 5.7. Strictly speaking, we are now no longer dealing with cultural trait primitives but with cultural traits which are addressed in most of the other sections in Chapters 4 and 5 as well – if not caused by nature, the objects of investigation in paralinguistics are cultural phenomena. (See also the difference between the 'extralinguistic' and 'paralinguistic' mentioned in Section 1.1.)

A well-known gender stereotype is that women are more talkative than men. This is not backed up by the data; Mehl *et al.* (2007) found no reliable sex difference in daily word use. Spender (1980), summarising work on the talkativity of men versus women, found no evidence for women doing the talking. It seems to be more a matter of power: the more powerful people are, the more they talk. For instance, in academia, professors talk more than students. To find the proportion of speech per speaker in dyadic or multi-party scenarios is a straightforward task; recently it has been addressed within speech activity detection and turn-taking detection (Laskowski 2011; Laskowski *et al.* 2008).

Studies on paralinguistics are often done with a sample of speakers who display similar characteristics – because homogeneous groups such as students are easier to recruit, and because possible intervening factors can be kept more or less constant. Age is such a factor, and age filtered by cultural developments (i.e., generation) is another factor. There is ample evidence that traditionally, high pitch of females and femininity correlate, and that pitch of females is higher than should be expected, based on physiological constraints (Spender 1980). This has changed in some societies: Pemberton *et al.* (1998) compared recordings of Australian women from 1945 and 1993, and found that women in 1993 had significantly deeper voices than women of the same age recorded in 1945. More evidence is given in Section 4.2.2, for instance, on the higher pitch of Japanese women and possible reasons for this. Determining pitch height and gender is again straightforward and can be done automatically within cross-linguistic studies.

5.3 Personality

Her voice was ever soft,
Gentle, and low, an excellent thing in woman.

(Shakespeare, *King Lear*, Act V, Scene 3)

Were I hard-favour'd, foul, or wrinkled-old,
Ill-nurtured, crooked, churlish, harsh in voice,...

(Shakespeare, *Venus and Adonis*, stanza 23)

Personality is a concept that has been conceived and developed outside of linguistics, within psychology; yet, the object of investigation can be similar to the object of investigation in sociolinguistics. Interestingly, personality research relies heavily on metalanguage, that is, on catalogues of terms denoting personality traits of different granularity, but speech and language are not necessarily employed as 'object languages', that is, as features characterising different personality traits; for instance, in Flett (2007), speech or language features are not even mentioned. Yet, Sapir (1927) envisaged 'speech as a personality trait' in the title of his short essay. In this section, we start with classic studies on speech and personality, then we present studies dealing with single traits, and conclude with studies on bundles of traits.

Classic Studies

In an early study of 'judging personality from voice', Allport and Cantril (1934) found small but consistent relationships with listeners' judgements and both physical characteristics (age, height, etc.) and personality traits (political preference, extraversion, etc.). In a survey on 'speech and personality', Sanford (1942) listed 106 studies dealing with different aspects of speech and language related to and indicating personality such as style, specific linguistic constructions, voice, and disorders of speech. The author concludes that '... the problems in this field are still more numerous than the facts' (Sanford 1942, p. 811). Sanford (1948) tried to establish 'an empirical psychology of language', dealing especially with the 'relation between linguistic behavior and personal adjustment', that is, with style, lexicon, syntax, speech, and disorders of speech. Several studies addressed judgements on personality from non-verbal properties of the voice (Addington 1968; Aronovitch 1976; Kramer 1963; Markel *et al.* 1972), or more specific relationships such as the effect of regional dialect on judgements of personality (Markel *et al.* 1967); cf. Section 5.2. Speakers with voice disorders such as harsh-breathy or hypernasal voice quality elicited more negative responses on judgements of personality than speakers with normal voice quality in Blood *et al.* (1979). The collection of articles by renowned scholars such as Abercrombie, Osgood, and Hymes in Markel (1969) conveys a broad conceptualisation of psycholinguistics as the study of speech and personality.

In a cross-cultural (cross-linguistic) approach, Scherer (1972) compared judgements of American and German listeners on personality traits, based on voice samples of American and German male and female speakers, and discussed the differences.

Feldstein and Sloan (1984) addressed the speech tempo of extroverts and introverts and stereotyped notions about it; they suggest that the stereotypes somewhat exaggerate the actual differences with respect to the speech rates of extraverts (rapid) and introverts (slow).

In a review of the literature, Scherer (1979) described the state of the art, in spite of a wealth of (mostly German) studies, as 'bleak', characterised so far by phenomenologically oriented hypotheses rather than by hard facts; instead, the author focused his review on empirical studies on aspects of 'normal' personality, discussing vocal aspects (pitch, intensity, and voice quality), fluency aspects (pauses, discontinuities), linguistic cues (morphology and syntax), and conversational behaviour (length of turns and total verbal output). He refers to a '... lamentable lack of research on the relationship between linguistic style and normal personality'. Nowadays, automatic text analysis can be employed for this, using large-scale databases; see Chung and Pennebaker (2007) and Section 4.3.1 above. A comparable evaluation can be found in Scherer and Scherer (1981). The development of research on speech and personality during the last two decades has been influenced by the advent of the OCEAN model (Section 3.1) and by the availability of massive computer power and larger-scale databases. We follow a straightforward, albeit simplifying, partition into studies dealing with single traits and studies dealing with bundles of traits (OCEAN or some other combination). The partition is simplifying because even single traits such as 'charismatic' are normally conceived of and modelled as consisting of several 'sub-traits'.

Single Traits

In this discussion of single trait studies, we concentrate on a few traits that are obviously attractive topics because they help identify the reasons for success, whether it be in gender relationships, in presidential elections, in advertisements, or in professional life in general.

Men's voices as *dominance* signals are addressed in Puts *et al.* (2006, 2007): in agreement with the stereotypes mentioned in Sections 4.2.2 and 4.2.3, (artificially) lowered F0 and formant dispersion (closer spacing of formant frequencies) were perceived as being produced by more dominant men than the same parameters, artificially raised. In Puts (2005), positive evidence is discussed for this evolutionary tendency that cannot be explained fully by differences in body size: 'Results indicate that low [voice pitch] is preferred mainly in short-term mating contexts rather than in long-term, committed ones, and this mating context effect is greatest when women are in the fertile phase of their ovulatory cycles. Moreover, lower male F0 correlated with higher self-reported mating success' (Puts 2005, p. 388). In contrast to these studies, (political) *charisma* and *leadership* seem to be positively correlated with higher pitch (Weninger *et al.* 2012). This alleged contradiction might be resolved when we consider the context: more intimate in face-to-face (gender) relationships, and more public when it is about charisma and leadership (see Section 4.2.3).

Gregory and Gallagher (2002) analysed voices in presidential debates during eight of the elections between 1960 (Kennedy–Nixon) and 2000 (Bush–Gore): they were able to employ frequency below 500 Hz to predict the popular vote outcomes in all eight elections. Rosenberg and Hirschberg (2005, 2009) identified the association of charisma, using speech samples from American politicians. They found a significant agreement amongst raters, and significant correlations between charisma rating and duration of linguistic units, the number of first personal pronouns used, the complexity of lexical items, pitch, intensity, and speaking rate; for all these parameters, higher values indicated higher charisma. Weninger *et al.* (2012) collected a corpus of YouTube speeches given by persons with leadership abilities such as executives of global players. The data were rated using ten labels from a leadership questionnaire which subsequently were mapped onto three cover dimensions – 'achiever', 'charismatic', and 'teamplayer' – which could be recognised automatically with up to 72.5% accuracy. Higher variety in speech in general, higher loudness, and higher median F0 characterised achievers; similar but less pronounced correlations were found for 'charismatic'. The same feature characteristics turned out to be relevant for related traits, namely for indicating confident speakers, in Kimble and Seidel (1991): the more confident speakers were, the faster and louder they responded to trivia questions.

Possible applications for systems that automatically determine emergence of leadership can be found in automatic voice coaching, in detecting leadership qualities in multimedia and entertainment, and in generating them in avatars (human–machine interaction, especially computer games). Cost-intensive observer rating (Bono and Judge 2004) could be replaced by automatic procedures. Similar applications can be envisioned for *vocal attractiveness*. Zuckerman and Driver (1989) investigated the mutual dependency of physical and vocal attractiveness, and elaborated on these traits in Zuckerman *et al.* (1990). The acoustic characteristics (especially voice quality) of attractive voices were examined in Zuckerman and Miyake (1993). Bruckert *et al.* (2010) showed that the well-established effect of facial averaging on attractiveness can be replicated for averaging voices via auditory morphing, irrespective of the speaker's or the listener's gender. This phenomenon could be largely explained by reduced aperiodicities and thus a smoother voice texture, and a reduced distance to the means of pitch and timbre. The nature of female vocal attractiveness is explored in Liu and Xu (2011). The most attractive female voice is that which projects a small body size, and the most effective acoustic cue was voice quality. Gender stereotyping is addressed in Ko *et al.* (2006) who show that measured acoustic characteristics that differ between the genders are also relevant for within-gender femininity.

A similar concept, namely *likeability* or *pleasantness*, is investigated by Burkhardt *et al.* (2011), using the aGender corpus from the 2010 Interspeech Paralinguistic Challenge (Schuller *et al.* 2010b); this task was later used in the Interspeech 2012 Speaker Trait Challenge (Schuller *et al.* 2012b). Auditory spectral features seem to be most important for an automatic likeability analysis.

Bundle of Traits

The 'big five' factor model of personality (OCEAN; see Section 3.1) has achieved a measure of consensus in the community; however, it is still 'just' a model, and research deals with either the full model, or parts of it, or some other bundle of traits.

Mairesse *et al.* (2007) report results for the recognition of the big five, both for conversation and text, and for observer and self ratings, using classification, regression, and ranking models. They present a catalogue of markers of the five traits, especially of extraversion, which displays the most visible characteristics (comparable to arousal in emotion modelling) such as higher loudness and faster speech rate, alongside a less controlled and less formal speaking style, compared with introverts. Findings on markers of the other traits are more scanty, most likely simply because the linguistic-acoustic markers of extraversion are more frequent and easier to model. Using large text and speech corpora, they use different types of features such as counts and characteristics of word categories, utterance type (commands, questions, assertions), and prosody (pitch, intensity, speech tempo). As expected and especially when using observer reports, extraversion turns out to be the easiest trait to model, characterised by pitch and variation of intensity, followed by emotional stability and conscientiousness. Openness to experience is better modelled using textual characteristics. Generally, the results suggest that observed personality may be easier to model than self-reports, at least in conversational data:

> *This may be due to objective observers using similar cues as our models, while self-reports of personality may be more influenced by factors such as the desirability of the trait... (Mairesse et al. 2007, p. 491)*

As for observer versus self ratings, they propose the hypothesis

> *... that traits with a high visibility (e.g., extraversion) are more accurately assessed using observer reports, as they tend to yield a higher inter-judge agreement..., while low visibility traits (e.g., emotional stability) are better assessed by oneself. A personality recogniser aiming to estimate the true personality would therefore have to switch from observer models to self-report models, depending on the trait under assessment. (Mairesse et al. 2007, p. 492)*

Mohammadi *et al.* (2010) present preliminary results on 'nonverbal behavioral cues', that is, prosodic features, for clips taken from Radio Swiss Romande (the French-speaking Swiss broadcasting service), annotated with the big five traits by judges who did not understand French. An automatic traits assignment yields significant results for all five traits, extraversion scoring highest with a 76% recognition rate. This task later featured in the Interspeech 2012 Speaker Trait Challenge (Schuller *et al.* 2012b). Gawda (2007) tests the associations between neuroticism, extraversion, and paralinguistic expressions, finding significant relations of introversion and neuroticism with speech fluency impediments. Ivanov *et al.* (2011) describe an automated system for speaker-independent personality prediction in the context of human–human spoken conversations (simulated tourist call centre, 24 speakers simulating users and agents, personality measured by a big five personality test, feature extraction with the open-source Speech and Music Interpretation by Large Space Extraction (openSMILE) toolkit (Eyben *et al.* 2010b)). Again, only performance for conscientiousness and extraversion was significantly above chance level. In the study of Oberlander and Nowson (2006), OCEAN scores for participants producing personal weblog text were obtained by self rating; openness scores were discarded, due to non-normal distribution. For binary text-based classification tasks, using a simple statistical classifier (naïve Bayes), results are promising for all four remaining traits but slightly worse for extraversion. This might illustrate the differences between purely text-based classification and a classification based on purely acoustic or both acoustic and linguistic information.

Concluding Remarks on Personality

Summing up, so far there seem to be fewer studies especially on the automatic modelling of personality traits conveyed via acoustics and text than on emotions, and fewer on the full big five model than on single traits. There might be several reasons for this state of affairs. Emotions are easier to elicit with some experimental manipulation, and they are easier to annotate because they stand out locally from a non-emotional (i.e., neutral) context. They are short-term, and can be investigated in a speaker-dependent way as well because speakers can produce different emotions even within a single recording. Within-subject studies are not possible for personality research – or only within a longitudinal design: personality is long-term. Personalities cannot be induced, pertinent information has to be obtained by time-consuming self ratings or observer ratings. Actors can be assigned to two basic types: those who can impersonate different personalities (Meryl Streep), and those who always sort of impersonate themselves (Humphrey Bogart, Woody Allen, and Bruce Lee). As far as we can see, actors impersonating different personalities have not really been employed in research on personality and speech/language, Polzehl *et al.* (2010) being an exception.

Single trait personality studies can focus on specific traits, relevant for and visible within specific application tasks. Basically, results for single traits are corroborated in studies on the big five. Traits such as extraversion are easier to model (observers' labelling) because they are correlated with 'higher, longer' characteristics; this can be compared with the modelling of arousal in emotion research. Other traits cannot be modelled that easily, and results are not unambiguous. It seems obvious that one has to employ text-based procedures for characterising personality traits.

5.4   Emotion and Affect

But some emotions don't make a lot of noise.

(Ernest Hemingway)

There is an abundance of studies on emotion/affect and speech; thus, we cannot aim at a full coverage of the literature but will restrict ourselves to mention some central aspects and exemplary studies, early, 'classic' ones, early 'computational' ones, and those representing the state of the art today. Some of the methodological differences between personality and emotion research are discussed in Section 5.3. Yet, personality is not only the ground for emotion as figure (Reisenzein and Weber 2009; Revelle and Scherer 2009). We will see that for similar classes such as extraversion as personality trait and activity/arousal as emotional dimension, speech features are employed in a similar way (cf. Section 5.3).

Early Studies

The study of speech and emotion can be traced back to the early decades of the twentieth century; see Scripture (1921) and Skinner (1935). The impressionistic study of Henry (1936) dealt with linguistic expressions relating to fear and anger in a few exotic languages. In early experiments on emotion and speech, Fairbanks and Pronovost (1939) and Fairbanks and Hoaglin (1941) described durational and pitch characteristics of simulated emotions (anger, contempt, fear, grief, and indifference), attributing a slower speech tempo to contempt and grief, and distinguishing different pitch characteristics. Osgood's semantic differential was used to measure listeners' attitudes to different intonational patterns in Uldall (1960) with respect to ten scales of the type 'bored versus interested' and 'polite versus rude'; the intonational features had particular weight with respect to the three factors 'pleasant versus unpleasant', 'interest versus lack of interest', and 'authoritative versus submissive'. Lieberman and Michaels (1962) extracted pitch values from eight neutral test sentences read by three male American English speakers in certain 'emotional' modes such as question, objective statement, fearful utterance, or happy utterance, and presented the original utterances and different synthesised varieties to listeners in forced judgement tests. Correct identification rates were 85% for unprocessed speech, and 44% if only pitch information was presented. Davitz (1964a) presented an overview of facial and vocal emotion research. Davitz (1964b) studied correlations between emotion dimensions and speech parameters; semantically neutral carrier sentences were produced in 14 'emotional tones'. Ratings were obtained for (1) a forced choice recognition task, (2) scales representing the auditory variables loudness, pitch level, timbre, and speech tempo, and (3) the three dimensions valence, strength, and activity from Osgood *et al.* (1957). Correlations between auditory ratings were strong for activity (more active indicated by 'higher, louder, faster') but weak for valence and strength. Williams and Stevens (1972) employed actors reading a dialogue containing different emotional situations, and subjected the recordings to quantitative and qualitative analyses. Anger, fear, and sorrow produced characteristic, albeit sometimes speaker-specific, differences with respect to, amongst others, pitch, spectrum, and tempo. Averill (1975) contained a list of 558 words with emotional connotations. A multi-dimensional scaling of similarity ratings for 11 acted emotions in Green and Cliff (1997) yielded a three-dimensional interpretation with the dimensions 'pleasant–unpleasant', 'excitement', and 'yielding–resisting' which was eventually related to dimensions obtained from similarity ratings for measures of 'tone-of-voice' such as high-low pitch or 'pleasant–unpleasant': 'Stimuli that were either highly pleasant or unpleasant were also excited, while stimuli unmarked in pleasantness were low in excitement' (Green and Cliff 1997, p. 429).

Frick (1985) reviewed prosodic markers of emotions in speech; as might be expected, 'activity or arousal seems to be signaled by increased pitch height, pitch range, loudness, and rate' (Frick 1985, p. 412). Scherer (1981) presents a survey on the state of the art of 'speech and emotional states', containing tables summarising the studies done so far and the results on vocal indicators of emotional states. Similar, updated overviews by Scherer and colleagues can be found in later work (Banse and Scherer 1996; Juslin and Scherer 2005; Johnstone and Scherer 2000; Scherer 2003; Scherer *et al.* 2003). Kehrein (2002) is one of the early studies using non-prompted speech dealing with emotional dimensions (activation, dominance, valence) and their prosodic characteristics.

Automatic Processing of Emotions

In the 1990s automatic speech processing began to address phenomena 'above the word chain', that is, beyond pure speech recognition, and towards dealing with more natural, spontaneous speech data. Researchers employed prosodic information within higher linguistic modules such as determination of stress (word accent), phrase/'sentence' accent, syntactic boundaries, islands of reliability (passages that are important to convey and are thus pronounced more clearly) and semantic focus, and dialogue act modelling. The view broadened towards all those paralinguistic phenomena that are covered in this book – emotion and speech being arguably one of the most attractive topics. Dellaert *et al.* (1996) explored different pattern recognition techniques and used prosodic features to classify the emotional content of utterances; the authors reported a classification performance close to human performance. Cowie and Douglas-Cowie (1996) pointed out that prosodic departures from a reference point corresponding to a well-controlled, neutral state not only characterise emotions but also deviant speech (central and sensory impairments caused by schizophrenia and deafness). Protopapas and Lieberman (1997) concentrated on the effect of pitch for distinguishing stressful from non-stressful conditions (re-synthesised pitch based on naturally occurring speech in highly stressful and non-stressful conditions) and concluded that maximum F0 constitutes the primary indicator.

Perhaps the first paper dealing with the automatic processing of 'natural(istic)' speech and affect (mother–child interaction) was Slaney and McRoberts (1998) (see also Slaney and McRoberts 2003). Here, adult-directed versus infant-directed speech were classified correctly with a Gaussian mixture-model more than 80% of the time. Mothers' speech was easier to classify than fathers' speech, and pitch was found to be an important measure. In a comparable study Batliner et al. (2008a) use a larger feature vector and additionally contrasted infant-directed and pet-robot-directed speech. Basic studies on mother–infant interaction are Rothman and Nowicki (2004) who describe a test of the ability to decode emotion of varying intensity in the voices of children, and Grossmann (2010) who reviews the literature on behavioural findings on infants' developing emotion-reading abilities and concludes that by the age of 7 months, infants reliably match and recognise emotional information across face and voice.

At the turn of the century, researchers began to use non-acted emotional data from human–human or human–machine interactions such as appointment scheduling or call-centre dialogues (Ang et al. 2002; Batliner et al. 2000; Lee et al. 2001). The role of the machine was sometimes played by a human Wizard of Oz. In a detailed review article, Cowie et al. (2001) address potential applications, present an overview on psychological traditions, offer lists of emotion words and of dimensional representations of emotions, list speech parameters and feature types, and give a comparable overview for facial gestures. Cowie and Cornelius (2003) can be seen as a sequel concentrating on the emotional states that are expressed in speech.

Batliner et al. (2000, 2003a) contrasted acted and naturalistic emotional speech from the same scenario and demonstrated a marked performance deterioration from one speaker to several speakers, and from acted to naturalistic data. Freely available and therefore widely used acted databases are the Danish Emotional Speech Database (DES: Engberg et al. 1997) and the Berlin Emotional Speech Database (BES: Burkhardt et al. 2005). DES and BES are representative of the 'early' databases in 1990s but still serve as exemplars for the generation of acted emotional databases in other languages. In the Linguistic Data Consortium (LDC) Emotional Prosody Speech and Transcripts corpus, actors produced 15 distinct emotional categories; this corpus was used, for instance, by Liscombe et al. (2003). Representatives in real-life settings are TV recordings (the Vera-Am-Mittag (VAM) Corpus: Grimm et al. 2008), call-centre interactions (Ang et al. 2002; Batliner et al. 2004; Devillers et al. 2005; Lee and Narayanan 2005; Liscombe et al. 2005b), multi-party interaction (the ICSI meeting corpus used in Laskowski 2009), and the Speech In Minimal Invasive Surgery (SIMIS) Database (Schuller et al. 2010a). Typical human–machine interactions in the laboratory are stress detection in a driving simulation (Fernandez and Picard 2003), tutoring dialogues (Ai et al. 2006; Liscombe et al. 2005a; Litman and Forbes 2003; Zhang et al. 2004), information systems (Batliner et al. 2003b), and human–robot communication (Batliner et al. 2008b); a still emerging field is interaction with virtual agents (Schröder et al. 2008). Other topics are the detection of aggressiveness and fear for homeland security, surveillance, and monitoring (Clavel et al. 2008; Schuller et al. 2008; Kwon et al. 2008), and estimates of excitability in sports videos (Boril et al. 2011). Naturally enough, the state 'interest' and its counterpart 'boredom' are of great interest for several applications (Gatica-Perez *et al.* 2005; Jeon *et al.* 2010; Schuller and Rigoll 2009; Schuller *et al.* 2006, 2007a, 2009a).

Crossing the Bridges

From the perspective of the present day, the automatic approaches towards emotion in speech from only a few years ago were mostly quite straightforward. Matters were kept simple – acted data, homogeneous groups of speakers, only a few prosodic features, and very pronounced emotion classes such as the big six were employed. This has changed over the last few years.

The focus on prosodic features, and sometimes mel frequency cepstral coefficients (MFCCs; see Chapter 8), in the early studies on the automatic processing of emotional speech has gradually been replaced by a plethora of different feature types; moreover, linguistic (part-of-speech, N-grams; see Chapter 9), and pragmatic (dialogue acts) modelling as well as disfluencies and non-verbals have been employed. Quasi-standards such as WEKA (Hall *et al.* 2009) for classification and openSMILE (Eyben *et al.* 2010b) for feature extraction make it possible to aim for standards that make comparisons across studies easier. Employing large feature vectors makes feature evaluation difficult; groups consisting of only one feature type can be evaluated separately in a sort of univariate approach (Batliner *et al.* 2006a; Schuller *et al.* 2007b), or in combination with other feature types in a sort of multivariate approach (Batliner *et al.* 2011b). If we are aiming at utmost reality, we should use all data available and not preselect good candidates (so-called *open microphone* setting); this means in turn that we have to cope with ambiguous and less pronounced items as well. However, prototypical realisations – be they acted or obtained from consensual annotations – can be used for demonstrating emotions, for instance to autistic children (Baron-Cohen *et al.* 2009) or in situations where we want to minimise the number of false alarms (Batliner *et al.* 2005; Seppi *et al.* 2008).

We can reduce the complexity of the task and make it manageable if we concentrate on a few consistent classes from the very beginning. If we want to address a multitude of emotions or several complex ones, we have to reduce the complexity of the tasks later on in the processing chain. The system of Sobol-Shikler and Robinson (2010) characterised over 500 affective state concepts from the MindReading database. The data were acted, and classification was pairwise. Devillers and Vidrascu (2006) used a 'multi-level' of granularity consisting of dominant (major) and secondary (minor) labels, that is, a coarse level and fine-grained level. The authors had to resort to the coarse-grained level when it came to classification, due to sparse data.

Another challenge, especially for classification performance, is 'within culture but across corpora' experiments (Eyben *et al.* 2010a). So far, there are not many studies addressing cross-cultural aspects; and existing ones tend to use acted data. This is due to the complexity of the task and to the necessity of comparing within a setting where everything else has to be kept constant ('other things being equal'). When comparing non-verbal emotional vocalisations such as screams and laughter across two dramatically different cultural groups, Sauter *et al.* (2010) and Sauter (2006) found that negative emotions most likely have vocalisations that can be recognised across cultures, whereas positive emotions are communicated by culture-specific means.

Initiatives such as Combining Efforts for Improving Automatic Classification of Emotion in Speech (CEICES: Batliner *et al.* 2006a), the Emotion Challenge at Interspeech 2009 (Schuller *et al.* 2009b) and the Audio/Visual Emotion Challenges (Schuller *et al.* 2011a, 2012a) will hopefully help in establishing standards. Further references on emotion processing can be found in Batliner *et al.* (2011a) and Schuller *et al.* (2011b).

5.5   Subjectivity and Sentiment Analysis

Maybe the best proof that the language is patriarchal is that it oversimplifies feeling. I'd like to have at my disposal complicated hybrid emotions, Germanic train-car constructions like, say, 'the happiness that attends disaster.'

(Jeffrey Eugenides, *Middlesex*)

The phenomena dealt with in *subjectivity analysis* and *sentiment analysis* (or *opinion mining*) have much in common with the dimension of *valence* and the concept of *appraisal* in emotion and affect processing. The difference lies mainly in the medium (text versus pure audio or speech). Subjectivity analysis classifies the content of a text into *objective* and *subjective*; sentiment analysis classifies the subjective content of a text into *positive* or *negative*. Pang and Lee (2008) trace back the 'sudden eruption of activity' that could be observed in this field at the beginning of this century to factors such as the rise of machine learning methods, the availability of data sets, and the intellectual challenges and commercial and intelligence sentiment-aware applications such as summarisation of reviews and recommendation systems.

Interestingly, so far there has not been much contact between the two sub-cultures of emotion/affect processing on the one hand, and sentiment analysis on the other hand, although the topics dealt with have much in common. A reason might be that the main academic disciplines are still separated from each other: on the one hand natural language processing which is rooted in artificial intelligence, and on the other hand speech processing as part of engineering, or phonetics and psychology as part of humanities. However, recent approaches try to unify speech and language for sentiment analysis (Cambria *et al.* 2013).

Wiebe *et al.* (2004) investigate several markers of subjectivity in language such as unique words (so called hapax legomena), collocational clues (context), and adjective and verb features. Unsupervised subjectivity classifiers are dealt with in Wiebe and Riloff (2005). Spertus (1997) describes some approaches towards flame (abusive message) recognition, based on the syntax and semantics of each sentence. Kennedy and Inkpen (2006) discuss methods for determining the sentiment expressed by a movie review, based on valence shifters (negations, intensifiers, and diminishers) such as counting positive and negative terms in a review, or using support vector machine classifiers with unigrams and bigrams as features; combining the two methods achieves better results than either method alone. Based on psycholinguistic and psychophysical experiments targeting customer reviews on the Internet, Becker and Aharonson (2010) argue for concentrating computational efforts on the final positions in texts. Kousta *et al.* (2009) attribute a general processing advantage over neutral words to words that express (negative or positive) valence '. . . due to the relevance of both negative and positive stimuli for survival and for the attainment of goals'. For multilingual subjectivity and sentiment analysis, Banea *et al.* (2011) identify and overview three types of methodologies: word and phrase level annotations, sentence-level annotations, and document-level annotations. A combination of methods from sentiment analysis and automatic speech analysis along the lines of (Mairesse *et al.* 2007) seems promising. Evaluation can also be influenced by subtle (seemingly 'innocent') linguistic structural means such as the use of transitive or non-transitive verbs (Fausey and Boroditsky 2010) or of anaphoric pronouns (Batliner 1984); cf. the weak linguistic relativity in Section 2.11. An extensive survey on 'Opinion Mining and Sentiment Analysis', pertinent techniques, approaches, and applications, is given in Pang and Lee (2008). A comprehensive introduction to the field is provided in Liu (2012).

5.6 Deviant Speech

> *Life is a foreign language; all men mispronounce it.*
>
> (Christopher Morley)

In this chapter, we want to deal with all kinds of deviant speech which can be seen as 'not normal', 'wide spectrum', or not *typically developed* speech. This can be *pathological speech* which is normally a long-term phenomenon that has to be treated in some way. Due to origin and cause, we can distinguish three sub-types of pathological speech: *voice pathology* describes impaired phonation, for example, in post-laryngectomy speech; *speech disorders* pertain to articulation problems, for example, due to congenital malformations (cleft lip and palate), stuttering or hypernasality; *language disorders* are due, for example, to hearing loss or aphasia (as a result of brain injury, a stroke, or dementia). Atypical speech can also be observed related to disorders such as depression or autism spectrum condition (ASC) – this is the preferred term in the UK; in the US, mostly autism spectrum disorder (ASD) is used. Normally, such speech has developed as a long-term trait that has to be diagnosed, monitored, and treated in some way.

Medium-term states which often are more or less self-induced and which do influence speech and language use are sleepiness, intoxication (often caused by alcohol consumption), a cold, or excessive voice strain. Even if this type of speech can make a very 'pathological' impression, it is due to a state that normally changes quite soon, often within hours, and does not need any therapeutic treatment; of course, habitual drinkers can display the characteristics of long-term deviant speech (trait), and a temporarily strained voice can turn into a pathological voice. Within-speaker comparisons are possible for such medium-term states, whereas for pathological speech, improvements are normally only observable over a longer period of time.

Non-native speech, that is, speaking in a second language (L2) that is not one's mother tongue, differs from the two types that we have mentioned so far: in many cases, the specific level of proficiency (and thus, the level of deviation from the 'correct' pronunciation and use of language) is something the speaker would like to overcome. At least, it is this population of L2 learners that computer-aided language learning (CALL) and computer-aided pronunciation training (CAPT) are aimed at. There are also speakers of an L2 who are more or less content with their level of proficiency because they can make themselves understood and do not need to impress with a close-to-native pronunciation; moreover, there are immigrant sub-cultures whose language can evolve into a vernacular, thus changing the type of speech from 'deviant' into 'cultural trait' (Section 5.2).

When dealing with all these constellations, we implicitly contrast the characteristics of deviant (i.e., atypical) speech with the typical speech as it should be – either after treatment/teaching, or after the user state has changed because the person is no longer intoxicated or sleepy, or her cold is over. Non-native speech is normally not subsumed under 'pathological' speech. A more generic term is *communication disorders* which, however, does not cover non-native speech, though this is also addressed in this section because the procedures for annotation, processing, and classifying are very similar or the same across all these types of deviant speech.

5.6.1 Pathological Speech

Sometimes I heard and understood and other times sounds and speech reached my brain like the unbearable noise of an onrushing freight train.

(Temple Grandin)

Damico *et al.* (2010) and Harrison (2010) give a broad overview of communication disorders, causes, assessment, and rehabilitation. Ruben (2000) estimates the cost of communication disorders, with a prevalence (frequency) of some 5–10%, at between $154 billion and $186 billion per year in the US, equal to 2.5–3% of gross national product. Fritzell (1996) reports data from hospital departments of phoniatrics in Sweden: *phonasthenia* (functional vocal fatigue, i.e., voice too high, loud or hard) was the most common diagnosis, and the teaching profession most affected, besides social workers, lawyers and clergymen. This tendency is corroborated for the US in Titze *et al.* (1997).

A basic protocol for the functional assessment of voice pathology, including a set of minimal basic measurements, is presented in Dejonckere *et al.* (2001): 'perception (grade, roughness, breathiness), videostroboscopy (closure, regularity, mucosal wave and symmetry), acoustics (jitter, shimmer, Fo-range and softest intensity), aerodynamics (phonation quotient), and subjective rating by the patient'. Using similar measurements and a multivariate regression model, Bhuta *et al.* (2004) established correlations between acoustic features and perceptual evaluation; three noise parameters turned out to be significantly correlated with the perceptual voice analysis. Kent (1996) discusses several limits to the auditory-perceptual assessment of speech and voice disorders and offers suggestions for their improvement. Kreiman *et al.* (1990) contrast assessments by naïve and expert listeners: while the naïve listeners used similar perceptual strategies, clinicians 'differed substantially in the parameters they considered important when judging similarity'. Averaging across them might obscure important aspects. Kent and Kim (2003) presents an acoustic typology of motor speech disorders from 'a parametric assessment of the speech subsystems (e.g., phonation, nasal resonance, vowel articulation, consonant articulation, intonation, and rhythm)' with respect to the global functions in speech such as voice quality, intelligibility, and prosody. Schoentgen (2006) gives an overview of vocal cues of disordered voices, proposes a classificatory framework, and warns against simply 'distilling general rules or comparing results obtained in different frameworks'. Kreiman *et al.* (1994) discuss the multidimensional nature of pathologic vocal quality, scrutinising the terms 'breathy' and 'rough'.

Listeners' ratings of speech disorders, especially of intelligibility, and their inter-rater agreement constitute a pivotal topic in this field, addressed by numerous studies. In Zenner (1986) on the so-called Post-Laryngectomy Telephone Test (PLTT) subjects had to read a random set of words and phrases, and naïve listeners had to write down what they heard. McColl *et al.* (1998) investigated the intelligibility of tracheoesophageal speech in noise: as the levels of background noise increased, listener ratings of intelligibility decreased. Bunton *et al.* (2007) studied auditory-perceptual ratings of dysarthria (a motor speech disorder due to neurological injury): '. . . auditory-perceptual ratings show promise during clinical assessment for identifying salient features of dysarthria for speakers with various etiologies'. Sheard *et al.* (1991) examined ratings of ataxic dysarthric speech samples with varying intelligibility: 'Judges agreed equally well in rating dysarthric speech across the range from low to high intelligibility.' Wolfe *et al.* (2000) investigated the perception of dysphonic voice quality by naive listeners: the most important perceptual dimension was 'degree of abnormality', followed by high-frequency noise for females, and 'breathy-overtight' for males. Hardin *et al.* (1992) examined the correspondence between nasalance scores and listener judgments of hypernasality and hyponasality, reporting a good overall relationship. Keuning *et al.* (1999) conducted a pilot study of the intra-judge reliability of the perceptual rating of cleft palate speech before and after pharyngeal flap surgery: they reported no differences in intra-judge reliability of experts versus non-experts.

Dejonckere *et al.* (1996) assess the perceptual evaluation of pathological voice quality with respect to reliability and correlations with acoustic measurement. Lohmander and Olsson (2004) critically review the literature on perceptual assessment of speech in patients with cleft palate, pointing out excessive variability in the experimental design.

We now turn to studies dealing with (specific aspects of) pathological speech. The prevalence of laryngeal carcinomas is 10 per 100 000 population (Zimmermann *et al.* 2003). Brown *et al.* (2003) review the history and state of the art of *post-laryngectomy voice rehabilitation*;[1] on the use of different types of voice prosthesis, see also Hilgers and Schouwenburg (1990), Schutte and Nieboer (2002), and Torrejano and Guimarães (2009). Speech impairment in oral cancer patients is dealt with in Pauloski *et al.* (1993, 1998). The effects on the phonetics of speech sounds of complete (replacement) dentures and dental prostheses in edentulous (toothless) patients are discussed in McCord *et al.* (1994), Petrović (1985), and Jacobs *et al.* (2001).

Vanderas (1987) reviews a number of studies reporting incidences (prevalence) of *cleft lip*, *cleft palate*, and *cleft lip and palate* by race; frequencies vary considerably. Sayetta *et al.* (1989) discuss methodological problems that might be responsible for such discrepancies. However, it seems safe to conclude that instances range roughly from 1 per 1000 to 2 per 1000 across nations; see also Derijcke *et al.* (1996). Van Lierde *et al.* (2002) report significant differences between cleft palate children and a typical control group for nasalance values and overall intelligibility; see also Van Lierde *et al.* (2003). Whitehill (2002) deals with problems related to speech intelligibility measured in speakers with cleft lip and palate. The psychophysical effects of laryngectomy are addressed in Devins *et al.* (1994), those of cleft lip and palate in the meta-analysis of Hunt *et al.* (2005).

For *Parkinson's disease* in Europe, Campenhausen *et al.* (2005) report in a systematic literature search crude and widely varying prevalence rate estimates from 65.6 per 100 000 to 12 500 per 100 000. Parkinson's disease affects over 1 million people in North America (Lang and Lozano 1998). Darkins *et al.* (1988) compared the prosody of typical, 'normal' subjects with that of patients with idiopathic Parkinson's disease and claimed that 'the striking disorder of prosody in Parkinson's disease relates to motor control, not to a loss of the linguistic knowledge required to make prosodic distinctions'. Holmes *et al.* (2000) examined the acoustic and perceptual voice characteristics of patients with Parkinson's disease according to disease

---

[1]Laryngectomy is the removal of the larynx, as may be done in cases of laryngeal cancer.

severity, finding limited pitch and loudness variability, breathiness, harshness, and reduced loudness. Speech treatment for Parkinson's disease is addressed in Ramig *et al.* (2008). Rusz *et al.* (2011) present quantitative acoustic measurements for characterising speech and voice disorders in early untreated Parkinson's disease.

ASC is reported to occur in up to 1% of the population (Simonoff *et al.* 2008). Wing and Potter (2002) discuss possible reasons for the increase in frequency reported in several studies in recent decades – whether these are genuine rises or due to changes in diagnostic criteria and increasing awareness and recognition of the condition. Children and adults with ASC are reported to have specific speech and prosody characteristics such as ' . . . articulation distortion errors, uncodable utterances due to discourse constraints, and utterances coded as inappropriate in the domains of phrasing, stress, and resonance' (Shriberg *et al.* 2001, p. 1097). Moreover, they have major difficulties in recognising and responding to others' feelings and emotional/mental states, that is, an impairment of *theory of mind* skills, both in facial expressions (Baron-Cohen *et al.* 2009) and in the voice (Chevallier *et al.* 2011; Golan *et al.* 2007; Van Lancker *et al.* 1989). Several aspects of this voice and language impairment, especially with respect to prosody, are addressed by a number of authors (Bonneh *et al.* 2011; Demouy *et al.* 2011; Grossman *et al.* 2010; McCann and Peppé 2003; Paul *et al.* 2005; Peppé *et al.* 2007, 2011; Ploog *et al.* 2009; Russo *et al.* 2008; Van Santen *et al.* 2010). Mower *et al.* (2011) describe an emotionally targeted interactive agent for children with ASC which should help them to effectively produce social conversational behaviour.

In addition, we want to mention a few other impairments and syndromes where speech and language are affected. People with *hearing loss* displaying deficiencies in speech production are arguably one of the most important target groups, simply due to the prevalence of this syndrome, amounting to millions of people in the US, and to the obvious fact that not only perception and comprehension of speech and language, but also production must be taken care of. Note that there are many degrees of hearing loss, and people can be equipped with different types of hearing aids and cochlear implants. Shargorodsky *et al.* (2010) substantiate the change in prevalence of hearing loss in US adolescents. Osberger and McGarr (1982) carried out an in-depth study of the speech production characteristics of the hearing impaired. Several aspects of providing patients with cochlear implants are addressed in Tobey *et al.* (2003), Connor *et al.* (2006), and Lowenstein (2012).

Depression is another common disorder with typical voice and language characteristics that can be detected and subsequently monitored in follow-up examinations (Chevrie-Muller *et al.* 1978; Ellgring and Scherer 1996; Lott *et al.* 2002; Low *et al.* 2011).

Attention deficit (hyperactivity) disorder (AD(H)D) is a frequent developmental disorder. Irrespective of its much debated prevalence and causes, adolescents diagnosed with ADHD may display problems with respect to social skills and emotion recognition skills similar to adolescents with ASC (Kats-Gold *et al.* 2007).

Automatic Processing

In clinical research on voice pathology, speakers often have to produce sustained vowels. Such a database was employed by Dibazar and Narayanan (2002) who used MFCCs and measures of pitch dynamics, modelled with Gaussian mixtures in a hidden Markov model (HMM) classifier, for the automatic classification of different speech pathologies. Lederman *et al.* (2008)

successfully used HMMs for classifying the cries of cleft lip and palate infants with or without an intra-oral plate. The automatic assessment of speech intelligibility (correlations between human ratings and automatic measures such as the accuracy of a speech recognition system) has been addressed by several studies: see Scipioni *et al.* (2009) and Bocklet *et al.* (2009) comparing the automatic assessment of children with cleft lip and palate in Italian and German; Stelzle *et al.* (2010) who evaluated an automatic computer-based speech assessment on edentulous patients with and without complete dentures; and Bocklet *et al.* (2012) using Gaussian mixture models and cepstral features for automatically evaluating pathological speech, correlating the automatic system and the expert listeners.

Maier *et al.* (2009b) present a novel web-based system for the automatic evaluation of speech and voice disorders, which has been evaluated for laryngectomised patients and for children with cleft lip and palate; further details are given in Maier *et al.* (2009a). Middag *et al.* (2009, 2011) utilise phonological features and language-independent procedures for predicting the intelligibility of pathological speech. Ringeval *et al.* (2011) describe an automatic prosodic assessment of language-impaired children with ASC, unspecified developmental disorders, and specific language impairments; they show that all these children had difficulties in reproducing intonation contours, compared with typically developed children, confirming the clinical descriptions of these subjects' communication impairments. Haderlein *et al.* (2011) demonstrate that an objective assessment of the intelligibility of pathological speech on the phone can be done, using support vector regression (see Section 11.2.3), based on the word accuracy and word correctness of a speech recognition system, and a set of prosodic features. Such procedures can be used for the (remote) monitoring of speech rehabilitation.

Knipfer *et al.* (2012) and Rouzbahani and Daliri (2011) try to detect automatically whether the speech/voice of a person is affected by Parkinson's disease. Voice and prosody seem to be very early indicators of Parkinson's disease; Bocklet *et al.* (2011) report promising performance when using prosodic features for differentiating automatically between typical speakers and speakers with early-stage Parkinson's disease. The most important prosodic features were based on energy, pauses, and pitch. Mahmoudi *et al.* (2011) aimed to develop and evaluate automated classifications of voice disorder in children with cochlear implants and hearing aids into four different levels of disorder, employing HMMs and neural networks, and one human expert as reference.

So far, satisfying results of automatic evaluation have only been achieved for cumulative evidence (see Section 3.7), that is, for averaged evaluations of longer passages, and not for single instance decisions, that is, for single segments.

The automatic processing and detection of chronic *cough* as a sign of an abnormal health condition is addressed in Matos *et al.* (2006) using HMMs, following a keyword-spotting approach; that is, coughs are modelled as words. Walmsley *et al.* (2006) attempt a cough/noncough classification using spectral coefficients and a probabilistic neural network, calculating total number of coughs and cough frequency as a function of time. Similar research is reported in Shin *et al.* (2009) and Drugman *et al.* (2011). Such procedures can be used for modelling not only human non-verbal behaviour but also animal behaviour; see Giesert *et al.* (2011) who analyse coughs in pigs to diagnose respiratory infections, motivated by the possibility of an early detection of infected animals to reduce costs for treatment and indirect costs caused by diminished mast and breeding results. Note that here, normal, short-term coughs (throat cleaning) have to be distinguished from medium-term (temporary deviant) coughs produced by infected pigs.

5.6.2   Temporarily Deviant Speech

Here's to plain speaking and clear understanding.
                                                                  (Kasper Gutman in *The Maltese Falcon*)

Habitual, constant traits in the voice and in the linguistic usage of a speaker can be employed to find out something about his or her biological and cultural trait primitives (Sections 5.1 and 5.2), and about long-term deviations from typical speech and language (Section 5.6.1). If deviations are not long-term but temporary (several hours or days), then they are caused by short-term events such as catching a cold, staying up late, or drinking too much. Coughing is a formal characteristic which can be segmented on the time axis; it can be a short-term symptom, elicited by some respiratory tract irritation, a medium-term symptom, caused by a cold, or a long-term symptom, caused by some long-lasting, abnormal health condition. We dealt with coughing in Section 5.6.1. Here, we want to address two typical medium-term, temporary functional traits or states: intoxication by alcohol consumption and sleepiness. They are attractive objects of investigation, important for both the health sector and for private and public safety. Note that there are also other states or traits such as stress (cognitive/emotional overload, not word accent) which is subsumed under emotion/affect in Section 5.4, and which can be short-term (a state), long-term (evolved into a trait), or medium-term (temporarily deviant).

Early studies on the effect of alcohol on the acoustic-phonetic properties of speech are reported in Trojan and Kryspin-Exner (1968), Sobell and Sobell (1972), and Pisoni and Martin (1989). Johnson *et al.* (1990) studied speech characteristics of the captain of the *Exxon Valdez* during the stranding of the oil tanker. Further results are given in Cooney *et al.* (1998) and Künzel and Braun (2003). Levit *et al.* (2001) may have been the first attempt to automatically detect alcohol intoxication; they obtained up to 69% recognition rate for the two-class problem 'below/above a blood alcohol concentration of 0.08%', using prosodic features and artificial neural networks. In-car alcohol detection is addressed in Schiel and Heinrich (2009). Possible forensic applications are faced with two problems. First, it is of course easier to distinguish between totally sober and fully drunk; this holds for both human and automatic assessment (Schuller *et al.* 2012b). The crucial threshold decision – whether blood alcohol concentration is above or below a legally defined limit – is, however, very difficult to make, based only on speech parameters. Second, in a realistic setting, both sober and alcoholised speech for the very same speaker may not be available but only speech that is one or the other.

Thorpy and Billiard (2011) is a compendium on the pathophysiological and clinical features of sleepiness. Earlier studies addressing the characteristics of sleepy speech are (Vollrath 1993), Bard *et al.* (1996), and Harrison and Horne (1997, 2000). Attempts to detect sleepiness from speech automatically are reported in Nwe *et al.* (2006), Greeley *et al.* (2007), Krajewski *et al.* (2009, 2012), and Zhang *et al.* (2010).

For both tasks so far, the performance for two-class problems in the studies reported on in Schuller *et al.* (2013a) has been well above chance level but still too low for use in 'critical' applications with single instance decisions (see Section 3.7). Moreover, the caveat set out in Schuller *et al.* (2013a) for the state of the art for sleepiness detection using speech holds more or less for intoxication detection as well, albeit to a lesser extent: '. . . small sample sizes, irrelevant high time-since-sleep values, speaker-dependent modelling, and non-comparable sleepiness reference values narrowed the generalisability of the results found so far'. Surveys of the speech processing of intoxication and sleepiness, as well as conditions and baselines for both the speaker intoxication and the speaker sleepiness challenges at Interspeech 2011 are given in Schuller *et al.* (2011c, 2013a). Speaker normalisation seems very promising, however, narrowing down the pool of potential applications if only seen speakers can be processed.

As far as we can see, the temporary characteristics of speech while the speaker is eating (e.g., hot potatoes) have not yet been researched; first attempts were reported in Vennemann (1979) from a phonological point of view, assuming a marked reduction of the phoneme (and thus phone) inventory.

5.6.3   Non-native Speech

Every American child should grow up knowing a second language, preferably English.
(Mignon McLaughlin)

Normally, non-native speech is – even if only implicitly – conceived as deviant, as something that has to be improved. In line with the other parts of this section dealing with pathological and temporarily deviant speech, we will concentrate on studies on speech dealing with the first part of the endeavour: to find the differences between the target – the second language – and the present state of proficiency. Trying to improve non-native speech draws heavily on pedagogics and teaching, and not on computational paralinguistics in its strict snese, even if we can easily imagine computational approaches such as computer-aided teaching or serious language games. Eskenazi (2009) provides an overview of spoken language technology for education and language learning that began in the 1980s, its history and main issues, pointing out that '. . . many of the techniques used in non-native pronunciation detection could be used for handicapped speech as well'. 'Handicapped speech' is a subset of 'pathological speech', covered in Section 5.6.1.

The main issues in the modelling and processing of non-native speech are *pronunciation* in the sense of correct or incorrect pronunciation of segments or words, and *prosody* in the sense of good or bad production of suprasegmental traits. Both non-native pronunciation and prosody have to be *detected, analysed, assessed,* and *corrected.* In most studies so far, there exist manual annotations of erroneous pronunciation and manual assessments of prosody, strictly localised (i.e., per segment or word) or global (e.g., per utterance or per speaker). Segments can be *substituted* by variants that are more or less similar to the target segment, incorrect segments can be *inserted*, segments can be *deleted*. Prosodic errors are incorrect placement of word accent (stress), and non-native intonation and *rhythm*. Whereas the degree of freedom for stress placement is low, there is, at least in non-tonal languages, a certain degree of freedom in the use of different types of phrasing (integrating or isolating) and different types of intonation (tonal configurations). A native rhythm helps structure what is perceived and contributes to a large extent to intelligibility.

As always, the task becomes more manageable if variability is reduced, for instance, by modelling only one source language L1, and not by trying to establish L1-independent procedures; moreover, text-dependent procedures (using only a limited set of test-sentences, i.e., few types and many tokens per type) are always easier to model than unlimited, free speech. In general, research first concentrated on different aspects of pronunciation; in recent years research on prosody has caught up. In the future, the specific motivation of the language learner might influence L2 modelling and teaching: whether we aim at pronunciation and prosody that sound as native as possible, or whether we are content with high intelligibility – and this in turn depends on the primary use of L2, that is, whether it is used for communicating with native speakers of the L2, or whether it is used as a third language (lingua franca), that is, not especially in a native L2 setting.

One of the earlier paradigms for the automatic assessment of pronunciation quality by machine is presented in Neumeyer *et al.* (2000). Piske *et al.* (2001) provide a thorough review of the literature on foreign accent in a second language, and detail the relevant variables such as motivation, age, and formal instruction; both age of L2 learning and amount of continued L1 use were found to affect degree of foreign accent; see also Flege *et al.* (2006). Omar and Pelecanos (2010) report on procedures for detecting non-native speakers and their native language, and utilise the detected speaker characteristics within a speaker recognition system to improve its performance. Speech rate seems to be a rather straightforward but reliable indicator of degree of non-nativeness (Cucchiarini *et al.* 2000). In Hönig *et al.* (2012), the impact of speech tempo (modelled by duration features) is demonstrated by regressions with and without features modelling tempo for a database with read non-native English speech; the difference in absolute terms is, depending on the constellation, between 4% and 20%.

Witt (2012) discusses the state of the art of research on computer-assisted pronunciation teaching and its major components, and gives an overview of existing commercial language learning software. ASR-based systems for language learning and therapy are exemplified in Strik (2012) and Pellom (2012); the latter describes an on-line solution specifically designed to address the shortcomings of more traditional learning methods and to improve conversational fluency.

So far we have sketched the main issues in modelling non-native speech; it can also be studied with respect to its role in other paralinguistic tasks. Graham *et al.* (2001) found major differences between native and non-native listeners in their ability to identify emotions expressed in voice. They suggest that, irrespective of the level of proficiency in L2, learners of an L2 can only interpret emotional meaning when they have been exposed extensively to pertinent utterances in the native L2 context. For similar experiments, see Chen (2009).

It is not only the ability to assess the emotional content of utterances in an L2, that is, the perception, that changes; so also do others' judgements, depending on speakers speaking either their L1 or an L2. Lev-Ari and Keysar (2010) demonstrate that lower intelligibility of a foreign accent causes non-native speakers to sound less credible. A comparable impact has been reported for bilingual speakers. Several studies showed that bilingual speakers partly changed their personality as perceived by the self and by others, depending on which of the two languages they were speaking; see Chen and Bond (2010) for Chinese versus English, Koven (2007) for French versus Portuguese, Ramírez-Esparza *et al.* (2006) for Spanish versus English, and Danziger and Ward (2010) for Hebrew versus Arabic. All these studies corroborate a weak form of linguistic relativity, as far as paralinguistic functions such as credibility or personality traits are concerned.

5.7 Social Signals

One cannot not communicate.

(Paul Watzlawick)

So far, the phenomena addressed can be observed devoid of communicative context, either in read speech or in public speech, without any distinct communication partner. In this section, we want to deal with paralinguistic functions that can (only) be observed and assessed within a communication setting (human–human or human–machine, dyadic or multi-party). In such situations, we can also try to detect and classify everything that has been mentioned in this chapter so far: age, gender, personality, emotion/affect, and atypical speech. All these states and traits, however, can basically also be observed without taking into account the interaction between the partners. Thus it might be possible to define phenomena that can be observed and analysed in isolation, those that only can be observed and analysed in some kind of interaction such as 'interest' (in what/whom) or 'intimacy' (with whom), and perhaps those that solely are constituted within a social setting such as back-channelling or entrainment (the mutual adaptation of communication partners with respect to phonetic parameters such as pitch, or to the use of words or syntactic constructions). In practice, it comes down to calling those phenomena 'social signals' that can be observed in – and have some relevance for – social interaction.

Among the 'classic' fields are *pragmatics, ethnography of communication/speaking, linguistic anthropology, sociology of language, sociolinguistics*. New names are *social signals* or *behavioural signals*. Each of these fields has, of course, specific traditional core topics and perspectives.

For the ethnography of communication, culture is the object of investigation, manifested and mirrored in language and speech. Typical topics addressed within this and neighbouring fields such as sociology of language are female and male varieties of a language (Haas 1979), forms of address (Brown and Gilman 1960), and taboo words or constellations, for instance, when males are not allowed to address females (Jay 2009).

For traditional pragmatics which is rooted in linguistics and not in anthropology, the main topics are speech act theory and dialogue acts; glancing through the topics addressed in, for example, the *Journal of Pragmatics*, one realises that nowadays the perspective is based on the humanities and not on artificial intelligence or engineering approaches, but the topics are similar and mostly all-encompassing across fields. The same holds for the field of 'social signals' that has been coined with a particular emphasis on machine analysis and synthesis of human social signals (Vinciarelli *et al.* 2009) – the topics are similar, albeit that the field of 'social signal processing' has been restricted explicitly to the non-verbal aspect by these authors. An early and renowned predecessor is Malinowski (1923) who introduced the term *phatic communication* into linguistics, especially for the introductory formulaic phrases in a human–human encounter such as *how are you* or *nice day, isn't it* – at first sight void of meaning, but pivotal for constituting social relationships; see Senft (2009).

The classic introduction to linguistic pragmatics dealing with deixis, implicature, presupposition, speech acts, and conversational structure, is Levinson (1983). The more recent introduction by Mey (2001) broadens the view onto topics such as pragmatics across cultures and social aspects. Krauss and Pardo (2006) argue that the phonological sound structure of speech provides information on social behaviour.

Pentland and Madan (2005) and Pentland (2008) focus on 'unconscious face, hand, and body gestures [forming] a visual motion texture that conveys social signals', that is, on rather subtle patterns of human–human interaction. Vinciarelli *et al.* (2009) and Vinciarelli and Mohammadi (2011) introduce 'social signal processing' as an emerging domain dealing with the ability 'to understand and manage social signals of a person we are communicating with [as] the core of social intelligence'. They argue that next-generation computers should possess this type of ability in order to become more effective.

A related term is 'behavioural signal processing' as described in Black *et al.* (2010, p. 2030): 'Human behavioral signal processing involves using signal processing methods and machine learning algorithms to extract human-centered information from audio-video signals, including social cues . . . , affect and emotions . . . , and intent . . . '. This term does not necessarily focus on social encounters alone; rather, studies within this domain deal with typical versus atypical and distressed human behaviour.

The fields of 'social signals' and 'honest signals' are a specific cross-section of paralinguistics, under the umbrella of social interaction/communication, for instance, encompassing phenomena that usually are listed as emotions or affect; social signal processing concentrates on non-verbals, and both fields put emphasis on the technological aspect. We might say that every paralinguistic event is a social signal in this sense when it is observed within a social context, intended to perform a social function, or perceived by the communication partner as a social signal. Being able to employ social signals has been described as *social intelligence*, which seems to be closely related to emotional intelligence; both are members of the set of *multiple intelligences*, with the close equivalent *interpersonal intelligence* (Gardner 1993). The theoretical foundation of these multiple intelligences seems to be less clear and has been widely questioned; yet, intuitively and in a pre-theoretical understanding, they are plausible.

Gregory *et al.* (2001, p. 37) found 'that the F0 band [below 0.5 kHz] plays an important role in transmission of social status and dominance information and that elimination of the F0 leads to lessened perceived quality of conversation'. Ogden (2006) argues, based on concepts developed within *conversation analysis* (Couper-Kuhlen and Selting 1996), that the phonetic form of utterances influences their interpretation as agreement or disagreement. Bousmalis *et al.* (2009) detail the multimodal formal means (non-verbal cues) for indicating agreement or disagreement and list a number of tools and databases that could be used to train automatic tools for the analysis of spontaneous, audiovisual instances of agreement and disagreement. The display of frustration in conversations is analysed in Yu (2011). Two types of frustration are established: combined verbal and non-verbal expressions, and non-verbal expressions alone. The latter is claimed to be a stronger emotional display. Gravano *et al.* (2011) describe acoustic/prosodic and lexical correlates of social variables found in a corpus of task-oriented spontaneous speech. The data are labelled with the help of the Amazon Mechanical Turk (see Section 6.1.2), for attempts of a speaker to be liked, to be actually likeable, or to plan what to say. Significant differences in behaviour between single and cross-gender parings were found for the realisations of correlates of the social variables. Mori (2009) shows that not only linguistic but also prosodic features, especially pauses, and non-verbal behaviours characterise turn-switching in expressive dialogues. Laskowski *et al.* (2008) explore the relationship of social dimensions such as assigned role or seniority with low-level features that characterise talkspurt deployment, that is, the alternating sequence of utterances produced by the participants in a conversation; a *talkspurt* is the ' . . . speech by one party, including his pauses, which is preceded and followed, with or without intervening pauses, by speech from the other party perceptible to the one producing the talkspurt' (Norwine and Murphy 1938, p. 282). A system for 'predicting, detecting and explaining the occurrence of vocal activity in multi-party conversation' is described in Laskowski (2011).

It can be expected that most – if not all – phonetic and linguistic parameters and means can be employed in *dyadic* or *multi-party* conversations to regulate the interaction between participants. A prototypical phenomenon is *back-channelling* – giving feedback that one is still paying attention. This can be done using facial or body gestures (head nodding, raised eyebrows), *non-verbals* or *conversational sounds* (unspecific grunts, particles such as *uhm, mhm, yes/yeah*), or more explicit linguistic means (*is that true?, oh, really?*). Default back-channelling is neutral, meaning 'I'm still listening' (Kießling *et al.* 1993); of course, more or less negative or positive back-channelling is possible as well. As in the case of laughter (Section 4.5), which can function as a back-channelling signal, other back-channelling signals normally only occur at specific positions in the conversation. The participant in the active, speaking role would be very irritated if the listener either did not produce any back-channelling signals at all, or did so in the wrong places. Grivičić and Nilep investigate the phonetic form of the token *yeah*; produced with creaky voice, it indicates '... passive recipiency and either a disprefer-ence to continue the current topic, or a disalignment with the primary speaker' (Grivičić and Nilep 2004, p. 1). Sometimes it indicates a shift from recipient to speaker. Truong and Heylen (2010) try to disambiguate automatically the functions of conversational sounds with the help of prosody, focusing on the different functions of *yeah* such as back-channelling/assessment, or signalling intention to take the floor. Ward (2006, p. 129) exemplifies the roles of lexical conversational sounds in American English such as '... low-overhead control of turn-taking, negotiation of agreement, signaling of recognition and comprehension, management of inter-personal relations such as control and affiliation, and the expression of emotion, attitude, and affect'.

Emotional colouring, interpersonal expressive behaviour, disambiguating ambiguous emo-tional expressions, and pertinent automatic procedures or conversational systems are addressed in Acosta and Ward (2011), Mower *et al.* (2009), Yu *et al.* (2004), and Ambady and Rosenthal (1992).

Entrainment is the term used for synchronising with and adapting to the interaction partner with respect to acoustic-prosodic or linguistic parameters such as pitch or energy. Lee *et al.* (2010, 2011) quantify prosodic entrainment in affective spontaneous spoken interactions of married couples. We can speculate that, ontogenetically and phylogenetically, entrainment leads to the register of *intimacy* that can be observed within a non-conflict-laden interaction within couples, for instance, when picking up the phone: for a third person present, it is often obvious that the spouse is calling and not anybody else. This register of intimacy can be observed within the basically symmetric interaction between couples, and within other, non-symmetric interactions; most prototypical is the interaction with a baby or toddler. There are several other terms denoting this specific speech register, with more or less slightly different connotations: *child-directed speech, infant-directed speech, baby-talk, motherese, fatherese,* and *parentese*, the last two being politically more correct versions; they are mostly not used to tell whether the mother, the father, or (one of the) parents are speaking. 'Intimacy' can be used as a over term or it can be distinguished from 'motherese'. There are other varieties which differ from each other in some respect, such as *pet-directed speech* or *pet-robot-directed speech* (Batliner *et al.* 2006b).

Kuhl (2004, 2007) advances the hypothesis that the earliest phases of language acquisition require social interaction which affects the learning of speech based on statistical and prosodic patterns – thus providing evidence for the pivotal role of motheresing in these phases. Zebrowitz *et al.* (1992) demonstrated that simply showing pictures with baby-faced instead of mature-faced children to adults before they had to give instructions to children over the phone was sufficient to elicit baby-talk. Several aspects of the phonetics of motherese are dealt with in the following studies: Fernald (2000) on infant-directed speech 'as a form of "hyperspeech" which facilities comprehension'; Kitamura and Burnham (1998) on differences in pitch height and range between mothers addressing either boys or girls; Liu *et al.* (2003) showing that '... mothers' vowel space area is significantly correlated with infants' speech discrimination performance'; Liu *et al.* (2003, p. F1) and Thiessen *et al.* (2005) on pitch contours in infant-directed speech that facilitate word segmentation; and Trainor *et al.* (2000) and Trainor and Desjardins (2002) on the more pronounced prosody of infant-directed speech and its role for expressing emotions and discriminating vowels.

In several studies, different varieties of this register are compared with each other: Burnham *et al.* (1998, 2002) analyse infant-directed, pet-directed, and adult-directed speech for similarities and differences. Reissland *et al.* illustrate that depressed mothers are less attuned in their 'maternal speech' to their infants during storybook reading than non-depressed mothers; this deficit '... might force the infant into self-regulatory patterns that eventually compromise the child's development' (Reissland *et al.* 2003, p. 255). Based on a comparison of mother–infant gestural and vocal interactions in chimpanzees and humans, Falk (2004) speculates about the role of motherese in the early development of foraging strategies and the prosodic and gestural conventionalisation of the meanings of certain utterances (words). Batliner *et al.* (2006b) investigate the prosody of children addressing Sony's pet robot dog Aibo and suggest that these children used a register that resembles mostly child-directed and pet-directed speech and to some extent computer-directed speech.

Shami and Verhelst (2007) explore different approaches towards feature extraction and three machine learning algorithms for classifying within and across databases with infant-directed speech versus adult-directed speech. Mahdhaoui and Chetouani (2009) and Mahdhaoui *et al.* (2010) compare different types of motherese detector systems. Batliner *et al.* (2008a) employ support vector machines (Section 11.2.1) and random forests (Section 11.2.1) to classify three different types of speaker, namely mothers addressing their own children or an unknown adult, women with no children addressing an imaginary child or an imaginary adult, and children addressing a pet robot using both intimate and neutral speech, and discuss the most important acoustic feature types.

5.8 Discrepant Communication

I'd kill for a Nobel Peace Prize.

(Steven Wright)

In contrast to deviant speech where the deviation from typical, normal speech is usually caused by some external factor, we define *discrepant speech* and *discrepant communication* as chosen intentionally by the speaker to serve specific purposes. For example, the speaker acts while using deviant speech (e.g., pretends to be alcoholised or to stammer), pretends to be polite, truthful, etc. while lying, or uses irony or sarcasm, or even specific, formulaic rhetorical figures such as an oxymoron – a figure of speech consisting of incongruent components such as 'heavy lightness'. Moreover, the register chosen and the manner of speaking can be fully 'normal' and typical, but the speaker 'speaks aside' to some other person or to herself, and the dialogue partner has to realise that the addressee has changed because otherwise misunderstandings might occur.

In the case of discrepant communication, we normally want to find the 'truth behind'. We use the term 'discrepant' communication/speech to denote communicative behaviour, partly manifested within speech, that cannot or should not simply be taken at face value; to put it another way, there is some mismatch between what is said or how it is said and what is meant. Another possible term would be *incongruent*.

Discrepant communication usually refers to atypical speech, at least in the more technical, frequency-oriented sense. Following Grice (1975, pp. 45ff.), we can assume that partners in a conversation normally follow specific conditions in order to ensure the felicity of conversation, the so-called *cooperative principle* and its four *conversational maxims*:

1. **Maxim of Quantity** – *(1) Make your contribution as informative as is required.... (2) Do not make your contribution more informative than is required.*
2. **Maxim of Quality** – *(1) Do not say what you believe to be false. (2) Do not say that for which you lack adequate evidence.*
3. **Maxim of Relation** – *Be relevant.*
4. **Maxim of Manner** – *(1) Avoid obscurity of expression. (2) Avoid ambiguity. (3) Be brief.... (4) Be orderly.*

In discrepant communication, this is not really the case: lying violates the maxim of quality, as the speaker knows that what he said is false; speaking aside violates the maxim of relation, as it is not immediately relevant for the communication with the dialogue partner; and irony and sarcasm violate the maxim of manner, because the message is intrinsically ambiguous. As far as we can see, a violation of the maxim of quantity might be a topic better addressed in linguistic semantics and in procedures such as topic spotting. In a general sense, these maxims can be seen as philosophical abstractions that cannot be fully observed in daily life. For instance, there is a delicate equilibrium between the requirements of Grice's maxims and the requirements of culturally adequate interactions that often are not very straightforward and require the use of some indirect speech, as in the well-known avoidance of a plain 'no' in Far Eastern cultures. A failure to master such subtle communication rules can as well be observed within cultures, for example, in the speech of non-native speakers, speakers with autism spectrum condition, Down's syndrome, or hard-of-hearing speakers who are sometimes conceived as being less polite than expected. In all these constellations, discrepant communication might take place. However, this is not intended by the speaker, thus we subsume them under the heading 'deviant speech' dealt with in Section 5.6.

Social signals dealt with in Section 5.7 are normally 'honest' signals (Pentland 2008); discrepant signals show up in social settings as well but are not exactly 'honest' – at least, they are indirect, and sometimes they are simply dishonest.

5.8.1   Indirect Speech, Irony, and Sarcasm

The week starts well.

(Attributed to Matthias Kneißl on his way to the scaffold)

Indirect speech is a conventionalised form of 'pseudo-discrepant' communication – it is pseudo because the 'direct' meaning is not what is conveyed but another metaphorical one. Pinker *et al.* (2008) mention *Would you like to come up and see my etchings?* functioning as sexual come-on, and *If you could pass the guacamole, that would be awesome* functioning as a polite request. Conflicts would only arise if one of the communication partners prefers the literal meaning of the utterance.

Rhetorical figures (e.g., an oxymoron such as *bitter sweet*) are another type of conventionalised, intrinsically discrepant speech. *Irony* is a not (fully) conventionalised indirect way of expressing facts or opinions, and often a play with words and their meaning. The statement is *incongruent* with the linguistic or situational context. By and large, the definition of irony given in Freud (1905) is still valid: 'Its essence lies in saying the opposite of what one intends to convey to the other person, but in sparing him contradiction by making him understand – by one's tone of voice, by some accompanying gesture, or (where writing is concerned) by some small stylistic indications – that one means the opposite of what one says.' However, the paradox essence of irony is that it is perfect if it is concealed perfectly – and thus prone to misunderstanding. A conventional example is *nice weather, isn't it?* when the rain is pouring down. In order to understand less conventional irony, the communication partner has to realise that there is a discrepancy between the literal meaning of the statement and its context; for instance, a favourable comment on some political action might be 'irony-prone' if the speaker is known not to be a follower of this political direction. The *paradox communication* known from family therapy (Watzlawick *et al.* 1967) is also discrepant; in some instances, it can be very close to irony. A specific case, the so-called *double-bind communication*, is illustrated by Watzlawick *et al.* (1967) with a story. A mother gives two ties to her son, and when he puts on one of them, she complains about him not liking the other one. This can be an indication of a pathological family system, or it can be just a joke, using the means normally employed in irony.

There is no easy distinction between irony and *sarcasm*; irony is indirect, sarcasm more direct, in word or in deed; sarcasm is a sort of intensification of irony, intended to wound, as in the statement of the main figure in Schiller's drama *Wilhelm Tell*, 'und mit der Axt hab ich ihm's Bad gesegnet' (with the axe I blessed his bath), after the bailiff wanted to take a bath with Tell's wife and was subsequently killed by him. In between irony and sarcasm is this famous statement from the Icelandic sagas. A great warrior, Atli Asmundson, was defending the door to his hall and he fought hard. Finally, he was struck down and run through by a spear. He looked down at the spear as he fell and said: 'Those broad spears are in fashion now.' (Those old Icelanders were (literally) dying for a good last sentence.) The use of 'blessing' in the sense of 'killing', and the mention of fashion in the context of one's own death are both discrepant; whereas Tell really intended to wound (lethally), Atli 'only' commented on his (lethal) wound in his laconic last statement, thus Tell might be conceived as being more sarcastic than Atli.

It is difficult to give any clear-cut definition for all these terms, thus we have followed the usual strategy and resorted to illustrative examples. In the literature, irony is often not clearly distinguished from – or is simply equated with – sarcasm. The vagueness of the non-linguistic, contextual factors makes it difficult to process and recognise these phenomena automatically, especially in real-life settings, because their modelling would presuppose an all-embracing ontology, that is, a shared conceptualisation of all consistent and discrepant constellations. Discrepant speech is of course also difficult to deal with within text-based sentiment analysis (Section 5.5). For a straightforward analysis which only takes into account single words (unigrams), even a negation of a negative word ('not bad', meaning 'pretty good'), can be problematic. This holds even more for ironic statements.

Colston and O'Brien (2000) contrast the slightly different function of verbal irony and understatement, and Bryant and Tree (2002) distinguish the different impacts of acoustic and textual information on the recognition of verbal irony in spontaneous speech. The vocal and especially prosodic cues of sarcasm are addressed in Rockwell (2000, 2007) and Cheang and Pell (2008).

Indirect speech, irony and sarcasm are of course interesting topics in themselves. They are of interest for an automatic processing of human–human and human–machine communication because we want to either ignore them in automatic summarisation, or model them in sentiment/affect analysis, for instance, when classifying negative or positive reviews. First attempts towards the automatic processing of irony and sarcasm in written language are reported in Davidov *et al.* (2010) and Filatova (2012).

5.8.2 Deceptive Speech

All Cretan are liars.

(Epimenides, a Cretan)

Deceptive speech (lying) is not indirect but intended to sound fully 'normal'; however, it is discrepant because it is incongruent with reality. We can imagine a plethora of promising constellations where a successful detection of deceptive speech could be beneficial: in any therapeutic scenario where patients have to be monitored, in a mediation scenario (business partners, divorce, conflicts between parents and children), and in forensic scenarios. All these scenarios normally lead to single instance decisions with far reaching consequences, and impose high demands on reliability and validity; they are critical in the sense of Table 1.1 on page 16. We can imagine less critical scenarios, for instance, when we want to find out whether interviewees are tending to hide the (complete) truth with respect to specific questions. Here, we are aiming at cumulative evidence (Section 3.7), thus it is only important that we can uncover some general tendencies.

In an early study, Fay and Middleton (1941) investigated the ability to judge truth-telling or lying and found an accuracy of the judgements slightly above chance. Kraut (1978, 1980) employed humans as 'lie detectors' and showed that observers were moderately accurate in detecting lies, and that actors were consistently good or bad liars. Vocal cues to deception have been dealt with by various authors: verbal and non-verbal cues (DePaulo *et al.* 1982); clear speech or filtered/inverted speech (Scherer *et al.* 1985); verbal and acoustic-prosodic variables (Anolli and Ciceri 1997); verbal cues (Reich 1981); and non-verbal strategies for decoding deception (Zuckerman *et al.* 1982). Ekman and colleagues have investigated the ability of experts and non-experts to detect lying (Ekman 1988; Ekman and O'Sullivan 1991; Ekman *et al.* 1991). Ekman *et al.* (1999) claim that not only laypersons but even professionals concerned with lying show poor performance in detecting lies, but some professional 'lie catchers' seem to be highly accurate. Vrij *et al.* (2000) found a higher percentage of detected lies with an approach that used non-verbal and verbal indicators of deception. First attempts at automatically distinguishing deceptive from non-deceptive speech are reported in Hirschberg *et al.* (2005). Voice acoustical correlates of feigned depression and feigned sleepiness are addressed in Reilly *et al.* (2004). Conceptualisations of non-cooperative and deceptive behaviour in (virtual) humans and robots are described in Nijholt *et al.* (2012).

Despite high expectations, the performance of lie detectors is poor. We refer again to the chapter in (Kreiman and Sidtis 2011) (recall Section 1.4.3 above) where the authors clearly illustrate these deficiencies. Thus, they must not be used for single instance decisions. It remains to be seen whether they can help in establishing cumulative evidence, for instance, in computer surveillance at airports.

5.8.3   Off-Talk

Enter Guildenstern and Rosencrantz....
Hamlet You were sent for...
Rosencrantz To what end, my lord?
Hamlet That you must teach me....
*Rosencrantz [**Aside** to Guildenstern]* What say you?
*Hamlet [**Aside**]* Nay then, I have an eye of you! *[Aloud]* If you love me, hold not off.
Guildenstern My lord, we were sent for.

(Shakespeare, *Hamlet*, Act II, Scene 2)

Speech is the primary means of interaction between humans in dyadic (dialogue) or multi-party conversations. This is the 'normal' use and thus 'typical' for computational paralinguistics. In this context, speech not addressed to any communication partner is atypical – not because it is awkward or deficient but because it is less frequent and outside of any interaction/communication. In line with Oppermann *et al.* (2001) and Siepmann *et al.* (2001), we want to employ the generic term *off-talk* for this atypical use, contrasting it with *on-talk* for the typical, frequent use of speech within conversations. The human dialogue partner(s) and any automatic system involved have to detect off-talk and process it accordingly; otherwise, there is a high risk of more or less fatal miscommunication. Again, the pivotal point is that this type of speech is not only sparse within a communicative setting but in some way 'deviant', that is, displaying characteristics differing from 'normal', typical speech – at least, this is our hope because we want to treat it differently from speech directed towards the dialogue partner. Of course, these characteristics are not only found in speech: we turn away from our communication partner (head movement, gaze direction) when addressing someone else who is present (cf. on-focus versus off-focus below), or we look down when addressing ourself (*self-talk*).

Pre-school children can often be observed talking to themselves; for them, *thinking aloud* and speech accompanying actions are an important means of self-regulation, learning, and becoming acquainted with a theory of mind by attributing mental states to themselves. The pivotal role of this *private speech* – also called *egocentric speech* by Piaget (1923) or *self-directed speech* – has been pointed out by Vygotski (1962) and in later developmental research studies. Thus, self-talk can be observed within an interpersonal communication; in this case, it is 'off-communication', that is, off-talk. When observed without any interpersonal communication, it is not exactly off-talk but private speech – most likely displaying some formal characteristics of off-talk. Overviews of different aspects of private speech can be found in two readers (Diaz and Berk 1992; Winsler *et al.* 2009). A generic description of private speech is given in Ahmed (1994). For multimodal human–computer interaction, see Lunsford (2004).

Self-talk can be observed in the second aside produced by Hamlet in the quotation above; this can be conceived of as 'real' private, self-directed speech, or as speech directed to the audience. The first *aside* of Rosencrantz is addressed to Guildenstern, another partner in this multi-party interaction, and not directly to Hamlet who has asked for an explanation. Most likely, both Hamlet and Rosencrantz turn away from each other when producing the aside. Note that the question produced by Rosencrantz could as well be addressed to Hamlet without being nonsensical, meaning: 'I am asking you, what do you think yourself?' This might, however, be rather improper, considering that it would be produced by a courtier towards the Prince of Denmark. The stage direction disambiguates.

In recent years, a new research topic has emerged, namely multimodal, multi-party interaction with other humans, for instance in meetings, or with both other humans and computers, for instance with information systems and/or embodied agents. In such scenarios, several speakers can overlap, and light and audio conditions are less favourable. Perhaps because of these additional factors, so far, often rather coarse parameters have been employed for distinguishing between *on-focus* and *off-focus* (the multimodal varieties of on-talk and off-talk) such as head orientation in the video channel, and a binary decision of speech versus non-speech in the audio channel. Gaze direction and/or head orientation in dyadic or multi-party conversations, especially as indicators of attention and addressee, are dealt with in Stiefelhagen *et al.* (2002), Stiefelhagen and Zhu (2002), Katzenmaier *et al.* (2004), Jovanovic and Op den Akker (2004), Turnhout *et al.* (2005), and Rehm and André (2005); further references are given in Heylen (2005). The fusion of gaze direction and/or head orientation with sound/speech is addressed in Stiefelhagen *et al.* (2002), Katzenmaier *et al.* (2004), and Turnhout *et al.* (2005). For the multi-party, human–human scenario of Stiefelhagen *et al.* (2002), a thorough analysis of gaze direction has been conducted. However, as it makes no prosodic difference whether the one or the other person is addressed, there is no detailed analysis of the audio channel. In Katzenmaier *et al.* (2004) additionally human–machine interaction occurs. The main differences observed in the audio channel are commands given to the machine versus conversation with the human partners — a consequence of a low-complexity dialogue system. The scenario in Turnhout *et al.* (2005) is similar to the triadic scenario in SmartWeb described below; from the audio channel the length of the speech segment is computed and combined with facial information. In a human–machine scenario, Batliner *et al.* (2008c) report up to 84.5% classification performance for distinguishing between on-focus and off-focus, when using prosodic, linguistic, and video information.

On-talk versus off-talk within the SmartWeb scenario is further dealt with in Oppermann *et al.* (2001), Siepmann *et al.* (2001), and Batliner *et al.* (2002). Off-talk as a special dialogue act has not yet been the object of much investigation (Alexandersson *et al.* 1998; Carletta *et al.* 1997), most likely because it could not be observed in those human–human communications which were analysed for dialogue act modelling. In a normal human–human dialogue setting, off-talk might be somewhat self-contradictory, because of the 'impossibility of not communicating' (Watzlawick *et al.* 1967). We can, however, easily imagine the use of off-talk if someone is speaking in a low voice not *to* but *about* a third person present who is very hard of hearing.

5.9   Common Traits of Functional Aspects

The common denominator of all the functional aspects we have addressed in this chapter is the focus on distinguishing different typical or typical/atypical voice, speech, language varieties, or phenomena that characterise different classes (categories) or differently graded peculiarities (continua, dimensions); these varieties are characterised by more or less different formal

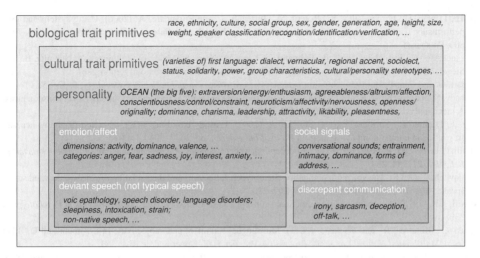

Figure 5.2   Layered figure–ground relationship of paralinguistic functions, with examples of states and traits

means. Nowadays, the methodologies employed in computational paralinguistics are largely independent of the specific phenomena – apart from different edge conditions such as speaker-dependent or independent modelling, two- or n-class problems, sparsity of data that impede the use of standard procedures, and of course, apart from the fact that different phenomena might be indicated by different (combinations of) features. In automatic processing, procedures adapted from ASR are nowadays often used in a sort of Swiss Army knife approach. This was different in the past, due to different research traditions; it is not clear yet whether in the future, methodologies will ramify into different methodologies, each specialised for specific tasks and applications.

Figure 5.2 displays especially those traits and states addressed in this chapter, in the same layered presentation as in Figure 5.1. The list is not complete; we invite the reader to consider and establish mutual relationships of the states and traits displayed, and of additional states and traits. To give one example: *age* is a biological trait primitive and related to *generation* (older/younger generation); generation is not only biologically but also culturally determined, and influences the expression of personality, emotion/affect, and social signals. It can be indicated by acoustic and linguistic means, and it can be investigated not only within computational paralinguistics but also within all the other disciplines mentioned (e.g., psychology, sociology, and socio-/psycholinguistics).

References

Acosta, J. C. and Ward, N. G. (2011). Achieving rapport with turn-by-turn, user-responsive emotional coloring. *Speech Communication*, **53**, 1137–1148.

Addington, D. (1968). The relationship of selected vocal characteristics to personality perception. *Speech Monographs*, **35**, 492–503.

Ahmed, M. K. (1994). Private speech: A cognitive tool in verbal communication. In S. Kimura and M. Leong (eds), *The Language Programs of the International University of Japan Working Papers*, volume 5. International University of Japan.

Ai, H., Litman, D., Forbes-Riley, K., Rotaru, M., Tetreault, J., and Purandare, A. (2006). Using system and user performance features to improve emotion detection in spoken tutoring dialogs. In *Proc. of Interspeech*, pp. 797–800, Pittsburgh.

Alexandersson, J., Buschbeck-Wolf, B., Fujinami, T., Kipp, M., Koch, S., Maier, E., Reithinger, N., Schmitz, B., and Siegel, M. (1998). Dialogue acts in VERBMOBIL-2 – second edition. Verbmobil Report 226.

Allport, G. and Cantril, H. (1934). Judging personality from voice. *Journal of Social Psychology*, **5**, 37–55.

Ambady, N. and Rosenthal, R. (1992). Thin slices of expressive behavior as predictors of interpersonal consequences: A meta-analysis. *Psychological Bulletin*, **111**, 256–274.

Ang, J., Dhillon, R., Shriberg, E., and Stolcke, A. (2002). Prosody-based automatic detection of annoyance and frustration in human-computer dialog. In *Proc. of Interspeech*, pp. 2037–2040, Denver.

Anolli, L. and Ciceri, R. (1997). The voice of deception: Vocal strategies of naive and able liars. *Journal of Nonverbal Behavior*, **21**, 259–284.

Aronovitch, C. (1976). The voice of personality: Stereotyped Judgments and their relation to voice quality and sex of speaker. *Journal of Social Psychology*, **99**, 207–220.

Averill, J. R. (1975). A semantic atlas of emotional concepts. *JSAS Catalog of Selected Documents in Psychology*, **5**, 330.

Banea, C., Mihalcea, R., and Wiebe, J. (2011). Multilingual sentiment and subjectivity. In I. Zitouni and D. Bikel (eds), *Multilingual Natural Language Processing*. Prentice Hall, Upper Saddle River, NJ.

Banse, R. and Scherer, K. R. (1996). Acoustic profiles in vocal emotion expression. *Journal of Personality and Social Psychology*, **70**, 614–636.

Bard, E. G., Sotillo, C., Anderson, A. H., Thompson, H. S., and Taylor, M. M. (1996). The DCIEM map task corpus: Spontaneous dialogue under SD and drug treatment. *Speech Communication*, **20**, 71–84.

Baron-Cohen, S., Golan, O., and Ashwin, E. (2009). Can emotion recognition be taught to children with autism spectrum conditions? *Philosophical Transactions of the Royal Society B*, **364**, 3567–3574.

Batliner, A. (1984). The comprehension of grammatical and natural gender: a cross-linguistic experiment. *Linguistics*, **22**, 831–856.

Batliner, A., Fischer, K., Huber, R., Spilker, J., and Nöth, E. (2000). Desperately seeking emotions: Actors, wizards, and human beings. In *Proc. of the ISCA Workshop on Speech and Emotion*, pp. 195–200, Newcastle, Co. Down.

Batliner, A., Zeissler, V., Nöth, E., and Niemann, H. (2002). Prosodic Classification of Offtalk: First Experiments. In *Proc. of Text, Speech and Dialogue (TSD)*, pp. 357–364, Berlin. Springer.

Batliner, A., Fischer, K., Huber, R., Spilker, J., and Nöth, E. (2003a). How to find trouble in communication. *Speech Communication*, **40**, 117–143.

Batliner, A., Zeissler, V., Frank, C., Adelhardt, J., Shi, R. P., and Nöth, E. (2003b). We are not amused – but how do you know? User states in a multi-modal dialogue system. In *Proc. of Interspeech*, pp. 733–736, Geneva.

Batliner, A., Hacker, C., Steidl, S., Nöth, E., and Haas, J. (2004). From emotion to interaction: Lessons from real human-machine dialogues. In *Affective Dialogue Systems, Proceedings of a Tutorial and Research Workshop*, pp. 1–12, Kloster Irsee.

Batliner, A., Steidl, S., Hacker, C., Nöth, E., and Niemann, H. (2005). Tales of tuning – prototyping for automatic classification of emotional user states. In *Proc. of Interspeech*, pp. 489–492, Lisbon.

Batliner, A., Steidl, S., Schuller, B., Seppi, D., Laskowski, K., Vogt, T., Devillers, L., Vidrascu, L., Amir, N., Kessous, L., and Aharonson, V. (2006a). Combining efforts for improving automatic classification of emotional user states. In *Proc. of IS-LTC 2006*, pp. 240–245, Ljubljana.

Batliner, A., Biersack, S., and Steidl, S. (2006b). The prosody of pet robot directed speech: Evidence from children. In *Proc. of Speech Prosody*, pp. 1–4, Dresden.

Batliner, A., Schuller, B., Schäffler, S., and Steidl, S. (2008a). Mothers, adults, children, pets — towards the acoustics of intimacy. In *Proc. of ICASSP*, pp. 4497–4500, Las Vegas.

Batliner, A., Steidl, S., Hacker, C., and Nöth, E. (2008b). Private emotions vs. social interaction — a data-driven approach towards analysing emotions in speech. *User Modeling and User-Adapted Interaction*, **18**, 175–206.

Batliner, A., Hacker, C., and Nöth, E. (2008c). To talk or not to talk with a computer – taking into account the user's focus of attention. *Journal on Multimodal User Interfaces*, **2**, 171–186.

Batliner, A., Schuller, B., Seppi, D., Steidl, S., Devillers, L., Vidrascu, L., Vogt, T., Aharonson, V., and Amir, N. (2011a). The automatic recognition of emotions in speech. In P. Petta, C. Pelachaud, and R. Cowie (eds), *Emotion-Oriented Systems: The Humaine Handbook*, Cognitive Technologies, pp. 71–99. Springer, Berlin.

Batliner, A., Steidl, S., Schuller, B., Seppi, D., Vogt, T., Wagner, J., Devillers, L., Vidrascu, L., Aharonson, V., and Amir, N. (2011b). Whodunnit: Searching for the most important feature types signalling emotional user states in speech. *Computer Speech and Language*, **25**, 4–28.

Becker, I. and Aharonson, V. (2010). Last but definitely not least: On the role of the last sentence in automatic polarity-classification. In *Proc. of the ACL 2010 Conference, Short Papers*, pp. 331–335, Uppsala, Sweden.

Beigi, H. (2011). *Fundamentals of Speaker Recognition*. Springer, New York.

Bhuta, T., Patrick, L., and Garnett, J. (2004). Perceptual evaluation of voice quality and its correlation with acoustic measurements. *Journal of Voice*, **18**, 299–304.

Black, M., Katsamanis, A., Lee, C.-C., Lammert, A. C., Baucom, B. R., Christensen, A., Georgiou, P. G., and Narayanan, S. (2010). Automatic classification of married couples' behavior using audio features. In *Proc. of Interspeech*, pp. 2030–2033, Makuhari.

Blood, G., Mahan, B., and Hyman, M. (1979). Judging personality and appearance from voice disorders. *Journal of Communication Disorders*, **12**, 63–67.

Bocklet, T., Maier, A., Riedhammer, K., and Nöth, E. (2009). Towards a language-independent intelligibility assessment of children with cleft lip and palate. In *Proc. of WOCCI*.

Bocklet, T., Stemmer, G., Zeissler, V., and Nöth, E. (2010). Age and gender recognition based on multiple systems – early vs. late fusion. In *Proc. of Interspeech*, pp. 2830–2833, Makuhari, Japan.

Bocklet, T., Nöth, E., Stemmer, G., Ruzickova, H., and Rusz, J. (2011). Detection of persons with Parkinson's disease by acoustic, vocal, and prosodic analysis. In *Proc. of the Automatic Speech Recognition and Understanding Workshop (ASRU)*, pp. 478–483, Big Island, Hawaii.

Bocklet, T., Riedhammer, K., Nöth, E., Eysholdt, U., and Haderlein, T. (2012). Automatic intelligibility assessment of speakers after laryngeal cancer by means of acoustic modeling. *Journal of Voice*, **26**, 390–397.

Bonneh, Y. S., Levanon, Y., Dean-Pardo, O., Lossos, L., and Adini, Y. (2011). Abnormal speech spectrum and increased pitch variability in young autistic children. *Frontiers in Human Neuroscience*, **4**, 237.

Bono, J. E. and Judge, T. A. (2004). Personality and transformational and transactional leadership: A meta-analysis. *Journal of Applied Psychology*, **89**, 901–910.

Boril, H., Sangwan, A., Hasan, T., and Hansen, J. (2011). Automatic excitement-level detection for sports highlights generation. In *Proc. of Interspeech*, pp. 2202–2205, Makuhari, Japan.

Bousmalis, K., Mehu, M., and Pantic, M. (2009). Spotting agreement and disagreement: A survey of nonverbal audiovisual cues and tools. In *Proc. of Affective Computing and Intelligent Interaction (ACII)*, Amsterdam.

Brown, D. H., Hilgers, F. J., Irish, J. C., and Balm, A. J. (2003). Postlaryngectomy voice rehabilitation: State of the art at the millennium. *World Journal of Surgery*, **27**, 824–831.

Brown, R. and Gilman, A. (1960). The pronouns of power and solidarity. In T. Sebeok (ed.), *Style in Language*, pp. 253–281. MIT Press, Cambridge, MA.

Bruckert, L., Lienard, J., Lacroix, A., Kreutzer, M., and Leboucher, G. (2006). Women use voice parameter to assess men's characteristics. *Proceedings of the Royal Society B*, **237**(1582), 83–89.

Bruckert, L., Bestelmeyer, P., Latinus, M., Rouger, J., Charest, I., Rousselet, G., Kawahara, H., and Belin, P. (2010). Vocal attractiveness increases by averaging. *Current Biology*, **20**, 116–120.

Bryant, G. and Tree, J. F. (2002). Recognizing verbal irony in spontaneous speech. *Metaphor and Symbol*, **17**, 99–117.

Bunton, K., Kent, R. D., Duffy, J. R., Rosenbek, J. C., and Kent, J. F. (2007). Listener Agreement for auditory-perceptual ratings of dysarthria. *Journal of Speech, Language, and Hearing Research*, **50**, 1481–1495.

Burkhardt, F., Paeschke, A., Rolfes, M., Sendlmeier, W., and Weiss, B. (2005). A database of German emotional speech. In *Proc. of Interspeech*, pp. 1517–1520, Lisbon.

Burkhardt, F., Eckert, M., Johannsen, W., and Stegmann, J. (2010). A database of age and gender annotated telephone speech. In *Proc. of LREC*, pp. 1562–1565, Valletta, Malta.

Burkhardt, F., Schuller, B., Weiss, B., and Weninger, F. (2011). 'Would you buy a car from me?' – On the likability of telephone voices. In *Proc. of Interspeech*, pp. 1557–1560, Florence.

Burnham, D., Francis, E., Vollmer-Conna, U., Kitamura, C., Averkiou, V., Olley, A., Nguyen, M., and Paterson, C. (1998). Are you my little pussy-cat? Acoustic, phonetic and affective qualities of infant- and pet-directed speech. In *Proc. of ICSLP*, pp. 4534–4556, Sydney.

Burnham, D., Kitamura, C., and Vollmer-Conna, U. (2002). What's new, pussycat? On talking to babies and animals. *Science*, **296**, 1435.

Cambria, E., Schuller, B., Xia, Y., and Havasi, C. (2013). New avenues in opinion mining and sentiment analysis. *IEEE Intelligent Systems Magazine, Special Issue on Concept-Level Opinion and Sentiment Analysis*, **28**(2).

Campenhausen, S. v., Bornschein, B., Wick, R., Bötzel, K., Sampaio, C., Poewe, W., Oertel, W., Siebert, U., Berger, K., and Dodel, R. (2005). Prevalence and incidence of Parkinson's disease in Europe. *European Neuropsychopharmacology*, **15**, 473–490.

Carletta, J., Dahlbäck, N., Reithinger, N., and Walker, M. (1997). Standards for Dialogue Coding in Natural Language Processing. Dagstuhl-Seminar-Report 167.

Cheang, H. S. and Pell, M. D. (2008). The sound of sarcasm. *Speech Communication*, **50**, 366–381.

Chen, A. (2009). Perception of paralinguistic intonational meaning in a second language. *Language Learning*, **59**(2), 367–409.

Chen, S. X. and Bond, M. H. (2010). Two languages, two personalities? Examining language effects on the expression of personality in a bilingual context. *Personality and Social Psychology Bulletin*, **36**(11), 1514–1528.

Chevallier, C., Noveck, I., Happé, F., and Wilson, D. (2011). What's in a voice? Prosody as a test case for the Theory of Mind account of autism. *Neuropsychologia*, **49**, 507–517.

Chevrie-Muller, C., Seguier, N., Spira, A., and Dordain, M. (1978). Recognition of psychiatric disorders from voice quality. *Language and Speech*, **21**, 87–111.

Childers, D. G. and Wu, K. (1991). Gender recognition from speech. Part II: Fine analysis. *Journal of the Acoustical Society of America*, **90**, 1841–1856.

Chung, C. and Pennebaker, J. (2007). The psychological functions of function words. In K. Fiedler (ed.), *Social Communication*, pp. 343–359. Psychology Press, New York.

Clavel, C., Vasilescu, I., Devillers, L., Richard, G., and Ehrette, T. (2008). Fear-type emotion recognition for future audio-based surveillance systems. *Speech Communication*, **50**(6), 487–503.

Clifford, B. (1980). Voice identification by human listeners: On earwitness reliability. *Law and Human Behavior*, **4**, 373–394.

Colston, H. L. and O'Brien, J. (2000). Contrast and pragmatics in figurative language: Anything understatement can do, irony can do better. *Journal of Pragmatics*, **32**, 1557–1583.

Connor, C. M., Craig, H. K., Raudenbush, S. W., Heavner, K., and Zwolan, T. A. (2006). The age at which young deaf children receive cochlear implants and their vocabulary and speech-production growth: Is there an added value for early implantation? *Ear & Hearing*, **27**, 628–644.

Cooney, O. M., McGuigan, K., Murphy, P., and Conroy, R. (1998). Acoustic analysis of the effects of alcohol on the human voice. *Journal of the Acoustical Society of America*, **103**, 2895.

Couper-Kuhlen, E. and Selting, M. (1996). Towards an interactional perspective on prosody and a prosodic perspective on interaction. In E. Couper-Kuhlen and M. Selting (eds), *Prosody in Conversation Interactional Studies*, pp. 11–56. Cambridge University Press, Cambridge.

Cowie, R. and Cornelius, R. R. (2003). Describing the emotional states that are expressed in speech. *Speech Communication*, **40**, 5–32.

Cowie, R. and Douglas-Cowie, E. (1996). Automatic statistical analysis of the signal and prosodic signs of emotion in speech. In *Proc. of ICSLP*, pp. 1989–1992, Philadelphia.

Cowie, R., Douglas-Cowie, E., Tsapatsoulis, N., Votsis, G., Kollias, S., Fellenz, W., and Taylor, J. (2001). Emotion recognition in human-computer interaction. *IEEE Signal Processing Magazine*, **18**, 32–80.

Cucchiarini, C., Strik, H., and Boves, L. (2000). Quantitative assessment of second language learners' fluency by means of automatic speech recognition technology. *Journal of the Acoustical Society of America*, **107**, 989–999.

Damico, J. S., Müller, N., and Ball, M. J. (eds) (2010). *The Handbook of Language and Speech Disorders*. Blackwell, Oxford.

Danziger, S. and Ward, R. (2010). Language changes implicit associations between ethnic groups and evaluation in bilinguals. *Psychological Science*, **21**, 6799–6800.

Darkins, A. W., Fromkin, V. A., and Benson, D. (1988). A characterization of the prosodic loss in Parkinson's disease. *Brain and Language*, **34**, 315–327.

Davidov, D., Tsur, O., and Rappoport, A. (2010). Semi-supervised recognition of sarcastic sentences in Twitter and Amazon. In *Proc. of the Fourteenth Conference on Computational Natural Language Learning*, pp. 107–116, Uppsala.

Davitz, J. R. (1964a). A review of research concerned with facial and vocal expressions of emotion. In J. R. Davitz (ed.), *The Communication of Emotional Meaning*, pp. 13–29. McGraw-Hill, New York.

Davitz, J. R. (1964b). Auditory correlates of vocal expressions of emotional feeling. In J. R. Davitz (ed.), *The Communication of Emotional Meaning*, pp. 101–112. McGraw-Hill, New York.

Dejonckere, P., Remacle, M., Fresnel-Elbaz, E., Woisard, V., Crevier-Buchman, L., and Millet, B. (1996). Differentiated perceptual evaluation of pathological voice quality: reliability and correlations with acoustic measurements. *Revue de Laryngologie - Otologie - Rhinologie*, **117**, 219–224.

Dejonckere, P. H., Bradley, P., Clemente, P., Cornut, G., Crevier-Buchman, L., Friedrich, G., Heyning, P. V. D., Remacle, M., and Woisard, V. (2001). A basic protocol for functional assessment of voice pathology, especially for investigating the efficacy of (phonosurgical) treatments and evaluating new assessment techniques. Guideline elaborated by the Committee on Phoniatrics of the European Laryngological Society (ELS). *European Archives of Oto-Rhino-Laryngology*, **258**, 77–82.

Dellaert, F., Polzin, T., and Waibel, A. (1996). Recognizing emotion in speech. In *Proc. of ICSLP*, pp. 1970–1973, Philadelphia.

Dellwo, V., Huckvale, M., and Ashby, M. (2007). How is individuality expressed in voice? An introduction to speech production and description for speaker classification. In C. Müller (ed.), *Speaker Classification I: Fundamentals, Features, and Methods*, pp. 1–20. Springer, Berlin.

Demouy, J., Plaza, M., Xavier, J., Ringeval, F., Chetouani, M., Périsse, D., Chauvin, D., Viaux, S., Golse, B., Cohen, D., and Robel, L. (2011). Differential language markers of pathology in autism, pervasive developmental disorder not otherwise specified and specific language impairment. *Research in Autism Spectrum Disorders*, **5**, 1402–1412.

DePaulo, B., Rosenthal, R., Rosenkrantz, J., and Green, C. (1982). Actual and perceived cues to deception: A closer look at speech. *Basic and Applied Social Psychology*, **3**, 291–312.

Derijcke, A., Eerens, A., and Carels, C. (1996). The incidence of oral clefts: a review. *British Journal of Oral and Maxillofacial Surgery*, **34**, 488–494.

Devillers, L. and Vidrascu, L. (2006). Real-life emotions detection with lexical and paralinguistic cues on human-human call center dialogs. In *Proc. of ICSLP*, pp. 801–804, Pittsburgh.

Devillers, L., Vidrascu, L., and Lamel, L. (2005). Challenges in real-life emotion annotation and machine learning based detection. *Neural Networks*, **18**, 407–422.

Devins, G. M., Stam, H., and Koopmans, J. P. (1994). Psychological impact of laryngectomy mediated by perceived stigma and illness intrusiveness. *Canadian Journal of Psychiatry*, **39**, 608–616.

Diaz, R. M. and Berk, L. E. (eds) (1992). *Private Speech. From Social Interaction to Self-Regulation*. Erlbaum, Hillsdale, NJ.

Dibazar, A. and Narayanan, S. (2002). A system for automatic detection of pathological speech. In *Proc. of Conference Signals, Systems, and Computers*, Asilomar, CA.

Drugman, T., Urbain, J., and Dutoit, T. (2011). Assessment of audio features for automatic cough detection. In *Proc. of the European Signal Processing Conference (EUSIPCO)*, pp. 1289–1293, Barcelona, Spain.

Ekman, P. (1988). Lying and nonverbal behavior: Theoretical issues and new findings. *Journal of Nonverbal Behavior*, **12**, 163–175.

Ekman, P. and O'Sullivan, M. (1991). Who can catch a liar? *American Psychologist*, **46**, 913–920.

Ekman, P., O'Sullivan, M., Friesen, W. V., and Scherer, K. R. (1991). Face, voice, and body in detecting deceit. *Journal of Nonverbal Behavior*, **15**, 125–135.

Ekman, P., O'Sullivan, M., and Frank, M. G. (1999). A few can catch a liar. *Psychological Science*, **10**, 263–266.

Ellgring, H. and Scherer, K. R. (1996). Vocal indicators of mood change in depression. *Journal of Nonverbal Behavior*, **20**, 83–110.

Engberg, I. S., Hansen, A. V., Andersen, O., and Dalsgaard, P. (1997). Design, recording and verification of a Danish emotional speech database. In *Proc. of Eurospeech*, pp. 1695–1698, Rhodes.

Eskenazi, M. (2009). An overview of spoken language technology for education. *Speech Communication*, **51**, 832–844.

Evans, S., Neave, N., and Wakelin, D. (2006). Relationships between vocal characteristics and body size and shape in human males: An evolutionary explanation for a deep male voice. *Biological Psychology*, **72**, 160–163.

Eyben, F., Batliner, A., Schuller, B., Seppi, D., and Steidl, S. (2010a). Cross-corpus classification of realistic emotions – some pilot experiments. In L. Devillers, B. Schuller, R. Cowie, E. Douglas-Cowie, and A. Batliner (eds), *Proc. of the 3rd International Workshop on EMOTION: Corpora for Research on Emotion and Affect, satellite of LREC 2010*, pp. 77–82, Valletta, Malta.

Eyben, F., Wöllmer, M., and Schuller, B. (2010b). openSMILE – The Munich Versatile and Fast Open-Source Audio Feature Extractor. In *Proc. of the 9th ACM International Conference on Multimedia, MM*, pp. 1459–1462, Florence.

Fairbanks, G. and Hoaglin, L. (1941). An experimental study of the durational characteristics of the voice during the expression of emotion. *Speech Monographs*, **8**, 85–91.

Fairbanks, G. and Pronovost, W. (1939). An experimental study of the pitch characteristics of the voice during the expression of emotion. *Speech Monographs*, **6**, 87–104.

Falk, D. (2004). Prelinguistic evolution in early hominins: Whence motherese? *Behavioral and Brain Sciences*, **27**, 491–503.

Fausey, C. M. and Boroditsky, L. (2010). Subtle linguistic cues influence perceived blame and financial liability. *Psychonomic Bulletin & Review*, **17**, 644–650.

Fay, P. and Middleton, W. (1941). The ability to judge truth-telling, or lying, from the voice as transmitted over a public address system. *Journal of General Psychology*, **24**, 211–215.

Feldstein, S. and Sloan, B. (1984). Actual and stereotyped speech tempos of extraverts and introverts. *Journal of Personality*, **52**, 188–204.

Fernald, A. (2000). Speech to infants as hyperspeech: Knowledge-driven processes in early word recognition. *Phonetica*, **57**, 242–254.

Fernandez, R. and Picard, R. W. (2003). Modeling drivers' speech under stress. *Speech Communication*, **40**, 145–159.

Filatova, E. (2012). Irony and sarcasm: Corpus generation and analysis using crowdsourcing. In *Proc. of LREC*, pp. 392–398, Istanbul.

Fitch, W. (1997). Vocal tract length and formant frequency dispersion correlate with body size in rhesus macaques. *Journal of the Acoustical Society of America*, **102**, 1213–1222.

Flege, J. E., Birdson, D., Bialystok, E., Mack, M., Sung, H., and Tsukada, K. (2006). Degree of foreign accent in English sentences produced by Korean children and adults. *Journal of Phonetics*, **34**, 153–175.

Flett, G. L. (2007). *Personality Theory & Research*. Wiley, Mississauga, ON.

Freud, S. (1905). *Der Witz und seine Beziehung zum Unbewussten*. Franz Deuticke, Leipzig and Vienna.

Frick, R. (1985). Communicating emotion: the role of prosodic features. *Psychological Bulletin*, **97**, 412–429.

Fritzell, B. (1996). Voice disorders and occupations. *Logopedics Phoniatrics Vocology*, **21**, 7–12.

Fuertes, J. N., Gottdiener, W. H., Martin, H., Gilbert, T. C., and Giles, H. (2012). A meta-analysis of the effects of speakers' accents on interpersonal evaluations. *European Journal of Social Psychology*, **42**, 120–133.

Gardner, H. (1993). *Multiple Intelligences: The Theory in Practice*. Basic Books, New York.

Gatica-Perez, D., McCowan, I., Zhang, D., and Bengio, S. (2005). Detecting group interest-level in meetings. In *Proc. of ICASSP*, pp. 489–492, Philadelphia.

Gawda, B. (2007). Neuroticism, extraversion, and paralinguistic expression. *Psychological REPORTS*, **100**, 721–726.

Ghai, S. and Sinha, R. (2010). Exploring the effect of differences in the acoustic correlates of adults' and children's speech in the context of automatic speech recognition. *EURASIP Journal on Audio, Speech, and Music Processing*, **2010**.

Giesert, A.-L., Balke, W.-T., and Jahns, G. (2011). Probabilistic analysis of coughs in pigs to diagnose respiratory infections. *Landbauforschung - vTI Agriculture and Forestry Research*, **3**, 237–242.

Gillick, D. (2010). Can conversational word usage be used to predict speaker demographics? In *Proc. of Interspeech*, pp. 1381–1384, Makuhari, Japan.

Golan, O., Baron-Cohen, S., Hill, J. J., and Rutherford, M. D. (2007). The 'Reading the Mind in the Voice' Test-Revised: A study of complex emotion recognition in adults with and without autism spectrum conditions. *Journal of Autism and Developmental Disorders*, **37**, 1096–1106.

Gonzalez, J. (2004). Formant frequencies and body size of speaker: a weak relationship in adult humans. *Journal of Phonetics*, **32**, 277–287.

Graham, C., Hamblin, A., and Feldstein, S. (2001). Recognition of emotion in English voices by speakers of Japanese, Spanish and English. *International Review of Applied Linguistics*, **39**, 19–37.

Gravano, A., Levitan, R., Willson, L., Beňuš, Š., Hirschberg, J., and Nenkova, A. (2011). Acoustic and prosodic correlates of social behavior. In *Proc. of Interspeech*, pp. 97–100, Florence.

Greeley, H. P., Berg, J., Friets, E., Wilson, J., Greenough, G., Picone, J., Whitmore, J., and Nesthus, T. (2007). Sleepiness estimation using voice analysis. *Behaviour Research Methods*, **39**, 610–619.

Green, R. and Cliff, N. (1997). Multidimensional comparisons of structures of vocally and facially expressed emotion. *Perception & Psychophysics*, **17**, 429–438.

Gregory, S. and Gallagher, T. (2002). Spectral analysis of candidates' nonverbal vocal communication: Predicting U.S. presidential election outcomes. *Social Psychology Quarterly*, **65**, 298–308.

Gregory, S. W., Green, B. E., Carrothers, R. M., Dagan, K. A., and Webster, S. W. (2001). Verifying the primacy of voice fundamental frequency in social status accommodation. *Language & Communication*, **21**, 37–60.

Grice, H. (1975). Logic and conversation. In P. Cole and J. Morgan (eds), *Syntax and Semantics, Volume 3, Speech Acts*, pp. 41–58. Academic Press, New York.

Grimm, M., Kroschel, K., and Narayanan, S. (2008). The Vera am Mittag German Audio-Visual Emotional Speech Database. In *Proc. of the IEEE International Conference on Multimedia and Expo (ICME)*, pp. 865–868, Hannover, Germany.

Grivičić, T. and Nilep, C. (2004). When phonation matters: The use and function of yeah and creaky voice. *Colorado Research in Linguistics*, **17**, 1–11.

Grossman, R. B., Bemis, R. H., Skwerer, D. P., and Tager-Flusberg, H. (2010). Lexical and affective prosody in children with high-functioning autism. *Journal of Speech, Language, and Hearing Research*, **53**, 778–793.

Grossmann, T. (2010). The development of emotion perception in face and voice during infancy. *Restorative Neurology and Neuroscience*, **28**, 219–236.

Haas, A. (1979). Male and female spoken language differences: Stereotypes and evidence. *Psychological Bulletin*, **86**, 616–626.

Haderlein, T., Nöth, E., Batliner, A., Eysholdt, U., and Rosanowski, F. (2011). Automatic intelligibility assessment of pathologic speech over the telephone. *Logopedics, Phoniatrics, Vocology*, **36**, 175–181.

Hall, M., Frank, E., Holmes, G., Pfahringer, B., Reutemann, P., and Witten, I. (2009). The WEKA data mining software: An update. *SIGKDD Explorations*, **11**.

Hanani, A., Russell, M., and Carey, M. (2013). Human and computer recognition of regional accents and ethnic groups from British English speech. *Computer Speech and Language*, **27**, 59–74.

Hao, J. (2002). *Cross-Racial Studies of Human Vocal Tract Dimensions and Formant Structures*. Ph.D. thesis, Ohio University.

Hardin, M. A., Demark, D.-R. V., Morris, H., and Payne, M. (1992). Correspondence between nasalance scores and listener judgments of hypernasality. *Cleft Palate-Craniofacial Journal*, **29**, 346–351.

Harrison, A. E. (ed.) (2010). *Speech Disorders: Causes, Treatment and Social Effects*. Nova Science Publishers, Hauppauge, NY.

Harrison, Y. and Horne, J. (1997). Sleep deprivation affects speech. *Sleep*, **20**, 871–877.

Harrison, Y. and Horne, J. (2000). The impact of sleep deprivation on decision making: A review. *Journal of Experimental Psychology: Applied*, **6**, 236–249.

Henry, J. (1936). The linguistic expression of emotion. *American Anthropologist*, **38**, 250–256.

Heylen, D. (2005). Challenges ahead. Head movements and other social acts in conversation. In *Proc. of AISB - Social Presence Cues for Virtual Humanoids*, pp. 45–52, Hatfield, UK.

Hilgers, F. J. M. and Schouwenburg, P. F. (1990). A new low-resistance, self-retaining prosthesis (Provox^{TM}) for voice rehabilitation after total laryngectomy. *The Laryngoscope*, **100**, 1202–1207.

Hirschberg, J., Beňuš, Š., Brenier, J. M., Enos, F., Friedman, S., Gilman, S., Girand, C., Graciarena, M., Kathol, A., Michaelis, L., Pellom, B., Shriberg, E., and Stolcke, A. (2005). Distinguishing deceptive from non-deceptive speech. In *Proc. of Interspeech*, pp. 1833–1836, Lisbon.

Holmes, R., Oates, J., Phyland, D., and Hughes, A. (2000). Voice characteristics in the progression of Parkinson's disease. *International Journal of Language & Communication Disorders*, **35**, 407–418.

Hönig, F., Batliner, A., and Nöth, E. (2012). Automatic assessment of non-native prosody annotation, modelling and evaluation. In *Proc. of the International Symposium on Automatic Detection of Errors in Pronunciation Training (ISADEPT)*, pp. 21–30, Stockholm.

Hunt, O., Burden, D., Hepper, P., and Johnston, C. (2005). The psychosocial effects of cleft lip and palate: A systematic review. *European Journal of Orthodontics*, **27**, 274–285.

Ivanov, A. V., Riccardi, G., Sporka, A. J., and Franc, J. (2011). Recognition of personality traits from human spoken conversations. In *Proc. of Interspeech*, pp. 1549–1552, Florence.

Jacobs, R., Van Steenberghe, D., Manders, E., Van Looy, C., Lembrechts, D., and Naert, I. (2001). Evaluation of speech in patients rehabilitated with various oral implant-supported prostheses. *Clinical Oral Implants Research*, **12**, 167–173.

James, M. (2011). Race. In E. N. Zalta (ed.), *The Stanford Encyclopedia of Philosophy*. http://plato.stanford.edu/archives/win2011/entries/race/.

Jay, T. (2009). The utility and ubiquity of taboo words. *Perspectives on Psychological Science*, **4**, 153–161.

Jeon, J. H., Xia, R., and Liu, Y. (2010). Level of interest sensing in spoken dialog using multi-level fusion of acoustic and lexical evidence. In *Proc. of Interspeech*, pp. 2802–2805, Makuhari, Japan.

Jessen, M. (2007). Speaker classification in forensic phonetics and acoustics. In C. Müller (ed.), *Speaker Classification I: Fundamentals, Features, and Methods*, pp. 180–204. Springer, Berlin.

Johnson, K. (2005). Speaker normalization in speech perception. In D. B. Pisoni and R. E. Remez (eds), *The Handbook of Speech Perception*, pp. 363–389. Blackwell, Malden, MA.

Johnson, K., Pisoni, D., and Bernacki, R. (1990). Do voice recordings reveal whether a person is intoxicated? A case study. *Phonetica*, **47**, 215–237.

Johnstone, T. and Scherer, K. R. (2000). Vocal communication of emotion. In M. Lewis and J. Haviland (eds), *Handbook of Emotions*, chapter 14, pp. 220–235. Guilford Press, New York, 2nd edition.

Jovanovic, N. and Op den Akker, R. (2004). Towards automatic addressee identification in multi-party dialogues. In M. Strube and C. Sidner (eds), *Proc. of the 5th SIGdial Workshop on Discourse and Dialogue*, pp. 89–92, Cambridge, MA. Association for Computational Linguistics.

Juslin, P. N. and Scherer, K. R. (2005). Vocal expression of affect. In J. A. Harrigan, R. Rosenthal, and K. R. Scherer (eds), *New Handbook of Methods in Nonverbal Behavior Research*, pp. 65–135. Oxford University Press, Oxford.

Kats-Gold, I., Besser, A., and Priel, B. (2007). The role of simple emotion recognition skills among school aged boys at risk of ADHD. *Journal of Abnormal Child Psychology*, **35**, 363–378.

Katzenmaier, M., Stiefelhagen, R., and Schultz, T. (2004). Identifying the addressee in human-human-robot interactions based on head pose and speech. In *Proc. of ICMI*, pp. 144–151, State College, PA.

Kehrein, R. (2002). The prosody of authentic emotions. In *Proc. of Speech Prosody*, pp. 423–426, Aix-in-Provence.

Kennedy, A. and Inkpen, D. (2006). Sentiment classification of movie reviews using contextual valence shifters. *Computational Intelligence*, **22**, 110–125.

Kent, R. D. (1996). Hearing and believing: Some limits to the auditory-perceptual assessment of speech and voice disorders. *American Journal of Speech-Language Pathology*, **5**, 7–23.

Kent, R. D. and Kim, Y.-J. (2003). Toward an acoustic typology of motor speech disorders. *Clinical Linguistics & Phonetics*, **17**, 427–445.

Keuning, K. H., Wieneke, G. H., and Dejonckere, P. H. (1999). The intrajudge reliability of the perceptual rating of cleft palate speech before and after pharyngeal flap surgery: the effect of judges and speech samples. *Cleft Palate-Craniofacial Journal*, **36**, 328–333.

Kießling, A., Kompe, R., Niemann, H., Nöth, E., and Batliner, A. (1993). 'Roger', 'Sorry', 'I'm still listening': Dialog guiding signals in information retrieval dialogs. In *Proc. of the ESCA Workshop on Prosody*, pp. 112–115, Lund.

Kimble, C. E. and Seidel, S. D. (1991). Vocal signs of confidence. *Journal of Nonverbal Behavior*, **15**, 99–105.

Kitamura, C. and Burnham, D. (1998). Acoustic and affective qualities of IDS in English. In *Proc. of ICSLP*, pp. 441–444, Sydney.

Knipfer, C., Bocklet, T., Nöth, E., Schuster, M., Sokol, B., Eitner, S., Nkenke, E., and Stelzle, F. (2012). Speech intelligibility enhancement through maxillary dental rehabilitation with telescopic prostheses and complete dentures: A prospective study using automatic, computer-based speech analysis. *The International Journal of Prosthodontics*, **1**, 24–32.

Ko, S., Judd, C., and Blair, I. (2006). What the voice reveals: Within- and between-category stereotyping on the basis of voice. *Personality and Social Psychology Bulletin*, **32**, 806–819.

Kockmann, M., Burget, L., and Černocký, J. (2010). Brno University of Technology System for Interspeech 2010 Paralinguistic Challenge. In *Proc. of Interspeech*, pp. 2822–2825, Makuhari, Japan.

Kousta, S.-T., Vinson, D. P., and Vigliocco, G. (2009). Emotion words, regardless of polarity, have a processing advantage over neutral words. *Cognition*, **112**, 473–481.

Koven, M. (2007). *Selves in Two Languages: Bilinguals' Verbal Enactments of Identity in French and Portuguese*. John Benjamins, Amsterdam.

Krajewski, J., Batliner, A., and Golz, M. (2009). Acoustic sleepiness detection: Framework and validation of a speech-adapted pattern recognition approach. *Behavior Research Methods*, **41**, 795–804.

Krajewski, J., Schnieder, S., Sommer, D., Batliner, A., and Schuller, B. (2012). Applying multiple classifiers and non-linear dynamics features for detecting sleepiness from speech. *Neurocomputing*, **84**, 65–75.

Kramer, E. (1963). Judgment of personal characteristics and emotions from nonverbal properties of speech. *Psychological Bulletin*, **60**, 408–420.

Krauss, R., Freyberg, R., and Morsella, E. (2002). Inferring speakers' physical attributes from their voices. *Journal of Experimental Social Psychology*, **38**, 618–625.

Krauss, R. M. and Pardo, J. S. (2006). Speaker perception and social behavior: Bridging social psychology and speech science. In P. A. V. Lange (ed.), *Bridging Social Psychology: Benefits of Transdisciplinary Approaches*, pp. 273–278. Routledge.

Kraut, R. (1978). Verbal and nonverbal cues in the perception of lying. *Journal of Personality and Social Psychology*, **36**, 380–391.

Kraut, R. (1980). Humans as lie detectors. *Journal of Communication*, **30**, 209–218.

Kreiman, J. and Sidtis, D. (2011). *Foundations of Voice Studies: An Interdisciplinary Approach to Voice Production and Perception*. Wiley-Blackwell, Malden, MA.

Kreiman, J., Gerratt, B., and Precoda, K. (1990). Listener experience and perception of voice quality. *Journal of Speech and Hearing Research*, **33**, 103–115.

Kreiman, J., Gerratt, B., and Berke, G. (1994). The multidimensional nature of pathologic vocal quality. *Journal of the Acoustical Society of America*, **96**, 1291–1302.

Kuhl, P. K. (2004). Early language acquisition: Cracking the speech code. *Neuroscience*, **5**, 831–843.

Kuhl, P. K. (2007). Is speech learning 'gated' by the social brain? *Developmental Science*, **10**, 110–120.

Künzel, H. J. and Braun, A. (2003). The effect of alcohol on speech prosody. In *Proc. of ICPhS*, pp. 2645–2648, Barcelona.

Kwon, H., Berisha, V., and Spanias, A. (2008). Real-time sensing and acoustic scene characterization for security applications. In *Proc. of the 3rd International Symposium on Wireless Pervasive Computing (ISWPC)*, pp. 755–758, Santorini, Greece.

Lang, A. E. and Lozano, A. M. (1998). Parkinson's disease. *New England Journal of Medicine*, **339**, 1044–1053.

Laskowski, K. (2009). Contrasting emotion-bearing laughter types in multiparticipant vocal activity detection for meetings. In *Proc. of ICASSP*, pp. 4765–4768, Taipei, Taiwan.

Laskowski, K. (2011). *Predicting, Detecting and Explaining the Occurrence of Vocal Activity in Multi-Party Conversation*. Ph.D. thesis, Language Technologies Institute, Carnegie Mellon University, Pittsburgh.

Laskowski, K., Ostendorf, M., and Schultz, T. (2008). Modeling vocal interaction for text-independent participant characterization in multi-party conversation. In *Proc. of the 9th SIGdial Workshop on Discourse and Dialogue*, pp. 148–155, Columbus, OH.

Lederman, D., Zmora, E., Hauschildt, S., Stellzig-Eisenhauer, A., and Wermke, K. (2008). Classification of cries of infants with cleft-palate using parallel hidden Markov models. *Medical & Biological Engineering & Computing*, **46**, 965–975.

Lee, C., Narayanan, S., and Pieraccini, R. (2001). Recognition of negative emotions from the speech signal. In *Proc. of ASRU*, pp. 240–243, Madonna di Campiglio, Italy.

Lee, C.-C., Black, M., Katsamanis, A., Lammert, A., Baucom, B., Christensen, A., Georgiou, P., and Narayanan, S. (2010). Quantification of prosodic entrainment in affective spontaneous spoken interactions of married couples. In *Proc. of Interspeech*, pp. 793–796, Makuhari, Japan.

Lee, C.-C., Katsamanis, A., Black, M., Baucom, B., Georgiou, P., and Narayanan, S. (2011). An analysis of PCA-based vocal entrainment measures in married couples' affective spoken interactions. In *Proc. of Interspeech*, pp. 3101–3104, Florence.

Lee, C. M. and Narayanan, S. S. (2005). Toward detecting emotions in spoken dialogs. *IEEE Transactions on Speech and Audio Processing*, **13**, 293–303.

Lev-Ari, S. and Keysar, B. (2010). Why don't we believe non-native speakers? The influence of accent on credibility. *Journal of Experimental Social Psychology*, **46**, 1093–1096.

Levinson, S. C. (1983). *Pragmatics*. Cambridge University Press, Cambridge.

Levit, M., Huber, R., Batliner, A., and Nöth, E. (2001). Use of prosodic speech characteristics for automated detection of alcohol intoxination. In *Proc. of the Workshop on Prosody and Speech Recognition*, pp. 103–106, Red Bank, NJ.

Li, M., Jung, C.-S., and Han, K. J. (2010). Combining five acoustic level modeling methods for automatic speaker age and gender recognition. In *Proc. of Interspeech*, pp. 2826–2829, Makuhari, Japan.

Li, Q. and Russell, M. J. (2001). Why is automatic recognition of children's speech difficult? In *Proc. of Eurospeech*, pp. 2671–2674, Aalborg.

Lieberman, P. and Michaels, S. B. (1962). Some aspects of fundamental frequency and envelope amplitude as related to the emotional content of speech. *Journal of the Acoustical Society of America*, **34**, 922–927.

Lingenfelser, F., Wagner, J., Vogt, T., Kim, J., and André, E. (2010). Age and gender classification from speech using decision level fusion and ensemble based techniques. In *Proc. of Interspeech*, pp. 2798–2801, Makuhari, Japan.

Liscombe, J., Venditti, J., and Hirschberg, J. (2003). Classifying subject ratings of emotional speech using acoustic features. In *Proc. of Eurospeech*, pp. 725–728, Geneva, Switzerland.

Liscombe, J., Hirschberg, J., and Venditti, J. (2005a). Detecting certainness in spoken tutorial dialogues. In *Proc. of Interspeech*, pp. 1837–1840, Lisbon, Portugal.

Liscombe, J., Riccardi, G., and Hakkani-Tür, D. (2005b). Using context to improve emotion detection in spoken dialog systems. In *Proc. of Interspeech*, pp. 1845–1848, Lisbon.

Litman, D. and Forbes, K. (2003). Recognizing emotions from student speech in tutoring dialogues. In *Proc. of ASRU*, pp. 25–30, Virgin Islands.

Liu, B. (2012). *Sentiment Analysis and Opinion Mining*. Morgan & Claypool Publishers, San Rafael, CA.

Liu, H.-M., Kuhl, P. K., and Tsao, F.-M. (2003). An association between mothers' speech clarity and infants' speech discrimination skills. *Developmental Science*, **6**, F1–F10.

Liu, X. and Xu, Y. (2011). What Makes a Female Voice Attractive? In *Proc. of ICPhS*, pp. 1274–1277, Hong Kong.

Lohmander, A. and Olsson, M. (2004). Methodology for perceptual assessment of speech in patients with cleft palate: A critical review of the literature. *Cleft Palate-Craniofacial Journal*, **41**, 64–70.

Lott, P., Guggenbühl, S., Schneeberger, A., Pulver, A., and Stassen, H. (2002). Linguistic analysis of the speech output of schizophrenic, bipolar, and depressive patients. *Psychopathology*, **35**, 220–227.

Low, L.-S., Maddage, M., Lech, M., Sheeber, L., and Allen, N. (2011). Detection of clinical depression in adolescents' speech during family interactions. *IEEE Transactions on Biomedical Engineering*, **58**, 574–586.

Lowenstein, J. H. (2012). *Artificial Hearing, Natural Speech: Cochlear Implants, Speech Production, and the Expectations of a High-Tech Society*. Routledge, New York.

Lunsford, R. (2004). Private speech during multimodal human-computer interaction. In *Proc. of ICMI*, page 346, Pennsylvania.

Mahdhaoui, A. and Chetouani, M. (2009). A new approach for motherese detection using a semi-supervised algorithm. In *Machine Learning for Signal Processing XIX – Proc. of the 2009 IEEE Signal Processing Society Workshop, MLSP*, pp. 1–6, Grenoble, France.

Mahdhaoui, A., Chetouani, M., and Kessous, L. (2010). Time-frequency features extraction for infant directed speech discrimination. In J. Sole-Casals and V. Zaiats (eds), *Advances in Non-Linear Speech Processing*, pp. 120–127. Springer, Berlin.

Mahmoudi, Z., Rahati, S., Ghasemi, M. M., Asadpour, V., Tayarani, H., and Rajati, M. (2011). Classification of voice disorder in children with cochlear implantation and hearing aid using multiple classifier fusion. *BioMedical Engineering OnLine*, **10**.

Maier, A., Hönig, F., Bocklet, T., Nöth, E., Stelzle, F., Nkenke, E., and Schuster, M. (2009a). Automatic detection of articulation disorders in children with cleft lip and palate. *Journal of the Acoustical Society of America*, **126**, 2589–2602.

Maier, A., Haderlein, T., Eysholdt, U., Rosanowski, F., Batliner, A., Schuster, M., and Nöth, E. (2009b). PEAKS: A system for the automatic evaluation of voice and speech disorders. *Speech Communication*, **51**, 425–437.

Mairesse, F., Walker, M. A., Mehl, M. R., and Moore, R. K. (2007). Using linguistic cues for the automatic recognition of personality in conversation and text. *Journal of Artificial Intelligence Research*, **30**, 457–500.

Majekodunmi, T. and Idachaba, F. (2011). A review of the fingerprint, speaker recognition, face recognition and iris recognition based biometric identification technologies. In *Proc. of the World Congress on Engineering*, London.

Malinowski, B. (1923). The problem of meaning in primitive languages. In C. K. Odgen and I. A. Richards (eds), *The Meaning of Meaning*, pp. 146–152. Kegan Paul, London.

Markel, N. (ed.) (1969). *Psycholinguistics; An Introduction to the Study of Speech and Personality*. Dorsey Press, Homewood, IL.

Markel, N., Eisler, R., and Reese, H. (1967). Judging personality from dialect. *Journal of Verbal Learning and Verbal Behavior*, **6**, 33–35.

Markel, N., Phillis, J., Vargas, R., and Howard, K. (1972). Personality traits associated with voice types. *Journal of Psycholinguistic Research*, **1**, 249–255.

Markley, E. D. (2000). Regional accent discrimination in hiring decisions: A language attitude study. Master's thesis, University of North Texas.

Matos, S., Birring, S., Pavord, I., and Evans, D. (2006). Detection of cough signals in continuous audio recordings using hidden Markov models. *IEEE Transactions on Biomedical Engineering*, **53**, 1078–1083.

McCann, J. and Peppé, S. (2003). Prosody in autism spectrum disorders: a critical review. *International Journal of Language & Communication Disorders*, **38**, 325–350.

McColl, D., Fucci, D., Petrosino, L., Martin, D. E., and McCaffrey, P. (1998). Listener ratings of the intelligibility of tracheoesophageal speech in noise. *Journal of Communication Disorders*, **31**, 279–289.

McCord, J., Firestone, H., and Grant, A. (1994). Phonetic determinants of tooth placement in complete dentures. *Quintessence International*, **25**, 341–345.

McGehee, F. (1937). The reliability of the identification of the human voice. *Journal of General Psychology*, **17**, 249–271.

Mehl, M., Vazire, S., Ramírez-Esparza, N., Slatcher, R., and Pennebaker, J. (2007). Are women really more talkative than men? *Science*, **317**, 82.

Meinedo, H. and Trancoso, I. (2010). Age and gender classification using fusion of acoustic and prosodic features. In *Proc. of Interspeech*, pp. 2818–2821, Makuhari, Japan.

Metze, F., Ajmera, J., Englert, R., Bub, U., Burkhardt, F., Stegmann, J., Müller, C., Huber, R., Andrassy, B., Bauer, J. G., and Littel, B. (2007). Comparison of four approaches to age and gender recognition for telephone applications. In *Proc. of ICASSP*, pp. 1089–1092, Honolulu.

Mey, J. L. (2001). *Pragmatics: An Introduction*. Blackwell, Oxford.

Middag, C., Martens, J.-P., Nuffelen, G. V., and Bodt, M. D. (2009). Automated intelligibility assessment of pathological speech using phonological features. *EURASIP Journal on Advances in Signal Processing*, **2009**.

Middag, C., Bocklet, T., Martens, J.-P., and Nöth, E. (2011). Combining phonological and acoustic ASR-free features for pathological speech intelligibility assessment. In *Proc. of Interspeech*, pp. 3005–3008, Florence.

Minematsu, N., Sekiguchi, M., and Hirose, K. (2002a). Automatic estimation of one's age with his/her speech based upon acoustic modeling techniques of speakers. In *Proc. of ICASSP*, pp. 137–140, Orlando, Florida.

Minematsu, N., Sekiguchi, M., and Hirose, K. (2002b). Performance improvement in estimating subjective agedness with prosodic features. In *Proc. of Speech Prosody*, pp. 507–510, Aix-en-Provence, France.

Minematsu, N., Yamauchi, K., and Hirose, K. (2003). Automatic estimation of perceptual age using speaker modeling techniques. In *Proc. of Eurospeech*, pp. 3005–3008, Geneva, Switzerland.

Mohammadi, G., Vinciarelli, A., and Mortillaro, M. (2010). The voice of personality: Mapping nonverbal vocal behavior into trait attributions. In *Proc. of SSPW'10*, pp. 17–20, Florence.

Mori, H. (2009). An analysis of switching pause duration as a paralinguistic feature in expressive dialogues. *Acoust. Sci. & Tech.*, **30**, 376–378.

Mower, E., Metallinou, A., Lee, C.-C., Kazemzadeh, A., Busso, C., Lee, S., and Narayanan, S. (2009). Interpreting ambiguous emotional expressions. In *Proc. of ACII*, pp. 662–669, Amsterdam.

Mower, E., Black, M., Flores, E., Williams, M., and Narayanan, S. (2011). Design of an emotionally targeted interactive agent for children with autism. In *Proc. of the IEEE International Conference on Multimedia and Expo (ICME)*, pp. 1–6, Barcelona, Spain.

Mporas, I. and Ganchev, T. (2009). Estimation of unknown speakers' height from speech. *International Journal of Speech Technology*, **12**, 149–160.

Müller, C. (ed.) (2007). *Speaker Classification I and II*. Springer, Berlin.

Muthusamy, Y., Barnard, E., and Cole, R. (1994). Reviewing automatic language identification. *Signal Processing Magazine, IEEE*, **11**, 33–41.

Mysak, E. D. (1959). Pitch and duration characteristics of older males. *Journal of Speech and Hearing Research*, **2**, 46–54.

Neumeyer, L., Franco, H., Digalakis, V., and Weintraub, M. (2000). Automatic scoring of pronunciation quality. *Speech Communication*, **30**, 83–93.

Nguyen, P., Le, T., Tran, D., Huang, X., and Sharma, D. (2010). Fuzzy support vector machines for age and gender classification. In *Proc. of Interspeech*, pp. 2806–2809, Makuhari, Japan.

Nijholt, A., Arkin, R. C., Brault, S., Kulpa, R., Multon, F., Bideau, B., Traum, D., Hung, H., Santos Jr., E., Li, D., Yu, F., Zhou, L., and Zhang, D. (2012). Computational deception and noncooperation. *IEEE Intelligent Systems*, **27**, 60–75.

Norwine, A. and Murphy, O. (1938). Characteristic time intervals in telephonic conversations. *Bell Systems Technical Journal*, **17**, 281–291.

Nwe, T. L., Li, H., and Dong, M. (2006). Analysis and detection of speech under sleep deprivation. In *Proc. of Interspeech*, pp. 17–21, Pittsburgh.

Oberlander, J. and Nowson, S. (2006). Whose thumb is it anyway? Classifying author personality from weblog text. In *Proc. of the COLING/ACL 2006 Main Conference Poster Sessions*, pp. 627–634, Sydney.

Ogden, R. (2006). Phonetics and social action in agreements and disagreements. *Journal of Pragmatics*, **38**, 1752–1775.

Omar, M. K. and Pelecanos, J. (2010). A novel approach to detecting non-native speakers and their native language. In *Proc. of ICASSP*, pp. 4398–4401, Dallas, Texas.

Oppermann, D., Schiel, F., Steininger, S., and Beringer, N. (2001). Off-talk – a problem for human-machine-interaction. In *Proc. of Eurospeech*, pp. 2197–2200, Aalborg.

Osberger, M. J. and McGarr, N. S. (1982). Speech production characteristics of the hearing impaired. Haskins Laborotories: Status Report on Speech Research SR-69, 227–288.

Osgood, C., Suci, G., and Tannenbaum, P. (1957). *The Measurement of Meaning*. University of Illinois Press, Urbana.

Pang, B. and Lee, L. (2008). Opinion mining and sentiment analysis. *Foundations and Trends in Information Retrieval*, **2**, 1–135.

Paul, R., Augustyn, A., Klin, A., and Volkmar, F. R. (2005). Perception and production of prosody by speakers with autism spectrum disorders. *Journal of Autism and Developmental Disorders*, **35**, 205–220.

Pauloski, B. R., Logemann, J. A., Rademaker, A. W., McConnel, F. M. S., Heiser, M. A., Cardinale, S., Shedd, D., Lewin, J., Baker, S. R., Graner, D., Cook, B., Milianti, F., Collins, S., and Baker, T. (1993). Speech and swallowing function after anterior tongue and floor of mouth resection with distal flap reconstruction. *Journal of Speech and Hearing Research*, **36**, 267–276.

Pauloski, B. R., Logemann, J. A., Colangelo, L. A., Rademaker, A. W., McConnel, F. M. S., Heiser, M. A., Cardinale, S., Shedd, D., Stein, D., Beery, Q., Myers, E., Lewin, J., Haxer, M., and Esclamado, R. (1998). Surgical variables affecting speech in treated patients with oral and oropharyngeal cancer. *The Laryngoscope*, **108**, 908–916.

Pellom, B. (2012). Rosetta Stone ReFLEX: Toward improving English conversational fluency in Asia. In *Proc. of the International Symposium on Automatic Detection of Errors in Pronunciation Training (isadept)*, pp. 15–20, Stockholm.

Pemberton, C., McCormack, P., and Russell, A. (1998). Have women's voices lowered across time? A cross sectional study of Australian women's voices. *Journal of Voice*, **12**, 208–213.

Pentland, A. (2008). *Honest Signals. How They Shape Our World*. MIT Press, Cambridge, MA.

Pentland, A. and Madan, A. (2005). Perception of social interest. In *Proc. of IEEE Int. Conf. on Computer Vision, Workshop on Modeling People and Human Interaction (ICCV-PHI)*, Beijing.

Peppé, S., McCann, J., and Gibbon, F. (2007). Receptive and expressive prosodic ability in children with high-functioning autism. *Journal of Speech, Language, and Hearing Research*, **50**, 1015–1028.

Peppé, S., Cleland, J., Gibbon, F., O'Hare, A., and Castilla, P. M. (2011). Expressive prosody in children with autism spectrum conditions. *Journal of Neurolinguistics*, **24**, 41–53.

Perrot, P., Aversano, G., and Chollet, G. (2007). Voice disguise and automatic detection: review and perspectives. In Y. Stylianou, M. Faundez-Zanuy, and A. Esposito (eds), *Progress in Nonlinear Speech Processing*, pp. 101–117. Springer, Berlin.

Petrović, A. (1985). Speech sound distortions caused by changes in complete denture morphology. *Journal of Oral Rehabilitation*, **12**, 69–79.

Piaget, J. (1923). *Le langage et la pensée chez l'enfant*. Delachaux & Niestlé, Neuchâtel.

Pinker, S., Nowak, M. A., and Lee, J. J. (2008). The logic of indirect speech. *Proceedings of the National Academy of Sciences of the United States of America*, **105**, 833–838.

Piske, T., McKay, I., and Flege, J. (2001). Factors affecting degree of foreign accent in an L2: a review. *Journal of Phonetics*, **29**, 191–215.

Pisoni, D. and Martin, C. (1989). Effects of alcohol on the acoustic-phonetic properties of speech: Perceptual and acoustic analyses. *Alcoholism: Clinical and Experimental Research*, **13**, 577–587.

Ploog, B. O., Banerjee, S., and Brooks, P. J. (2009). Attention to prosody (intonation) and content in children with autism and in typical children using spoken sentences in a computer game. *Research in Autism Spectrum Disorders*, **3**, 743–758.

Polzehl, T., Möller, S., and Metze, F. (2010). Automatically assessing personality from speech. In *Proc. of the IEEE Fourth International Conference on Semantic Computing (ICSC '10)*, pp. 134–140, Washington, DC.

Popp, D., Donovan, R. A., Crawford, M., Marsh, K. L., and Peele, M. (2003). Gender, race, and speech style stereotypes. *Sex Roles*, **48**, 317–325.

Porat, R., Lange, D., and Zigel, Y. (2010). Age recognition based on speech signals using weights supervector. In *Proc. of Interspeech*, pp. 2814–2817, Makuhari, Japan.

Protopapas, A. and Lieberman, P. (1997). Fundamental frequency of phonation and perceived emotional stress. *Journal of the Acoustical Society of America*, **101**, 2267–2277.

Puts, D. (2005). Mating context and menstrual phase affect women's preferences for male voice pitch. *Evolution and Human Behavior*, **26**, 388–397.

Puts, D., Gaulin, S., and Verdolini, K. (2006). Dominance and the evolution of sexual dimorphism in human voice pitch. *Evolution and Human Behavior*, **27**, 283–296.

Puts, D., Hodges, C., Cárdenas, R., and Gaulin, S. (2007). Men's voices as dominance signals: vocal fundamental and formant frequencies influence dominance attributions among men. *Evolution and Human Behavior*, **28**, 340–344.

Rakić, T., Steffens, M. C., and Mummendey, A. (2011). When it matters how you pronounce it: The influence of regional accents on job interview outcome. *British Journal of Psychology*, **102**, 868–883.

Ramig, L. O., Fox, C., and Sapir, S. (2008). Speech treatment for Parkinson's disease. *Expert Review of Neurotherapeutics*, **8**, 299–311.

Ramírez-Esparza, N., Gosling, S. D., Benet-Martínez, V., Potter, J. P., and Pennebaker, J. W. (2006). Do bilinguals have two personalities? A special case of cultural frame switching. *Journal of Research in Personality*, **40**, 99–120.

Rehm, M. and André, E. (2005). Where do they look? Gaze behaviors of multiple users interacting with an ECA. In *Intelligent Virtual Agents: 5th International Working Conference, IVA 2005*, pp. 241–252, Berlin, New York. Springer.

Reich, A. (1981). Detecting the presence of vocal disguise in the male voice. *Journal of the Acoustical Society of America*, **69**, 1458–1461.

Reilly, N., Cannizzaro, M. S., Harel, B. T., and Snyder, P. J. (2004). Feigned depression and feigned sleepiness: A voice acoustical analysis. *Brain and Cognition*, **55**, 383–386.

Reisenzein, R. and Weber, H. (2009). Personality and emotion. In P. J. Corr and G. Matthews (eds), *The Cambridge Handbook of Personality Psychology*, pp. 54–71. Cambridge University Press, Cambridge.

Reissland, N., Shepherd, J., and Herrera, E. (2003). The pitch of maternal voice: a comparison of mothers suffering from depressed mood and non-depressed mothers reading books to their infants. *Journal of Child Psychology and Psychiatry*, **44**, 255–261.

Rendall, D., Vokey, J., and Nemeth, C. (2007). Lifting the curtain on the Wizard of Oz: Biased voice-based impressions of speaker size. *Journal of Experimental Psychology: Human Perception and Performance*, **33**, 1208–1219.

Revelle, W. and Scherer, K. (2009). Personality and emotion. In *Oxford Companion to the Affective Sciences*, pp. 1–4. Oxford University Press, Oxford.

Ringeval, F., Demouy, J., Szaszák, G., Chetouani, M., Robel, L., Xavier, J., Cohen, D., and Plaza, M. (2011). Automatic intonation recognition for the prosodic assessment of language-impaired children. *IEEE Transactions on Audio, Speech and Language Processing*, **19**, 1328–1342.

Rockwell, P. (2000). Lower, slower, louder: Vocal cues of sarcasm. *Journal of Psycholinguistic Research*, **29**, 483–495.

Rockwell, P. (2007). Vocal features of conversational sarcasm: A comparison of methods. *Journal of Psycholinguistic Research*, **36**, 361–369.

Rosenberg, A. and Hirschberg, J. (2005). Acoustic/prosodic and lexical correlates of charismatic speech. In *Proc. of Interspeech*, pp. 513–516, Lisbon.

Rosenberg, A. and Hirschberg, J. (2009). Charisma perception from text and speech. *Speech Communication*, **51**, 640–655.

Rothman, A. D. and Nowicki, S. (2004). A measure of the Ability to identify emotion in children's tone of voice. *Journal of Nonverbal Behavior*, **28**, 67–92.

Rouzbahani, K. H. and Daliri, M. R. (2011). Diagnosis of Parkinson's disease in human using voice signals. *Basic and Clinical Neuroscience*, **2**, 12–20.

Ruben, R. J. (2000). Redefining the survival of the fittest: Communication disorders in the 21st century. *The Laryngoscope*, **110**, 241.

Russo, N., Larson, C., and Kraus, N. (2008). Audio-vocal system regulation in children with autism spectrum disorders. *Experimental Brain Research*, **188**, 111–124.

Rusz, J., Cmejla, R., Ruzickova, H., and Ruzicka, E. (2011). Quantitative acoustic measurements for characterization of speech and voice disorders in early untreated Parkinson's disease. *Journal of the Acoustical Society of America*, **129**, 350–367.

Sanford, F. (1942). Speech and personality. *Psychological Bulletin*, **39**, 811–845.

Sanford, F. H. (1948). Speech and personality. In L. A. Pennington and I. A. Berg (eds), *An Introduction to Clinical Psychology*, pp. 157–177. Ronald Press, Oxford.

Sapir, E. (1927). Speech as a personality trait. *American Journal of Sociology*, **32**, 892–905.

Sauter, D. (2006). *An investigation into vocal expressions of emotions: The roles of valence, culture, and acoustic factors*. Ph.D. thesis, University College London.

Sauter, D. A., Eisner, F., Ekman, P., and Scott, S. K. (2010). Cross-cultural recognition of basic emotions through nonverbal emotional vocalizations. *Proceedings of the National Academy of Sciences of the United States of America*, **107**, 2408–2412.

Sayetta, R., Weinrich, M., and Coston, G. (1989). Incidence and prevalence of cleft lip and palate: what we think we know. *Cleft Palate Journal*, **26**, 242–247.

Scherer, K. (1972). Judging personality from voice: A cross-cultural approach to an old issue in interpersonal perception. *Journal of Personality*, **40**, 191–210.

Scherer, K. R. (1979). Personality markers in speech. In K. R. Scherer and H. Giles (eds), *Social Markers in Speech*, pp. 147–209. Cambridge University Press, Cambridge.

Scherer, K. R. (1981). Speech and emotional states. In J. Darby (ed.), *Speech Evaluation in Psychiatry*, pp. 115–135. Grune & Stratton, New York.

Scherer, K. R. (2003). Vocal communication of emotion: A review of research paradigms. *Speech Communication*, **40**, 227–256.

Scherer, K. R. and Scherer, U. (1981). Speech behavior and personality. In J. Darby (ed.), *Speech Evaluation in Psychiatry*, pp. 115–135. Grune & Stratton, New York.

Scherer, K. R., Feldstein, S., Bond, R. N., and Rosenthal, R. (1985). Vocal cues to deception: a comparative channel approach. *Journal of Psycholinguistic Research*, **14**, 409–425.

Scherer, K. R., Johnstone, T., and Klasmeyer, G. (2003). Vocal expression of emotion. In R. J. Davidson, K. R. Scherer, and H. H. Goldsmith (eds), *Handbook of Affective Sciences*, pp. 433–456. Oxford University Press, Oxford.

Schiel, F. and Heinrich, C. (2009). Laying the foundation for in-car alcohol detection by speech. In *Proc. of Interspeech*, pp. 983–986, Brighton, UK.

Schoentgen, J. (2006). Vocal cues of disordered voices: An overview. *Acta Acustica united with Acustica*, **92**, 667–680.

Schötz, S. (2006). *Perception, Analysis and Synthesis of Speaker Age*. Ph.D. thesis, University of Lund, Sweden.

Schötz, S. (2007). Acoustic analysis of adult speaker age. In C. Müller (ed.), *Speaker Classification I*, pp. 88–107. Springer, Berlin.

Schröder, M., Cowie, R., Heylen, D., Pantic, M., Pelachaud, C., and Schuller, B. (2008). Towards responsive sensitive artificial listeners. In *Proc. of the 4th Intern. Workshop on Human-Computer Conversation*, Bellagio.

Schuller, B. and Rigoll, G. (2009). Recognising interest in conversational speech – comparing bag of frames and supra-segmental features. In *Proc. of Interspeech*, pp. 1999–2002, Brighton, UK.

Schuller, B., Köhler, N., Müller, R., and Rigoll, G. (2006). Recognition of interest in human conversational speech. In *Proc. of Interspeech*, pp. 793–796, Pittsburgh, PA.

Schuller, B., Müller, R., Hörnler, B., Höthker, A., Konosu, H., and Rigoll, G. (2007a). Audiovisual recognition of spontaneous interest within conversations. In *Proc. of the 9th ACM International Conference on Multimodal Interfaces, ICMI*, pp. 30–37, Nagoya, Japan.

Schuller, B., Batliner, A., Seppi, D., Steidl, S., Vogt, T., Wagner, J., Devillers, L., Vidrascu, L., Amir, N., Kessous, L., and Aharonson, V. (2007b). The relevance of feature type for the automatic classification of emotional user states: Low level descriptors and functionals. In *Proc. of Interspeech*, pp. 2253–2256, Antwerp, Belgium.

Schuller, B., Wimmer, M., Arsić, D., Moosmayr, T., and Rigoll, G. (2008). Detection of security related affect and behaviour in passenger transport. In *Proc. of Interspeech*, pp. 265–268, Brisbane, Australia.

Schuller, B., Müller, R., Eyben, F., Gast, J., Hörnler, B., Wöllmer, M., Rigoll, G., Höthker, A., and Konosu, H. (2009a). Being bored? Recognising natural interest by extensive audiovisual integration for real-life application. *Image and Vision Computing*, **27**(12), 1760–1774.

Schuller, B., Steidl, S., and Batliner, A. (2009b). The Interspeech 2009 Emotion Challenge. In *Proc. of Interspeech*, pp. 312–315, Brighton, UK.

Schuller, B., Eyben, F., Can, S., and Feussner, H. (2010a). Speech in minimal invasive surgery – towards an affective language resource of real-life medical operations. In *Proc. of the 3rd International Workshop on EMOTION: Corpora for Research on Emotion and Affect, satellite of LREC 2010*, pp. 5–9, Valletta, Malta.

Schuller, B., Steidl, S., Batliner, A., Burkhardt, F., Devillers, L., Müller, C., and Narayanan, S. (2010b). The Interspeech 2010 Paralinguistic Challenge. In *Proc. of Interspeech*, pp. 2794–2797, Makuhari, Japan.

Schuller, B., Valstar, M., Eyben, F., McKeown, G., Cowie, R., and Pantic, M. (2011a). AVEC 2011 – The First International Audio/Visual Emotion Challenge. In B. Schuller, M. Valstar, R. Cowie, and M. Pantic (eds), *Proceedings of the First International Audio/Visual Emotion Challenge and Workshop, AVEC 2011, held in conjunction with the International HUMAINE Association Conference on Affective Computing and Intelligent Interaction 2011, ACII 2011, Memphis, TN*, volume II, pp. 415–424. Springer, Berlin.

Schuller, B., Batliner, A., Steidl, S., and Seppi, D. (2011b). Recognising realistic emotions and affect in speech: State of the art and lessons learnt from the first challenge. *Speech Communication*, **53**, 1062–1087.

Schuller, B., Batliner, A., Steidl, S., Schiel, F., and Krajewski, J. (2011c). The Interspeech 2011 Speaker State Challenge. In *Proc. of Interspeech*, pp. 3201–3204, Florence.

Schuller, B., Valstar, M., Cowie, R., and Pantic, M. (2012a). AVEC 2012 – The Continuous Audio/Visual Emotion Challenge. In *Proc. of the 2nd International Audio/Visual Emotion Challenge and Workshop, AVEC, Grand Challenge and Satellite of ACM ICMI 2012*, Santa Monica, CA.

Schuller, B., Steidl, S., Batliner, A., Nöth, E., Vinciarelli, A., Burkhardt, F., Son, R. v., Weninger, F., Eyben, F., Bocklet, T., Mohammadi, G., and Weiss, B. (2012b). The Interspeech 2012 Speaker Trait Challenge. In *Proc. of Interspeech*, Portland, OR.

Schuller, B., Steidl, S., Batliner, A., Schiel, F., Krajewski, J., Weninger, F., and Eyben, F. (2013a). Medium-term speaker states – a review on intoxication, sleepiness and the first challenge. *Computer Speech and Language*.

Schuller, B., Steidl, S., Batliner, A., Burkhardt, F., Devillers, L., Müller, C., and Narayanan, S. (2013b). Paralinguistics in speech and language – state-of-the-art and the challenge. *Computer Speech and Language*, **27**, 4–39.

Schultz, T. (2007). Speaker characteristics. In C. Müller (ed.), *Speaker Classification I*, pp. 47–74. Springer, Berlin.

Schutte, H. K. and Nieboer, G. (2002). Aerodynamics of esophageal voice production with and without a Groningen voice prosthesis. *Folia Phoniatrica et Logopaedica*, **54**, 8–18.

Scipioni, M., Gerosa, M., Giuliani, D., Nöth, E., and Maier, A. (2009). Intelligibility assessment in children with cleft lip and palate in Italian and German. In *Proc. of Interspeech*, pp. 967–970, Brighton.

Scripture, E. (1921). A study of emotions by speech transcription. *Vox*, **31**, 179–183.

Senft, G. (2009). Phatic communion. In G. Senft, J.-O. Östman, and J. Verschueren (eds), *Culture and Language Use*, pp. 226–233. John Benjamins, Amsterdam.

Seppi, D., Batliner, A., Schuller, B., Steidl, S., Vogt, T., Wagner, J., Devillers, L., Vidrascu, L., Amir, N., and Aharonson, V. (2008). Patterns, prototypes, performance: Classifying emotional user states. In *Proc. of Interspeech*, pp. 601–604, Brisbane, Australia.

Shami, M. and Verhelst, W. (2007). Automatic classification of expressiveness in speech: a multi-corpus study. In C. Müller (ed.), *Speaker Classification II*, pp. 43–56. Springer, Berlin.

Shargorodsky, J., Curhan, S. G., Curhan, G. C., and Eavey, R. (2010). Change in prevalence of hearing loss in US adolescents. *Journal of the American Medical Association*, **304**, 772–778.

Sheard, C., Adams, R. D., and Davis, P. J. (1991). Reliability and agreement of ratings of ataxic dysarthric speech samples with varying intelligibility. *Journal of Speech and Hearing Research*, **34**, 285–293.

Shin, S.-H., Hashimoto, T., and Hatano, S. (2009). Automatic detection system for cough sounds as a symptom of abnormal health condition. *IEEE Transactions on Information Technology in Biomedicine*, **13**, 486–493.

Shriberg, L., Paul, R., McSweeny, J., Klin, A., Cohen, D., and Volkmar, F. (2001). Speech and prosody characteristics of adolescents and adults with high-functioning autism and Asperger syndrome. *Journal of Speech, Language, and Hearing Research*, **44**, 1097–1115.

Siepmann, R., Batliner, A., and Oppermann, D. (2001). Using prosodic features to characterize off-talk in human-computer interaction. In *Proc. of the Workshop on Prosody and Speech Recognition 2001*, pp. 147–150, Red Bank, NJ.

Simonoff, E., Pickles, A., Chairman, T., Chandler, S., Loucas, T., and Baird, G. (2008). Psychiatric disorders in children with autism spectrum disorders: Prevalence, comorbidity, and associated factors in a population-derived sample. *Journal of the American Academy of Child and Adolescent Psychiatry*, **47**, 921–929.

Skinner, E. (1935). A calibrated recording and analysis of the pitch, force, and quality of vocal tones expressing happiness and sadness. *Speech Monographs*, **2**, 81–137.

Slaney, M. and McRoberts, G. (1998). Baby Ears: A recognition system for affective vocalizations. In *Proc. of ICASSP*, pp. 985–988, Seattle.

Slaney, M. and McRoberts, G. (2003). BabyEars: A recognition system for affective vocalizations. *Speech Communication*, **39**, 367–384.

Sobell, L. and Sobell, M. (1972). Effects of alcohol on the speech of alcoholics. *Journal of Speech and Hearing Research*, **15**, 861–868.

Sobol-Shikler, T. and Robinson, P. (2010). Classification of complex information: Inference of co-occurring affective states from their expressions in speech. *IEEE Transactions on Pattern Analysis and Machine Intelligence*, **32**, 1284–1297.

Solan, L. and Tiersma, P. (2003). Hearing voices: Speaker identification in court. *Hastings Law Journal*, **54**, 373.

Spender, D. (1980). *Man Made Language*. Routledge, London.

Spertus, E. (1997). Smokey: Automatic recognition of hostile messages. In *Proc. of Innovative Applications of Artificial Intelligence (IAAI) '97*, pp. 1058–1065.

Stelzle, F., Ugrinovic, B., Knipfer, C., Bocklet, T., Nöth, E., Schuster, M., Eitner, S., Seiss, M., and Nkenke, E. (2010). Automatic, computer-based analysis of speech intelligibility on edentulous patients with and without complete dentures – preliminary results. *Journal of Oral Rehabilitation*, 37(3), 209–216.

Stiefelhagen, R. and Zhu, J. (2002). Head orientation and gaze direction in meetings. In *Proc. of: CHI '02 Extended Abstracts on Human Factors in Computing Systems*, pp. 858–859, New York.

Stiefelhagen, R., Yang, J., and Waibel, A. (2002). Modeling focus of attention for meeting indexing based on multiple cues. *IEEE Transactions on Neural Networks*, 13, 928–938.

Strik, H. (2012). ASR-based systems for language learning and therapy. In *Proc. of the International Symposium on Automatic Detection of Errors in Pronunciation Training (isadept)*, pp. 9–14, Stockholm.

Thiessen, E. D., Hill, E. A., and Saffran, J. R. (2005). Infant-directed speech facilitates word segmentation. *Infancy*, 7, 53–71.

Thorpy, M. J. and Billiard, M. (eds) (2011). *Sleepiness: Causes, Consequences and Treatment*. Cambridge University Press, Cambridge.

Timoshenko, E. (2012). *Rhythm Information for Automated Spoken Language Identification*. Ph.D. thesis, Technische Universität München.

Titze, I. R., Lemke, J., and Montequin, D. (1997). Population in the U.S. workforce who rely on voice as a primary tool of trade: a preliminary report. *Journal of Voice*, 11, 254–259.

Tobey, E. A., Geers, A. E., Brenner, C., Altuna, D., and Gabbert, G. (2003). Factors associated with development of speech production skills in children implanted by age five. *Ear & Hearing*, 24, 36S–45S.

Torre, P. and Barlow, J. (2009). Age-related changes in acoustic characteristics of adult speech. *Journal of Communication Disorders*, 42, 324–333.

Torrejano, G. and Guimarães, I. (2009). Voice quality after supracricoid laryngectomy and total laryngectomy with insertion of voice prosthesis. *Journal of Voice*, 23, 240–246.

Trainor, L., Austin, C., and Desjardins, R. (2000). Is infant-directed speech prosody a result of the vocal expression of emotion? *Psychological Science*, 11, 188–195.

Trainor, L. J. and Desjardins, R. N. (2002). Pitch characteristics of infant-directed speech affect infants' ability to discriminate vowels. *Psychonomic Bulletin & Review*, 9, 335–340.

Trojan, F. and Kryspin-Exner, K. (1968). The decay of articulation under the influence of alcohol and paraldehyde. *Folia Phoniatrica et Logopaedica*, 20, 217–238.

Truong, K. P. and Heylen, D. (2010). Disambiguating the functions of conversational sounds with prosody: the case of 'yeah'. In *Proc. of Interspeech*, pp. 2554–2557, Makuhari.

Turnhout, K. v., Terken, J., Bakx, I., and Eggen, B. (2005). Identifying the intended addressee in mixed human-human and human-computer interaction from non-verbal features. In *Proc. of ICMI*, pp. 175–182, New York.

Uldall, E. (1960). Attitudinal meanings conveyed by intonation contours. *Language and Speech*, 3, 223–234.

Van Lancker, D., Cornelius, C., and Kreiman, J. (1989). Recognition of emotional-prosodic meanings in speech by autistic, schizophrenic, and normal children. *Developmental Neuropsychology*, 5, 207–226.

Van Lierde, K., De Bodt, M., Van Borsel, J., Wuyts, F., and Van Cauwenberge, P. (2002). Effect of cleft type on overall speech intelligibility and resonance. *Folia Phoniatrica et Logopaedica*, 54, 158–168.

Van Lierde, K., De Bodt, M., Baetens, I., Schrauwen, V., and Van Cauwenberge, P. (2003). Outcome of treatment regarding articulation, resonance and voice in Flemish adults with unilateral and bilateral cleft palate. *Folia Phoniatrica et Logopaedica*, 55, 80–90.

Van Santen, J. P. H., Prud'hommeaux, E. T., Black, L. M., and Mitchell, M. (2010). Computational prosodic markers for autism. *Autism*, 14, 215–236.

Vanderas, A. P. (1987). Incidence of cleft lip, cleft palate, and cleft lip palate among races: A review. *Cleft Palate Journal*, 24, 216–225.

Vennemann, T. (1979). Was ißt und was soll die Phonologie? In *Nordica et Mystica. Festschrift für Kurt Schier*, pp. 64–83. Institut für Nordische Philologie, München.

Vinciarelli, A. and Mohammadi, G. (2011). Towards a technology of nonverbal communication: Vocal behavior in social and affective phenomena. In D. Gökçay and G. Yildirim (eds), *Affective Computing and Interaction: Psychological, Cognitive and Neuroscientific Perspectives*, pp. 133–156. IGI Global.

Vinciarelli, A., Pantic, M., and Bourlard, H. (2009). Social signal processing: Survey of an emerging domain. *Image and Vision Computing*, 27, 1743–1759.

Vipperla, R., Renals, S., and Frankel, J. (2010). Ageing voices: The effect of changes in voice parameters on ASR performance. *EURASIP Journal on Audio, Speech, and Music Processing*, **2010**.

Vollrath, M. (1993). *Mikropausen im Sprechen*. Peter Lang, Frankfurt.

Vrij, A., Edward, K., Roberts, K. P., and Bull, R. (2000). Detecting deceit via analysis of verbal and nonverbal behavior. *Journal of Nonverbal Behavior*, **24**, 239–263.

Vygotski, L. (1962). *Thought and language*. MIT Press, Cambridge, MA.

Walmsley, A. D., Morice, A. H., Dane, A. D., and Barry, S. J. (2006). The automatic recognition and counting of cough. *Cough*, **2**.

Ward, N. (2006). Non-lexical conversational sounds in American English. *Pragmatics and Cognition*, **14**, 129–182.

Watzlawick, P., Beavin, J., and Jackson, D. D. (1967). *Pragmatics of Human Communications*. W.W. Norton, New York.

Wells, J. C. (1982). *Accents of English*. Cambridge University Press.

Weninger, F., Krajewski, J., Batliner, A., and Schuller, B. (2012). The voice of leadership: Models and performances of automatic analysis in online speeches. *IEEE Transactions on Affective Computing*, **3**, 496–508.

Whitehill, T. L. (2002). Assessing Intelligibility in speakers with cleft palate: A critical review of the literature. *Cleft Palate-Craniofacial Journal*, **39**, 50–58.

Wiebe, J. M. and Riloff, E. (2005). Creating subjective and objective sentence classifiers from unannotated texts. In *Proc. of the Conference on Computational Linguistics and Intelligent Text Processing (CICLing)*, number 3406 in Lecture Notes in Computer Science, pp. 486–497. Springer, Berlin.

Wiebe, J. M., Wilson, T., Bruce, R., Bell, M., and Martin, M. (2004). Learning subjective language. *Computational Linguistics*, **30**(3), 277–308.

Williams, C. and Stevens, K. (1972). Emotions and speech: some acoustic correlates. *Journal of the Acoustical Society of America*, **52**, 1238–1250.

Wing, L. and Potter, D. (2002). The epidemiology of autistic spectrum disorders: is the prevalence rising? *Mental Retardation and Developmental Disabilities Research Reviews*, **8**, 151–161.

Winsler, A., Fernyhough, C., and Montero, I. (eds) (2009). *Private Speech, Executive Functioning, and the Development of Verbal Self-Regulation*. Cambridge University Press, Cambridge.

Witt, S. M. (2012). Automatic error detection in pronunciation training: Where we are and where we need to go. In *Proc. of the International Symposium on Automatic Detection of Errors in Pronunciation Training (isadept)*, pp. 1–8, Stockholm.

Woehrling, C. and Mareüil, P. B. d. (2006). Identification of regional accents in French: perception and categorization. In *Proc. of Interspeech*, pp. 1511–1514, Pittsburgh.

Wolfe, V. I., Martin, D. P., and Palmer, C. I. (2000). Perception of dysphonic voice quality by naive listeners. *Journal of Speech, Language, and Hearing Research*, **43**, 697–705.

Wöllmer, M., Weninger, F., Knaup, T., Schuller, B., Sun, C., Sagae, K., and Morency, L.-P. (2013). YouTube movie reviews: In, cross, and open-domain sentiment analysis in an audiovisual context. *IEEE Intelligent Systems Magazine, Special Issue on Concept-Level Opinion and Sentiment Analysis*, **28**(2).

Wu, K. and Childers, D. G. (1991). Gender recognition from speech. Part I: Coarse analysis. *Journal of the Acoustical Society of America*, **90**, 1828–1840.

Yu, C. (2011). The display of frustration in arguments: A multimodal analysis. *Journal of Pragmatics*, **43**, 2964–2981.

Yu, C., Aoki, P., and Woodruf, A. (2004). Detecting user engagement in everyday conversations. In *Proc. of Interspeech*, pp. 1329–1332, Jeju Island, Korea.

Zebrowitz, L. A., Brownlow, S., and Olson, K. (1992). Baby talk to the babyfaced. *Journal of Nonverbal Behavior*, **16**, 143–158.

Zenner, H. (1986). The Post-Laryngectomy Telephone Intelligibility Test (PLTT). In I. Hermann (ed.), *Speech Restoration via Voice Prosthesis*, pp. 148–152. Springer, Berlin.

Zhang, T., Hasegawa-Johnson, M., and Levinson, S. E. (2004). Children's emotion recognition in an intelligent tutoring scenario. In *Proc. of Interspeech*, pp. 1441–1444, Jeju Island, Korea.

Zhang, X.-J., Gu, J. H., and Tao, Z. (2010). Research of detecting fatigue from speech by PNN. In *Proc. of 2010 International Conference on Information Networking and Automation (ICINA)*, pp. 278–281, Kunming, China.

Zimmermann, R., Budach, W., Dammann, F., Einsele, H., Ohle, C. v., Preßler, H., Ruck, P., Wehrmann, M., and Zenner, H.-P. (2003). *Pharynx- und Larynxtumoren*. Interdisziplinäres Tumorzentrum Tübingen, Tübingen.

Zuckerman, M. and Driver, R. (1989). What sounds beautiful is good: The vocal attractiveness stereotype. *Journal of Nonverbal Behavior*, **13**, 67–82.

Zuckerman, M. and Miyake, K. (1993). The attractive voice: What makes it so? *Journal of Nonverbal Behavior*, **17**, 119–135.

Zuckerman, M., Spiegel, N., DePaulo, B., and Rosenthal, R. (1982). Nonverbal strategies for decoding deception. *Journal of Nonverbal Behavior*, **6**, 171–187.

Zuckerman, M., Hodgins, H., and Miyake, K. (1990). The vocal attractiveness stereotype: Replication and elaboration. *Journal of Nonverbal Behavior*, **14**, 97–112.

6

Corpus Engineering

The temptation to form premature theories upon insufficient data is the bane of our profession.

(Sir Arthur Conan Doyle)

Human beings, for all their pretensions, have a remarkable propensity for lending themselves to classification somewhere within neatly labeled categories. Even the outrageous exceptions may be classified as outrageous exceptions!

(William John Reichmann)

What one would need in almost 'any case' dealing with computational paralinguistics is speech and language data alongside label information. This requires *corpus engineering* which, as used here, means all the steps that are necessary before speech or language data can be processed in classification, regression, or other procedures, whether stand-alone or within applications. This involves basic, technical questions such as sample rate or type of microphone, and the following steps (adapted from Batliner *et al.* (2011) and expanded):

1. deciding on existing recordings if appropriate (TV, broadcast, Internet, other), or
2. the design of an (application-oriented) recording scenario – this can mean simply ensuring a quiet office environment or establishing an elaborate scenario such as multi-party with/without virtual agents or robots
3. deciding on the type of speech (isolated vowels, read, prompted, acted, elicited, realistic/natural(istic), other)
4. deciding on the type of recordings (close talk microphone, room microphone, video recordings for documentation or, within a multimodal setting, for later processing)
5. recruitment of the necessary personnel such as subjects and supervisors (especially in a Wizard of Oz setting), based on appropriate ethical considerations and privacy regulations
6. the recordings and – if necessary – subsequent transfer onto storage media with/without resampling of the audio signal
7. transliteration, that is, orthographic transcription of the data, sometimes including the annotation of extra- or non-linguistic events such as breathing or noise

Computational Paralinguistics: Emotion, Affect and Personality in Speech and Language Processing, First Edition.
Björn W. Schuller and Anton M. Batliner. © 2014 John Wiley & Sons, Ltd. Published 2014 by John Wiley & Sons, Ltd.

8. definition and extraction of appropriate units of analysis such as words, chunks, turns, dialogue moves with appropriate criteria (intuitive or based on prosodic, linguistic, or pragmatic criteria), or based on 'physical' criteria such as speech pauses, number of frames, partitioning of speech files into n segments of equal length
9. (data-driven, iterative) establishing of phenomena to be annotated and processed
10. annotation of states or traits on a categorical or continuous basis, possibly with subsequent mapping onto fewer main classes
11. establishing a gold standard for the annotations, based – if possible – on several annotators, and evaluating the quality of these annotations by applying some measures of correlation/correspondence
12. other pre-processing steps such as manual processing or correction of automatically processed feature values, documentation
13. detailed documentation of recording conditions, room acoustics, all other details that could be relevant
14. additional perception tests, if appropriate
15. defining and applying exchange formats, partitioning the data into training/development/ test sets
16. if possible, free release of the data (adequate licensing agreement, taking into account privacy considerations).

Procedures should meet the standards if possible, and the number of speakers, recordings per speaker, and annotators should be as high as possible, and transcriptions and annotations should be as diverse and detailed as possible. Gibbon *et al.* (1997, 2000) give a broad and extensive overview of all aspects of 'standards and resources for spoken language systems' and of 'multimodal and spoken dialogue systems'. Schiel and Draxler (2004) and Schiel *et al.* (2004) are intended ' . . . to be used as "cookbooks" providing practical help and ready-to-use solutions for the production or validation of spoken language corpora'. Cowie *et al.* (2011a,b) discuss principles, history, and basic issues in the collection of emotional databases; Cowie *et al.* (2011a) deal with the labelling of such emotional databases.

Several aspects of the list given above have already been dealt with in this book, such as the decision whether to employ acted or 'realistic' speech, units of analysis, and detailed versus coarse categorical or continuous modelling. In this chapter, we will concentrate on annotations and their evaluation, and on exemplars of databases.

6.1 Annotation

Automatic speech recognition (ASR) 'traditionally' needs transliterations (orthographic transcriptions), that is, the spoken word chain. Everything else – technical noise, filled pauses, other non-verbals – is basically not relevant. In the early days, these events were sometimes not annotated at all; nowadays, they most often are. For computational paralinguistics we need more than – even enriched – transliterations. When we analyse written language, this normally involves further processing and clustering of words and phrases (see Chapter 9). For speech, we can harness extralinguistic and linguistic context; besides, annotation is performed by assigning labels to speech units (see Section 3.3). This can be done *model-driven*, *data-driven*, or *application-driven*; mostly, a mixture of these three approaches is used. Models of 'big n'

type prevail for personality and emotion; notwithstanding the type of data, 'big *n*' annotation is possible for personality traits simply because it is based on a generic model of long-lasting characteristics. By and large, the same holds for biological and cultural trait primitives. It is normally a bad choice for emotions if we are dealing with realistic, non-acted data because subjects are free to 'choose their own' emotional state; thus, emotion annotation should normally be data-driven: we only can annotate what we find. Moreover, we can select only those states or traits for annotation we are interested in. In a call-centre scenario, we might only be interested in finding out whether and when a customer gets angry; in the case of non-native or pathological speech, it might arguably be most important to assess intelligibility. Of course, the more additional information we have, the better are our chances to employ this information for processing the phenomena we actually are interested in: knowing the sex, the first language, the age, the previous mental or emotional state – all this can be harnessed for modelling and processing other states or traits. In other words, the more detailed annotations we have, the better; however, we could also do with the labels for those phenomena we especially are interested in.

Annotations can be binary-categorical, 'yes–no' attributions, or *n*-ary categorical, 'multiple-choice' attributions. The concept of rating scales can be traced back to the 'attitude scale' proposed in Likert (1932), and can be seen as an extension of a yes–no attribution on an ordinal (lower–higher) scale. Related to this is the semantic differential (Osgood *et al.* 1957; Osgood 1964), intended to measure connotative meanings on a rating scale. Markel (1965) makes an early attempt to prove that formal paralinguistic parameters can be annotated reliably. In the same vein, Huttar (1968) evaluated functional distinctions (emotional states) on nine seven-point semantic differential scales. Dimensions can be annotated with such rating scales or quasi-continuously, for example, when scales from 1 to 100 or tools such as feeltrace (Cowie 2000) are used.

The pros and cons of categorical versus dimensional/continuous modelling and annotations have been widely discussed, and the same holds for the granularity of annotation (coarse-grained or fine-grained, complex/mixed or simple); more details are given in Sections 2.3 and 2.5. The meaningful number of annotators is addressed in Section 3.10. From the point of view of basic research, we might favour the most detailed type of annotations; from the point of application-oriented research, we might favour just the labels for those phenomena we are most interested in.

6.1.1   Assessment of Annotations

As already mentioned in Section 2.4, the terms 'ground truth' and 'gold standard' are often used more or less synonymously in the literature – here, we want to define 'ground truth' as the actual truth as measured on the ground, as compared to the 'gold standard' that might ideally be identical to the ground truth but might also be the (slightly) error-prone labelling as seen from the 'sky above'. Often in computational paralinguistics, the gold standard is not reliable, that is, the training and testing label itself may be erroneous. This strongly depends on the task: for example, the age of a speaker is usually known, but the emotion of a speaker is usually difficult to assess. When interpreting results, one thus has to bear in mind that the reference is usually the gold standard and not necessarily the ground truth. This has a double impact: on the one hand, the learnt models are error-prone; on the other hand, the test results might be over- or under-interpretations.

Thus, in order to achieve a reliable gold standard close to the ground truth, usually several annotators (labellers, raters) are used – the less certain the task is, the greater the number of annotators. There are a couple of measures to identify the agreement among labellers – *inter-rater reliability* — in the usual case where two or more labellers are involved.

If the task is modelled continuously, such as likeability of a speaker on a continuous scale, the (mean) correlation coefficient and the (average) mean linear error (MLE), mean absolute error (MAE) or mean square error (MSE), and standard deviation among labellers are frequently used. If one is to be preferred reported in isolation, it may be the correlation in the case of subjective tasks, as it is usually more informative in the given case of a gold standard without reliable reference point. On the other hand, the MSE or similar may be preferred if the task has a well-defined reference point and solid ground truth, as in the case of speaker age or height determination, the reason being that this may be more intuitive to interpret.

In the case of categorical modelling, a variety of measures can be employed for agreement evaluation such as Krippendorff's alpha, or Cohen's or Fleiss's kappa. As a continuum can be discretised, the latter statistics can also be used in this case, often with a linear or quadratic weighting. Pearson's correlation coefficient (Pearson 1901) and Spearman's rank correlation coefficient rho (Spearman 1904) can be used for such ranked intervals. Cohen's kappa (Cohen 1960) is defined for two raters as follows:

$$\kappa = \frac{p_0 - p_c}{1 - p_c}, \tag{6.1}$$

where p_0 is the measured agreement among two labellers and p_c is the chance level of agreement. If labellers agree throughout, $\kappa = 1$. If they agree only on the same level as chance, then $\kappa = 0$. Negative values indicate systematic disagreement. Fleiss's κ (Fleiss 1971, 1981) – a generalisation of Scott's π – is related to Cohen's κ and suitable for more than two raters. It is one of the most frequently encountered measures in the field. It requires all raters to rate all data, and is particularly suitable for larger data sets. Linear and quadratic weighting are commonly used in the case of ordinal-scaled class properties.

The values should be interpreted with care and, as such, do not necessarily imply any quality assessment. According to Landis and Koch (1977), values of 0.4–0.6, for example, indicate moderate agreement, while higher values are considered as good to excellent agreement (see also Section 3.10). However, the number of categories and subjects has an impact on this value, and there are easier and more difficult tasks. Thus, an excellent kappa value can also mean that the task has been defined in such a way that the outcome is very homogeneous. To give a straightforward example, when 'anger' is operationalised as 'high pitched' or the occurrence of the word 'angry', then this procedure might result in a very good labeller agreement and subsequent classification procedure; yet the validity of such an operationalisation may be rather doubtful. At the other end of the scale, a very low kappa value normally goes together with poor classification performance and indicates that something has gone wrong. In the middle of the scale, values can indicate poor planning and instructions of the annotators; however, it can also illustrate the difficulty of the task, which might result in different labelling strategies which cannot be said to be right or wrong. Careful reasoning is required to distinguish these two possibilities.

Carletta (1996) introduced the kappa statistics into the area of discourse and dialogue analysis. Several reliability coefficients are critically assessed in Krippendorff (2004), Hayes and Krippendorff (2007), and Gwet (2008).

Further, a weight can be computed for each labeller. This can help to achieve a more consistent gold standard. The justification is that labellers may suffer lapses in concentration if they have to label huge amounts of data, or stop taking the task seriously at some time. The evaluator weighted estimator (EWE) as described by Grimm and Kroschel (2005) provides an elegant model to achieve a weighted gold standard $y_{\mathrm{EWE},n}$:

$$
y_{\mathrm{EWE},n} = \frac{1}{\sum_{k=1}^{K} r_k} \sum_{k=1}^{K} r_k y_{n,k}, \tag{6.2}
$$

where the subscript $k = 1, \ldots, K$ denotes the rater, $y_{n,k}$ is the label of rater k for instance n, and r_k is an evaluator-dependent weight. The average of the individual evaluators' responses thus takes the fact that each evaluator is subject to an individual amount of disturbance during evaluation into account:

$$
r_k = \frac{\sum_{n=1}^{N} \left(y_{n,k} - \frac{1}{N} \sum_{n'=1}^{N} y_{n',k} \right) \left(\bar{y}_n - \frac{1}{N} \sum_{n'=1}^{N} \bar{y}_{n'} \right)}{\sqrt{\sum_{n=1}^{N} \left(y_{n,k} - \frac{1}{N} \sum_{n'=1}^{N} y_{n',k} \right)^2} \sqrt{\sum_{n=1}^{N} \left(\bar{y}_n - \frac{1}{N} \sum_{n'=1}^{N} \bar{y}_{n'} \right)^2}}. \tag{6.3}
$$

These weights measure the correlation between the listener's estimations $y_{n,k}$ and the average ratings of all evaluators, $\bar{y}_{n,k}$, where

$$
\bar{y}_n = \frac{1}{K} \sum_{k=1}^{K} y_{n,k}. \tag{6.4}
$$

The inter-evaluator agreement can be described by the correlation coefficients r_k using equation (6.3) and by the standard deviations σ_n of the assessments,

$$
\sigma_n = \sqrt{\frac{1}{K-1} \sum_{k=1}^{K} \left(y_{n,k} - y_{\mathrm{EWE},n} \right)^2}. \tag{6.5}
$$

The standard deviation indicates how similarly a speech or text instance is perceived by the human judge in terms of the target problem. The inter-evaluator correlation measures the agreement among the individual evaluators and thus focuses on the more general evaluation performance (Grimm and Kroschel 2005). If the weights are chosen constant among raters, the gold standard is the simple mean of the raters' continuous labels $y_{n,k}$.

Different multi-labeller evaluations have been conducted by Mower et al. (2009) who claim that 'the acoustic properties of emotional speech are better captured using models formed from averaged evaluations rather than from individual-specific evaluations'; Steidl et al. (2005) who proposed an entropy-based method for evaluating systematic confusions in the annotations of emotional speech; and Hönig *et al.* (2010a, b) and Hönig *et al.* (2011) who evaluated pairwise and averaged annotations for the assessment of non-native prosody.

6.1.2   New Trends

To achieve annotations with labels y_n for instance n of the computational paralinguistics task of interest with reduced cost, new methods for community or distributed annotation such as crowd sourcing, for example, by Amazon Mechanical Turk[1] will be of interest. If one further wants to reduce the amount of speech and language data prior to the labelling to those instances that will likely result in the best gain for the system, the field of active learning provides solutions to this end (Riccardi and Hakkani-Tur 2005); see Section 12.2 below. In addition, to obtain even larger amounts of data without the usual amount of annotation effort, merging databases for training (Schuller *et al.* 2011b) and semi- or even unsupervised learning techniques have recently been shown beneficial (Zhang and Schuller 2012; Zhang *et al.* 2011). In particular, the latter allows for exploitation of practically infinite amounts of data, such as text, audio, and audiovisual video streams available on-line. A more complex, yet also very promising alternative was shown in Schuller and Burkhardt (2010), where synthesised training material was shown to be highly beneficial in cross-corpus testing, that is, using a different database for training than for testing.

6.2   Corpora and Benchmarks: Some Examples

To illustrate the typical procedures of collection and preparation of data and later give some results on typical tasks, we will now provide some examples of typical corpora in the field.

Fifteen years ago, a survey of paralinguistic corpora that both are well suited to computational processing and publicly available would have produced a reasonably manageable list. This has changed, partly because researchers themselves have realised that isolated efforts will not suffice, partly because the large governmental bodies increasingly require, as a condition for funding, that databases collected within projects be made publicly available. Nowadays, it seems not to be possible to provide an exhaustive list. Thus, we will restrict ourselves to those corpora that have been employed within the four challenges held at the ISCA annual conference, Interspeech 2009–2012 (Schuller *et al.* 2009b, 2010, 2011a, 2012), which cover some of the main topics in paralinguistics as described in Chapter 5: biological and cultural traits primitives, personality, emotion, and deviant (here, pathological) speech. We will also describe TIMIT, which is a good example of how one can harness a rich transcription and annotation for different paralinguistic tasks. Basically, all these corpora are available from the owners; however, different conditions such as licensing fees, restrictions on commercial use, or privacy requirements may apply. All the corpora have been widely used – especially by the participants in the challenges, thus they can tell us a great deal in terms of do's and don'ts, promising and less promising approaches, and lacunas to be addressed in the future.

---

[1] https://www.mturk.com/mturk/

6.2.1   FAU Aibo Emotion Corpus

The Friedrich Alexander University (FAU) Aibo Emotion Corpus (AEC) (Steidl 2009) used in the 2009 Emotion Challenge (Schuller *et al.* 2009b) is a corpus with recordings of children interacting with Sony's pet robot Aibo; see also Section 2.13.1 above. The corpus consists of spontaneous German speech that is emotionally coloured. The speech is spontaneous because the children were not told to use specific instructions but to talk to Aibo as they would talk to a friend. The children were led to believe that Aibo was responding to their commands, whereas the robot was actually controlled by a human operator. The wizard caused Aibo to perform a fixed, predetermined sequence of actions; sometimes Aibo behaved disobediently, provoking emotional reactions. The data were collected at two different schools, Mont and Ohm, from 51 children (aged 10–13, 21 boys, 30 girls; about 9.2 hours of speech without pauses). Speech was transmitted with a high-quality wireless headset (UT 14/20 TP SHURE UHF series with microphone WH20TQG) and recorded with a DAT recorder (sampling rate 48 kHz, quantisation 16 bit, 48 kHz down-sampled to 16 kHz). The recordings were segmented automatically into 'turns' using a pause threshold of 1 s. Five labellers (advanced students of linguistics) listened to the turns in sequential order and annotated each word independently as neutral (default) or as belonging to one of ten other classes. Since many utterances are only short commands and rather long pauses can occur between words due to Aibo's reaction time, the emotional/emotion-related state of the child can change within turns. Hence, the data were labelled at the word level. If three or more labellers agreed, the label was attributed to the word; all in all, there were 48 401 words.

Classification experiments on a subset of the corpus (Steidl 2009) showed that the best unit of analysis is neither the word nor the turn, but some intermediate chunk being the best compromise between the length of the unit of analysis and the homogeneity of the different emotional/emotion-related states within one unit. Hence, manually defined chunks based on syntactic-prosodic criteria (Steidl 2009) are used here (see also Batliner *et al.* 2010). The whole corpus consisting of 18 216 chunks was used in the 2009 Challenge.

Two problems are formulated: a five-class and a two-class problem. The latter was chosen subsuming the first by introducing the cover classes NEGative (subsuming *angry*, *touchy*, *reprimanding*, and *emphatic*) and IDLe (consisting of all non-negative states). A heuristic approach similar to that applied in Steidl (2009) is used to map the labels of the five labellers at the word level onto one label for the whole chunk. Since the whole corpus is used, the classes are highly unbalanced. Speaker independence is guaranteed by using the data of one school (Ohm, 13 boys, 13 girls) for training and the data of the other school (Mont, 8 boys, 17 girls) for testing. In the training set, the chunks are given in sequential order and the chunk ID says which child the chunk belongs to. In the test set, the chunks are presented in random order without any information about the speaker. Additionally, the transliteration of the spoken word chain of the training set and the vocabulary of the whole corpus is provided, allowing for training of ASR and linguistic feature computation.

6.2.2   aGender Corpus

For the recording of the aGender corpus used in the 2010 Challenge, an external company was employed to identify possible speakers of the targeted age and gender groups (Burkhardt *et al.* 2010; Schuller *et al.* 2010, 2013). The subjects received written instructions on the procedure and a financial reward, the calls were free of charge. They were asked to call the recording system six times with a mobile phone, alternating indoors and outdoors to obtain different recording environments. They were prompted by an automated interactive voice response system to repeat given utterances or produce free content. Between each session a break of one day was scheduled to ensure more variations of the voices. The utterances were stored on the application server as 8 bit, 8 kHz, A-law. To validate the data, the associated age cluster was compared with a manual transcription of the self-stated date of birth. Four age groups – child, youth, adult, and senior – were defined. This choice was originally partially motivated for usage in call-centre dialogue systems that can address these four different age groups in different ways. For gender, children are not subdivided into female and male, thus giving three 'genders' and overall seven age/gender classes.

The content of the database was designed as follows. Each of the six recording sessions contains 18 utterances taken from a set of utterances listed in detail in Burkhardt *et al.* (2010). The topics of these were command words, embedded commands, month, week day, relative time description, public holiday, birth date, time, date, telephone number, postal code, first name, last name, and yes/no. The spoken contents were either free or restricted and 'eliciting' questions were used such as 'Please tell us any date, for example the birthday of a family member'. In total, 47 hours of speech in 65 364 single utterances of 954 speakers were collected. Note that not all volunteers completed all six calls, and some called more than six times, resulting in different numbers of utterances per speaker. The mean utterance length was 2.58 s. Twenty-five speakers were selected randomly for each of the seven classes as a fixed test partition (17 332 utterances, 12.45 hours) and the other 770 speakers as a training partition (53 076 utterances, 38.16 hours), which was further subdivided into train (32 527 utterances, 23.43 hours, 471 speakers) and develop (20 549 utterances, 14.73 hours, 299 speakers) partitions. Overall, this random speaker-based partitioning results in a roughly 40%–30%–30% train–develop–test distribution.

The age group can be handled either as combined age/gender task by classes as indicated above or as age group task independent of gender by classes.

6.2.3 TUM AVIC Corpus

For the the second task in the 2010 Challenge, the Technische Universität München (TUM) Audiovisual Interest Corpus (AVIC) database (Schuller *et al.* 2009a) was used. It features 2 hours of human conversational speech recording, annotated for different levels of interest. The corpus further features a uniquely detailed transcription of spoken content with word boundaries by forced alignment, non-linguistic vocalisations, single annotator tracks, and the sequence of (sub-)speaker turns.

An experimenter and a subject sat on opposite sides of a desk. The experimenter played the role of a product presenter and led the subject through a commercial presentation. The subject's role was to listen to the experimenter's explanations and topic presentations, ask several questions about anything of interest, and actively interact with the experimenter considering his/her interest in the addressed topics. The subject was explicitly asked not to worry about being polite to the experimenter, for example, by always showing a certain level of 'polite' attention.

Voice data were recorded by two microphones – one headset and one far-field microphone. Recordings were stored at 44.1 kHz, 16 bit. Twenty-one subjects took part in the recordings, three of them Asian, the remaining European. The language throughout the experiments is English, and all subjects are non-native, but very experienced English speakers.

To acquire reliable labels of a subject's 'level of interest', the collected material was first segmented into speaker and sub-speaker turns. Then it was labelled by four male annotators, independently of each other. The annotators were undergraduate students of psychology. The intention was to annotate observed interest in the 'common sense'. A speaker turn was thus defined as a continuous speech segment produced solely by one speaker. Back-channel interjections (*mhm*, etc.) were ignored: every time there is a speaker change, a new speaker turn begins. This is in accordance with the common understanding of 'turn-taking'. Thus, speaker turns can contain multiple and long sentences. In order to provide level of interest analysis at a finer time scale, the speaker turns were additionally segmented at grammatical phrase boundaries. A turn lasting longer than 2 s is split by punctuation and syntactical and grammatical rules, until each remaining segment is shorter than 2 s. The segments resulting from this 'chunking' are referred to as 'sub-speaker turns'.

The level of interest is annotated by sub-speaker turn. To familiarise the annotators with a subject's character and behaviour patterns prior to the actual annotation task, the annotators first had to watch approximately 5 minutes of video of a subject. Each sub-speaker turn had to be viewed at least once to label the level of interest displayed by the subject. Five levels of interest were distinguished as follows:

- *disinterest* (level -2) – the subject is tired of listening and talking about the topic, is totally passive, and disengaged;
- *indifference* (level -1) – the subject is passive, does not give much feedback to the experimenter's explanations, and asks unmotivated questions, if any;
- *neutrality* (level 0) – the subject follows and participates in the discourse, though it cannot be recognised whether she/he is interested in or indifferent to the topic;
- *interest* (level $+1$) – the subject wants to discuss the topic, closely follows the explanations, and asks questions;
- *curiosity* (level $+2$) – the subject shows a strong desire to talk and learn more about the topic.

In addition to the levels of interest annotation, the spoken content was transcribed by one annotator and checked by another. In this process, long pauses, short pauses, and further types of non-linguistic vocalisations were labelled. These vocalisations are breathing (452), consent (325), hesitation (1147), laughter (261), and coughing, other human noise (716). There are in total 18 581 spoken words and 23 084 word-like units, including 2901 non-linguistic vocalisations (19.5%). The overall annotation thus contains information by sub-speaker turn on the spoken content, non-linguistic vocalisations, individual level of interest annotator tracks, and the mean level of interest across annotators.

The gold standard is established either by majority vote on discrete ordinal classes or by shifting to a continuous scale obtained by averaging over the single annotators' level of interest leading to a continuous representation. The subjects had a tendency to be quite polite: almost no negative average level of interest was annotated. Apart from a higher resolution, the continuous representation form allows for subtraction of a subject's long-term interest profile to adapt to the mood or personality of the individual.

The speakers (and 3880 sub-speaker turns) were divided into three partitions for training (1512 sub-speaker turns, 51 min 44 s of speech, 4 female and 4 male speakers), system development (1161 sub-speaker turns, 43 min 7 s of speech, 3 female and 3 male speakers), and testing (1207 sub-speaker turns, 42 min 44 s of speech, 3 female and 4 male speakers). This was done speaker-independently to provide the best achievable balance with priority accorded to gender, then next age, and finally ethnicity.

6.2.4 Alcohol Language Corpus

A brief description of the Alcohol Language Corpus (ALC) project is now given. Details can be found in Schiel and Heinrich (2009) and Schiel et al. (2012). ALC as used in the 2011 Challenge (Schuller et al. 2011a) comprises 162 speakers (84 men, 78 women) aged 21–75 years, mean age 31.0 years and standard deviation 9.5 years, from five different locations in Germany. Non-native speakers, speakers with a strong dialect as well as non-cooperative speakers were excluded from participation. To obtain a gender-balanced set, 154 speakers (77 men, 77 women) were selected randomly; these were further randomly partitioned into gender-balanced training, development and test sets.

Speakers voluntarily underwent a systematic intoxication test supervised by the staff of the Institute of Legal Medicine, Munich. Before the test, each speaker chose the blood alcohol content (BAC) she/he wanted to reach during the intoxication test. Using both Watson and Widmark formulas (Schiel et al. 2012), the amount of required alcohol for each person was estimated and handed to the subject. After consumption, the speaker waited another 20 minutes before undergoing a breath alcohol concentration test (BRAC, not considered) and a blood sample test (for BAC). The possible range is between 0.028% and 0.175%. Immediately after the tests, the speaker was asked to undergo the ALC speech test, which lasted no longer than 15 minutes, to avoid significant changes caused by fatigue or saturation/decomposition of the measured blood alcohol level. At least two weeks later the speaker was required to undergo a second test while sober, which took about 30 minutes. Both tests took place in the same acoustic environment and were supervised by the same member of staff, who also acted as the conversation partner for dialogue recordings.

The speech signal was recorded with two different microphones, of which the Beyerdynamic Opus 54.16/3 headset was used for the Challenge. It was connected to an M-AUDIO MobilePre audio interface where the analogue signal was converted to digital and transferred to a laptop. Signals were down-sampled to 16 kHz. All speakers were prompted with the same material. Three different speech styles are part of each ALC recording: read speech, spontaneous speech, and command and control.

A two-class task was constituted by division into 'non-alcoholised' (BAC in the range [0, 0.05] per cent) and 'alcoholised' (BAC in the range]0.05, 0.175] per cent) recordings. Further, the data were partitioned into training, development, and testing sets.

6.2.5 Sleepy Language Corpus

Ninety-nine participants took part in six partial sleep deprivation studies for the recording of the Sleepy Language Corpus (SLC) (Krajewski et al. 2012; Schuller et al. 2010) used for the second task in the 2011 Challenge. The mean age of subjects was 24.9 years, with a standard deviation of 4.2 years and a range of 20–52 years. The recordings took place in a realistic car environment or in lecture rooms (sampling rate 44.1 kHz, down-sampled to

16 kHz, quantisation 16 bit, microphone-to-mouth distance 0.3 m). The speech data consisted of different tasks as follows: isolated vowels, including sustained vowel phonation, sustained loud vowel phonation, and sustained smiling vowel phonation; read speech, from 'Die Sonne und der Nordwind' (the story of 'The North Wind and the Sun', widely used within phonetics, speech pathology, etc.); commands/requests, consisting of ten simulated driver assistance system commands/requests in German, such as 'Ich suche die Friesenstrasse' ('I am looking for Friesen Street'); four simulated pilot to air traffic controller communication statements; and a description of a picture and a regular lecture.

A well-established, standardised subjective sleepiness questionnaire measure, the Karolinska Sleepiness Scale, was used by the subjects (self-assessment) and additionally by the two experimental assistants (observer assessment, given by assessors who had been formally trained to apply a standardised set of criteria). In the version used, scores ranged from 1 to 10: extremely alert (1), very alert (2), alert (3), rather alert (4), neither alert nor sleepy (5), some signs of sleepiness (6), sleepy, but no effort to stay awake (7), sleepy, some effort to stay awake (8), very sleepy, great effort to stay awake, struggling against sleep (9), extremely sleepy, cannot stay awake (10). Given these verbal descriptions, scores greater than 7.5 appear to be most relevant from a practical perspective as they describe a state in which the subject feels unable to stay awake. For training and classification purposes, the recordings (mean 5.9, standard deviation 2.2) were thus divided into two classes: not sleepy and sleepy samples with a threshold of 7.5 (ca. 94 samples per subject; in total 9277 samples). A more detailed description of the data can be found in Krajewski and Kroeger (2007) and Krajewski et al. (2009, 2012).

The available turns were divided into male and female speakers for each study. Then the turns from male and female subjects were split speaker-independently, in ascending order of subject ID, into training (roughly 40%), development (30%), and test (30%) instances. This subdivision not only ensures speaker-independent partitions but also provides for stratification by gender and study set-up (environment and degree of sleep deprivation). Of the 99 subjects, 36 (20 female, 16 male) were assigned to the training set, 30 (17 female, 13 male) to the development set, and 33 (19 female, 14 male) to the test set. All turns including linguistic cues on the sleepiness level (e.g., 'Ich bin sehr müde' – 'I'm very tired') were removed from the test set – 188 in total.

6.2.6 Speaker Personality Corpus

The Speaker Personality Corpus (SPC) (Mohammadi et al. 2010) was used as one of three corpora in the 2012 Challenge (Schuller et al. 2012). It includes 640 clips (one person per clip, 322 individuals in total) randomly extracted from the French news bulletins transmitted by Radio Suisse Romande (the Swiss national broadcast service) during February 2005. The most frequent speaker appears in 16 clips, while 61.0% of the subjects talk in one clip and 20.2% in two. The average length of the clips is 10 seconds (roughly 1 hour 40 minutes in total).

Eleven labellers performed the personality assessment via an on-line application. Each labeller listened to all clips and, for each one completed the BFI-10, a personality assessment questionnaire commonly applied in the literature (Rammstedt and John 2007) and aimed at calculating a score for each of the big five OCEAN dimensions (Wiggins 1996): openness to experience (artistic, curious, imaginative, insightful, original, wide interests);

conscientiousness (efficient, organised, planful, reliable, responsible, thorough); extraversion (active, assertive, energetic, outgoing, talkative); agreeableness (appreciative, forgiving, generous, kind, sympathetic, trusting); neuroticism (anxious, self-pitying, tense, touchy, unstable, worrying). The labellers had no understanding of French; thus only non-verbal cues could be considered. They were allowed to work no more than 60 minutes per day (split into two 30-minute sessions) to ensure an appropriate level of concentration during the entire assessment. Furthermore, the clips were presented in a different order to each labeller to avoid tiredness effects in the last clips of a session. Attention was paid to avoiding clips containing words that might be understood by non-French speakers (e.g., names of places or famous persons) and might have a priming effect. For a given labeller, the assessment of each clip yields five scores corresponding to the OCEAN personality traits. The scores for each clip and each dimension were averaged over the different labellers. Then, based on the mean for each personality dimension, the two classes high or low on trait were obtained: each clip is labelled above average for a given trait if at least six labellers (the majority) assign it a score higher than their average for the same trait; otherwise, it is labelled below average. Training, development and test sets were defined by speaker-independent subdivision of the SPC, stratifying by speaker gender.

6.2.7 Speaker Likability Database

In the 2012 Challenge the Speaker Likability Database (SLD) was also used (Burkhardt *et al.* 2011), an age- and gender-balanced subset comprising 800 speakers from the German aGender database (Burkhardt *et al.* 2010) used in the 2010 Challenge (see Section 6.2.2). For each speaker, the longest sentence consisting of a command embedded in a free sentence was used, in order to keep the effort involved in judging the data as low as possible. Thirty-two labellers (17 male, 15 female, aged 20–42 years, mean 28.6 years, standard deviation 5.4 years) were employed and paid for their work. To control for effects of gender and age group on the likeability ratings, the stimuli were presented in six blocks with a single gender/age group. To mitigate the effects of fatigue or boredom, each of the 32 labellers rated only three out of the six blocks in randomised order with a short break between each block. The order of stimuli within each block was randomised for each labeller as well. The stimuli had to be rated on a seven-point Likert scale as for likeability, without taking into account sentence content or transmission quality. A preliminary analysis of the data shows no significant impact of participants' age or gender on the ratings, whereas the samples rated are significantly different (mixed effects model, $p < 0.0001$). To establish a consensus from the individual likeability ratings (16 per instance), the EWE (Grimm and Kroschel 2005) was used, as described in Section 6.1.1. For each rater, this cross-correlation is computed only on the block of stimuli which she/he rated. In general, the raters exhibit varying 'reliability' with a cross-correlation ranging from 0.057 to 0.697.

The EWE rating was discretised into the two classes 'likeable' and 'non-likeable' based on the median EWE rating of all stimuli in the SLD. For the challenge, the data were partitioned into a training, development, and test set based on the subdivision for the 2010 Challenge. Roughly 30% of the development speakers were shifted to the test set in a stratified way in order to increase its size.

SLD thus provides an interesting example of a corpus that was enriched after its original release (aGender) by further annotation for a paralinguistic speaker trait. In this way, analysis of interdependence and its potential exploitation in automatic analysis systems becomes feasible.

6.2.8   NKI CCRT Speech Corpus

The last corpus in the 2012 Challenge was the NKI CCRT Speech Corpus (NCSC) recorded at the Department of Head and Neck Oncology and Surgery of the Netherlands Cancer Institute as described in Van der Molen *et al.* (2009). The corpus contains recordings and perceptual evaluations of 55 speakers (10 female, 45 male, average speaker age 57) who underwent concomitant chemoradiation treatment (CCRT) for inoperable tumours of the head and neck. Recordings and evaluations in the corpus were made before and after CCRT: before CCRT (54 speakers), 10 weeks after CCRT (48 speakers) and 12 months after CCRT (39 speakers). Not all speakers were Dutch native speakers, yet all speakers read a Dutch text of neutral content.

Recordings were made in a sound-treated room with a Sennheiser MD421 dynamic microphone and a portable 24-bit digital wave recorder (Edirol Roland R-1). The sampling frequency was 44.1 kHz; the mouth to microphone distance was 30 cm.

Thirteen recently graduated or about to graduate speech pathologists (all female, native Dutch speakers, average age 23.7 years) evaluated the speech recordings in an on-line experiment on an intelligibility scale from 1 to 7. They were requested to do this in a quiet environment. All annotaters completed an on-line familarisation module.

The samples were manually transcribed and an automatic phoneme alignment was generated by a speech recogniser trained on Dutch speech using the Spoken Dutch Corpus (CGN). Transcription and phonemisation were provided for the participants. For the Challenge, the original samples were segmented at the sentence boundaries. The training, development, and test partitions were obtained by stratifying according to age, gender, and nativeness of the speakers, roughly following a 40%–30%–30% split. The average rank correlation (Spearman's rho) of the individual ratings with the mean rating is 0.783. In accordance with the Likability Sub-Challenge, the EWE was calculated and discretised into binary class labels (intelligible, non-intelligible), dividing at the median of the distribution. Note that the class labels of the speech segments are not exactly balanced (1200/1186) since the median was taken from the ratings of the non-segmented original speech.

6.2.9   TIMIT Database

TIMIT (Fisher *et al.* 1986) is a well-known and popular database named after its origins: Texas Instruments (TI) and the Massachusetts Institute of Technology (MIT). Originally, it was intended to provide speech data for the development and evaluation of ASR systems. The rich transcription, annotation, and documentation, however, make it well suited to computational paralinguistic experiments as well. Thus it is well suited to speaker trait assessments such as height determination experiments because it contains a sufficiently large number of speakers (630 in total). Each of them produced ten phonetically rich sentences. In addition to featuring sufficient different speakers, TIMIT provides a rich amount of metadata on its speakers' traits: their age, gender, height, dialect (one out of eight major American English ones), level of education, and race. All TIMIT recordings are in 16 bit, 16 kHz. In particular, TIMIT was repeatedly used for speaker height determination (Mporas and Ganchev 2009).

6.2.10   Final Remarks on Databases

An overview of the databases introduced above is given in Table 6.1. One can see that once speaker traits are targeted, the number of subjects is usually considerably higher. When the state or trait is of more subjective nature, the number of labellers is usually higher, as in the case of the SLD.

Further, one can see that the majority of the data is prompted rather than spontaneous. In particular, for less researched paralinguistic tasks, databases featuring Germanic and Latin languages prevail. These last two issues – dominance of prompted and thus phonetically limited material and imbalance of languages featured – highlight the urgent need for more diversity on the linguistic side.

Similarly, lab conditions prevail during recording, apart from data taken from broadcast media and the few corpora recorded over the telephone.

Database size is strikingly small: often just a few hundred up to a few thousand instances or a few hours of speech material are contained in the sets. This holds, in particular, for those databases featuring more subjective states and traits. In fact, such a larger number of labellers per speech clip can be even more cost-intensive than requiring a large number of diverse speakers for the recordings in the first place. This is in stark contrast to related fields such as speech recognition, where up to several years of speech material are used for training and testing of systems.

To provide an impression of the state of the art in fully automatic paralinguistic recognition performance, Table 6.2 gives results from the series of comparative challenges held at Interspeech since 2009. The databases shown served as the basis for comparison and the table shows the baseline results as given by the challenge organisers, the best result by individual

Table 6.1   Overview of statistics from the examplary corpora: total time (hours), number of 'chunks' (units of analysis), subjects (subs), labellers (labs), and type (spontaneous, S, or prompted, P), language (lang) by ISO 3166-1 two-letter country code, audio quality (lab condition, LAB, telephone, TEL, or broadcast, FM), as well as bandwidth (kilohertz)

Corpus	[h]	# chunks	# subs	# labs	Type	Lang	Audio	[kHz]
FAU AEC	8.9	18 216	51	5	S	DE	LAB	16
TUM AVIC	2.3	3 880	21	4	S	UK	LAB	44
aGender	50.6	65 364	945	–	P	DE	TEL	8
ALC	43.8	12 360	162	–	P	DE	LAB	16
SLC	21.3	9 089	99	3	P	DE	LAB	16
SPC	1.7	640	322	11	S	FR	FM	8
SLD	0.7	800	800	32	P	DE	TEL	8
NCSC	2.0	2 386	55	13	P	NL	LAB	16
TIMIT	4.4	6 300	630	–	P	US	LAB	16

Table 6.2 Results of the Interspeech 2009–2012 challenges. Given are baseline results of the challenges (base), the winner's result (best), and the optimal result by majority vote over N best participants. Vote results are not available (n. a.) for the continuous regression task of level of interest recognition and the individual personality dimensions. UAR = Unweighted Average Recall, CC = correlation coefficient (see Section 11.3.3)

Year	Task	Corpus	Classes #	Base	Best UAR[%]/*CC	Vote(N)
2009	Emotion	FAU AEC	5	38.2	41.7	44.0 (5)
	Negativity	FAU AEC	2	67.7	70.3	71.2 (7)
2010	Age	aGender	4	48.9	52.4	53.6 (4)
	Gender	aGender	3	81.2	84.3	85.7 (5)
	Interest	TUM AVIC	[−1, 1]	0.421*	0.428*	n. a.
2011	Intoxication	ALC	2	65.9	70.5	72.2 (3)
	Sleepiness	SLC	2	70.3	71.7	72.5 (3)
2012	Openness	SPC	2	59.0	n. a.	n. a.
	Conscientiousness	SPC	2	79.1	n. a.	n. a.
	Extroversion	SPC	2	75.3	n. a.	n. a.
	Agreeableness	SPC	2	64.2	n. a.	n. a.
	Neuroticism	SPC	2	64.0	n. a.	n. a.
	Personality	SPC	5 x 2	68.3	69.0	70.4 (5)
	Likeability	SLD	2	59.0	65.8	68.7 (3)
	Intelligibility	NCSC	2	68.9	76.8	76.8 (1)

participating groups (but not for each individual personality dimension), and the usually overall best result to the present day as obtained by voting over the N best participants' results. Looking at these upper benchmarks, a 'magic number' seems to be a value of around 70% UAR for two-class tasks along the time continuum (e.g., negativity, personality), and intelligibility. For a higher number of classes, this number falls, except for gender which possesses a solid ground truth. This seems to hold for the other results as well: the more subjective a task is, the more challenging it seems to be. The baselines have been calculated with different feature sets (cf. Appendix A.1), and partly with different classifiers. They should thus not be compared directly across tasks – rather, all these results are meant to give a basic impression on the obtainable and typical benchmarks for the named corpora.

References

Batliner, A., Seppi, D., Steidl, S., and Schuller, B. (2010). Segmenting into adequate units for automatic recognition of emotion-related episodes: a speech-based approach. *Advances in Human-Computer Interaction*. 15 pp.

Batliner, A., Schuller, B., Seppi, D., Steidl, S., Devillers, L., Vidrascu, L., Vogt, T., Aharonson, V., and Amir, N. (2011). The automatic recognition of emotions in speech. In P. Petta, C. Pelachaud, and R. Cowie (eds), *Emotion-Oriented Systems: The Humaine Handbook*, Cognitive Technologies, pp. 71–99. Springer, Berlin.

Burkhardt, F., Eckert, M., Johannsen, W., and Stegmann, J. (2010). A database of age and gender annotated telephone speech. In *Proc. of LREC*, pp. 1562–1565, Valletta, Malta.

Burkhardt, F., Schuller, B., Weiss, B., and Weninger, F. (2011). 'Would you buy a car from me?' – On the likability of telephone voices. In *Proc. of Interspeech*, pp. 1557–1560, Florence.

Carletta, J. (1996). Assessing agreement on classification tasks: the kappa statistic. *Computational Linguistics*, **22**, 1–6.

Cohen, J. (1960). A coefficient of agreement for nominal scales. *Educational and Psychological Measurement*, **20**, 37–46.

Cowie, R. (2000). Describing the emotional states expressed in speech. In *Proc. of the ISCA Workshop on Speech and Emotion*, pp. 11–18, Newcastle, Co. Down.

Cowie, R., Douglas-Cowie, E., McRorie, M., Sneddon, I., Devillers, L., and Amir, N. (2011a). Issues in data collection. In P. Petta, C. Pelachaud, and R. Cowie (eds), *Emotion-Oriented Systems: The Humaine Handbook*, Cognitive Technologies, pp. 197–212. Springer, Berlin.

Cowie, R., Douglas-Cowie, E., Sneddon, I., Batliner, A., and Pelachaud, C. (2011b). Principles and history. In P. Petta, C. Pelachaud, and R. Cowie (eds), *Emotion-Oriented Systems: The Humaine Handbook*, Cognitive Technologies, pp. 167–196. Springer, Berlin.

Fisher, W., Doddington, G., and Goudie-Marshall, K. (1986). The DARPA Speech Recognition Research Database: Specifications and status. In *Proc. of the DARPA Workshop on Speech Recognition*, pp. 93–99.

Fleiss, J. L. (1971). Measuring nominal scale agreement among many raters. *Psychological Bulletin*, **76**, 378–382.

Fleiss, J. L. (1981). The measurement of interrater agreement. In *Statistical Methods for Rates and Proportions*, chapter 13, pp. 212–236. John Wiley & Sons, New York, 2nd edition.

Gibbon, D., Moore, R., and Winski, R. (eds) (1997). *Handbook of Standards and Resources for Spoken Language Systems*. Mouton de Gruyter, Berlin.

Gibbon, D., Mertins, I., and Moore, R. K. (eds) (2000). *Handbook of Multimodal and Spoken Dialogue Systems*. Kluwer Academic, Boston.

Grimm, M. and Kroschel, K. (2005). Evaluation of natural emotions using self assessment manikins. In *Proc. of ASRU*, pp. 381–385, Cancún, Mexico.

Gwet, K. L. (2008). Computing inter-rater reliability and its variance in the presence of high agreement. *British Journal of Mathematical and Statistical Psychology*, **61**, 29–48.

Hayes, A. F. and Krippendorff, K. (2007). Answering the call for a standard reliability measure for coding data. *Communication Methods and Measures*, **1**, 77–89.

Hönig, F., Batliner, A., Weilhammer, K., and Nöth, E. (2010a). Automatic assessment of non-native prosody for English as L2. In *Proc. of Speech Prosody,* Chicago IL.

Hönig, F., Batliner, A., Weilhammer, K., and Nöth, E. (2010b). How many labellers? Modelling inter-labeller agreement and system performance for the automatic assessment of non-native prosody. In *Proc. of SLATE*, Tokyo.

Hönig, F., Batliner, A., and Nöth, E. (2011). How many labellers revisited – naïves, experts and real experts. In *Proc. of SLATE*, Venice, Italy.

Huttar, G. L. (1968). Relations between prosodic variables and emotions in normal American English utterances. *Journal of Speech and Hearing Research*, **11**, 481–487.

Krajewski, J. and Kroeger, B. (2007). Using prosodic and spectral characteristics for sleepiness detection. In *Proc. of Interspeech*, pp. 1841–1844, Antwerp.

Krajewski, J., Batliner, A., and Golz, M. (2009). Acoustic sleepiness detection: Framework and validation of a speech-adapted pattern recognition approach. *Behavior Research Methods*, **41**, 795–804.

Krajewski, J., Schnieder, S., Sommer, D., Batliner, A., and Schuller, B. (2012). Applying multiple classifiers and non-linear dynamics features for detecting sleepiness from speech. *Neurocomputing*, **84**, 65–75.

Krippendorff, K. (2004). Reliability in content analysis. Some common misconceptions and recommendations. *Human Communication Research*, **30**, 411–433.

Landis, J. and Koch, G. (1977). The measurement of observer agreement for categorical data. *Biometrics*, **33**, 159–174.

Likert, R. (1932). A technique for the measurement of attitudes. *Archives of Psychology*, **22**, 1–55.

Markel, N. (1965). The reliability of coding paralanguage: Pitch, loudness, and tempo. *Journal of Verbal Learning and Verbal Behavior*, **4**, 306–308.

Mohammadi, G., Vinciarelli, A., and Mortillaro, M. (2010). The voice of personality: Mapping nonverbal vocal behavior into trait attributions. In *Proc. of SSPW'10*, pp. 17–20, Florence.

Mower, E., Mataric, M. J., and Narayanan, S. S. (2009). Evaluating evaluators: a case study in understanding the benefits and pitfalls of multi-evaluator modeling. In *Proc. of Interspeech*, pp. 1583–1586, Brighton.

Mporas, I. and Ganchev, T. (2009). Estimation of unknown speakers' height from speech. *International Journal of Speech Technology*, **12**, 149–160.

Osgood, C., Suci, G., and Tannenbaum, P. (1957). *The Measurement of Meaning*. University of Illinois Press, Urbana.

Osgood, C. E. (1964). Semantic differential technique in the comparative study of cultures. *American Anthropologist*, **66**, 171–200.

Pearson, K. (1901). On lines and planes of closest fit to systems of points in space. *Philosophical Magazine*, **2**, 559–572.

Rammstedt, B. and John, O. P. (2007). Measuring personality in one minute or less: A 10-item short version of the Big Five Inventory in English and German. *Journal of Research in Personality*, **41**, 203–212.

Riccardi, G. and Hakkani-Tur, D. (2005). Active learning: theory and applications to automatic speech recognition. *IEEE Transactions on Speech and Audio Processing*, **13**, 504–511.

Schiel, F. and Draxler, C. (2004). The production of speech corpora. http://www.phonetik.uni-muenchen.de/forschung/BITS/TP1/Cookbook/Tp1.html.

Schiel, F. and Heinrich, C. (2009). Laying the foundation for in-car alcohol detection by speech. In *Proc. of Interspeech*, pp. 983–986, Brighton, UK.

Schiel, F., Baumann, A., Draxler, C., Ellbogen, T., Hoole, P., and Steffen, A. (2004). The validation of speech corpora. http://www.phonetik.uni-muenchen.de/forschung/BITS/TP2/Cookbook/.

Schiel, F., Heinrich, C., and Barfuesser, S. (2012). Alcohol Language Corpus – The first public corpus of alcoholized German speech. *Language Resources and Evaluation*, **46**, 503–521.

Schuller, B. and Burkhardt, F. (2010). Learning with synthesized speech for automatic emotion recognition. In *Proc. of ICASSP*, pp. 5150–5515, Dallas, TX.

Schuller, B., Müller, R., Eyben, F., Gast, J., Hörnler, B., Wöllmer, M., Rigoll, G., Höthker, A., and Konosu, H. (2009a). Being bored? Recognising natural interest by extensive audiovisual integration for real-life application. *Image and Vision Computing*, **27**, 1760–1774.

Schuller, B., Steidl, S., and Batliner, A. (2009b). The Interspeech 2009 Emotion Challenge. In *Proc. of Interspeech*, pp. 312–315, Brighton, UK.

Schuller, B., Steidl, S., Batliner, A., Burkhardt, F., Devillers, L., Müller, C., and Narayanan, S. (2010). The Interspeech 2010 Paralinguistic Challenge. In *Proc. of Interspeech*, pp. 2794–2797, Makuhari, Japan.

Schuller, B., Batliner, A., Steidl, S., Schiel, F., and Krajewski, J. (2011a). The Interspeech 2011 Speaker State Challenge. In *Proc. of Interspeech*, pp. 3201–3204, Florence.

Schuller, B., Zhang, Z., Weninger, F., and Rigoll, G. (2011b). Using multiple databases for training in emotion recognition: To unite or to vote? In *Proc. of Interspeech*, pp. 1553–1556, Florence.

Schuller, B., Steidl, S., Batliner, A., Nöth, E., Vinciarelli, A., Burkhardt, F., Son, R. v., Weninger, F., Eyben, F., Bocklet, T., Mohammadi, G., and Weiss, B. (2012). The Interspeech 2012 Speaker Trait Challenge. In *Proc. of Interspeech*, Portland, OR.

Schuller, B., Steidl, S., Batliner, A., Burkhardt, F., Devillers, L., Müller, C., and Narayanan, S. (2013). Paralinguistics in speech and language – state-of-the-art and the challenge. *Computer Speech and Language*, **27**, 4–39.

Spearman, C. (1904). The proof and measurement of association between two things. *American Journal of Psychology*, **15**, 72–101.

Steidl, S. (2009). *Automatic Classification of Emotion-Related User States in Spontaneous Children's Speech*. Logos Verlag, Berlin.

Steidl, S., Levit, M., Batliner, A., Nöth, E., and Niemann, H. (2005). 'Of all things the measure is man': Automatic classification of emotions and inter-labeler consistency. In *Proc. of ICASSP*, pp. 317–320, Philadelphia.

Van der Molen, L., Van Rossum, M. A., Ackerstaff, A. H., Smeele, L. E., Rasch, C. R. N., and Hilgers, F. J. M. (2009). Pretreatment organ function in patients with advanced head and neck cancer: clinical outcome measures and patients' views. *BMC Ear Nose Throat Disorders*, **9**(10).

Wiggins, J. S. (1996). *The Five-Factor Model of Personality*. Guilford Press, New York.

Zhang, Z. and Schuller, B. (2012). Semi-supervised learning helps in sound event classification. In *Proc. of ICASSP*, pp. 333–336, Kyoto, Japan.

Zhang, Z., Weninger, F., Wöllmer, M., and Schuller, B. (2011). Unsupervised learning in cross-corpus acoustic emotion recognition. In *Proceedings 12th Biannual IEEE Automatic Speech Recognition and Understanding Workshop, ASRU 2011*, pp. 523–528, Big Island, HI.

Part II
Modelling

7

Computational Modelling of Paralinguistics: Overview

To manage a system effectively, you might focus on the interactions of the parts rather than their behavior taken separately.

(Russell L. Ackoff)

To deal with the actual computational modelling of paralinguistics, we first give an overview on the usual steps and building blocks involved. Then, in the subsequent chapters, we will highlight these in detail, focusing on those methods and aspects most frequently encountered in the field today (Schuller *et al.* 2011, 2013a,b).

A unified overview of a typical computational paralinguistics analysis system is given in Figure 7.1. Each component is described below.:

Preprocessing. After digitalisation of the speech waveform, preprocessing takes place. The aim of this step is usually to enhance the speech signal of interest (de-noising and de-reverberation) or to separate frequency patterns which stem from independent sources (source separation). In terms of enhancement, de-noising (reducing background noise) is dealt with more frequently than de-reverberation (reducing the influence of varying room impulse responses). Popular source separation algorithms include independent component analysis (ICA) in the case of multiple microphones/arrays, and non-negative matrix factorisation (NMF: Schmidt and Olsson 2006) in the case of single microphones. Additionally, automatic gain control is performed in some cases to mitigate the influences of varying recording levels and microphone gains.

Low-level descriptor extraction. Next, audio features are extracted – at approximately 100 frames per second with typical window sizes of 10–30 ms. Windowing functions are usually rectangular for extraction of a low-level descriptor (LLD) in the time domain and smooth (e.g., Hamming or Hann windows) for extraction in the frequency or time-frequency domains. Many systems process features on this level directly, either to provide frame-by-frame results, or to provide supra-segmental results by bagging of feature frames with a sliding window approach or by using dynamical approaches which provide some sort of temporal alignment and warping such as hidden Markov models or general dynamic Bayesian networks.

Computational Paralinguistics: Emotion, Affect and Personality in Speech and Language Processing, First Edition.
Björn W. Schuller and Anton M. Batliner. © 2014 John Wiley & Sons, Ltd. Published 2014 by John Wiley & Sons, Ltd.

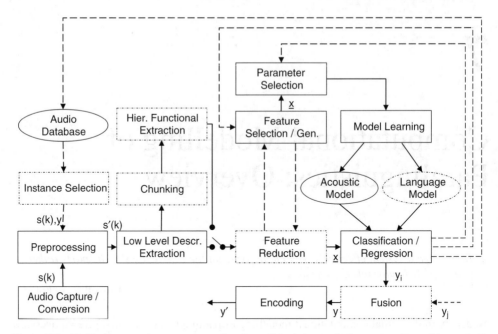

Figure 7.1 Unified perspective on computational paralinguistics analysis systems. Dotted boxes indicate optional components. Dashed lines show steps carried out only during system training or adaptation phases, where $s(k)$, $\underline{x}$, y are the speech signal, feature vector, and target (vector), respectively, a prime indicates an altered version and subscripts indicate diverse vectors. The connection from classification/regression back to the database indicates active and semi-supervised or unsupervised learning. The fusion block supports integration of other signals by late 'semantic' fusion

Typical audio LLDs cover: intensity (energy, loudness, etc.), intonation (pitch, etc.), linear prediction cepstral coefficients (LPCCs), perceptual linear prediction (PLP), cepstral coefficients (MFCCs, PLP-CCs etc.), formants (amplitude, position, width, etc.), magnitude spectra (mel frequency bands, NMF based components), harmonicity (harmonics-to-noise ratio, noise-to-harmonics ratio, etc.), vocal chord perturbation (jitter, shimmer, etc.), spectral statistics (MPEG-7 standard, roll-off points, flux, variance, slope, etc.), and many more. These are often augmented by further descriptors derived from the raw LLDs such as delta coefficients or regression coefficients. Further, diverse filtering techniques (smoothing with moving average filter, mean/variance normalisation, etc.) are often applied.

In addition to audio features, textual features, derived from the transcripts of the spoken words, can be used. These are called linguistic LLDs. Typical features are: linguistic strings (phoneme sequences, word sequences, etc.), non-linguistic strings (laughter, sighs, etc.), and disfluencies (pauses, hesitations, etc.). Again, deriving further LLDs may be considered (stemmed, part-of-speech tagged, semantically tagged, etc.). Finally, these can be tokenised in different ways, such as word (back-off) N-grams, character N-grams, etc. Note that their extraction usually requires automatic speech recognition, which allows the posteriors from the speech recogniser to be used as an additional speech recognition confidence feature.

Chunking (optional). In paralinguistics analysis most phenomena (emotion and speaker states and traits, etc.) are expressed by the evolution of certain LLDs over time. Thus, analysing the short LLD frames directly is not feasible. In the chunking stage LLD frames are grouped into meaningful temporal units of analysis. Typically, these units are between a few hundred milliseconds and a few seconds. Different types of such units have been investigated in prior studies: a fixed number of frames, acoustic chunking (e.g., by Bayesian information criterion), voiced/unvoiced parts of speech, phonemes, syllables, words, or phrases in the sense of syntactically or semantically motivated chunkings below the sentence level, or complete sentences (Batliner *et al.* 2010). Obviously, high level chunking requires pre-analysis such as speech activity detection, voicing analysis, or structural analysis based on (automatic) speech transcripts.

(Hierarchical) functional extraction (optional). In this stage, functionals are applied to each LLD within the analysis window (see above paragraph) (Eyben *et al.* 2010; Pachet and Roy 2009). The intention is a further information reduction and projection of the LLD time series of potentially unknown and variable length to a scalar value for each applied functional and LLD. This leads to what is called 'supra-segmental' analysis.

Frequently encountered functionals for speech chunks are: extremes (minimum, maximum, range, etc.), means (arithmetic, absolute, etc.), percentiles (quartiles, ranges, etc.), standard deviation, higher moments (skewness, kurtosis, etc.), peaks (number, distances, etc.), distinct segments (number, duration, etc.), regression (coefficients, error, etc.), spectrum (discrete cosine transformation coefficients, etc.), and tempo (durations, positions, etc.).

For linguistic LLDs, the following functionals can be computed for each chunk: vector space modelling (bag-of-words, etc.), look-up (word lists, concepts, etc.), statistical and information theoretic measures (salience, information gain, etc.). Also at this level, further and altered features can be obtained from the raw functionals. That is to say, by going to longer temporal units, hierarchical functionals can be computed (functionals of functionals). Functionals between LLDs, so-called cross-LLD functionals, are another option. Finally, another stage of filtering (smoothing, normalising, etc.) is sometimes applied.

Feature reduction. The aim of this step is to remove redundant information from the features and keep information related to the target of interest (e.g., emotion). The feature space is thus usually transformed – typically by a translation into the origin of the feature space and a rotation to reduce covariance between features in the transformed space. This is typically achieved by principal component analysis (PCA) (Jolliffe 2002). Linear discriminant analysis (LDA) additionally uses target information (usually discrete, i.e., class labels) to maximise the distance between class centres and to minimise dispersion of classes. Next, a feature reduction in the new space takes place by selecting a limited number of features. In the case of PCA and LDA, this is done by choosing the components with the largest eigenvalues. These features – principle components, for example – still require all the features of the original space to be computed, because the features in the new space are linear (or other) combinations of all the features in the original space.

Feature selection/generation (training/adaptation phase). It is now decided which features to discard from the feature space, that is, which features are not correlated with the task of interest. This may be of interest if a new task is not well known – for example, estimation of a speaker's weight, race or heart rate, or malfunction of a technical system from acoustic properties. In such a case, a large number of features can be 'brute-forced'. Only those well correlated with the task at hand are kept. Typically, a target function is defined in a first step.

In the case of 'open loop' selection, typical target functions are information gain or correlation among features and correlation of features with the target of the task at hand. For 'closed loop' selection, the target function is the accuracy of the learning algorithm, which is to be maximised. Usually, an efficient search function is required in addition as a fully exhaustive search is computationally hardly feasible in most cases. Such a search could start with an empty set adding features in the 'forward' direction, with the full set deleting features in the 'backward' direction, or bi-directional starting 'somewhere in the middle' and then adding and/or deleting features.

Often a random element is injected or the search is based entirely on an initial random selection guided by principles such as evolutionary search (genetic algorithms). As the search methods are usually based on the assumption of accepting a sub-optimal solution in order to reduce computational effort, the principle of 'floating' is often added to overcome nesting effects (Pudil *et al.* 1994; Ververidis and Kotropoulos 2006). That is, in the case of forward search, a (limited) number of backward steps is added to avoid too 'greedy' a search. This so-called 'sequential forward floating search' (SFFS) is among the most popular methods in the field, as we typically desire a small number of final features out of a large set of features – which is exactly what this algorithm provides. In addition, generating new feature variants can be considered during the feature selection step, for example, by applying single feature or multiple feature mathematical operations such as logarithm or division.

Parameter selection (training/adaptation phase). Parameter selection is the 'fine tuning' of the learning algorithm and the models. This can include optimisation of a model's topology, initialisation values, type of transfer or kernel functions used, or step sizes and number of iterations in the learning phase. In reality, the performance of a machine learning algorithm can indeed be significantly influenced by the choice of an optimal or sub-optimal parametrisation. While this step is seldom carried out systematically, the most popular approach is a grid search. As for feature selection, it is crucial not to use instances already used for evaluation during the parameter tuning, as obviously this would lead to an overestimation of performance.

Model learning (training/adaptation phase). This is the phase of supervised learning in which the classifier or regressor model is built based on given labelled data. There are also classifiers or regressors which do not need this phase (so-called 'lazy learners') as they decide at run-time which class to choose for a test sample, for example, by the class label of an example (training) instance with shortest distance in the feature space to the test sample. However, these are not used often because they typically do not lead to sufficient accuracy in the rather complex task of computational paralinguistics and are slow and memory-hungry at run-time for complex tasks (with many training samples).

Classification/regression. In this step a target label is assigned to an unknown test instance. In the case of classification, these are discrete labels. In the case of regression, the output is a continuous value. In general, a high diversity of classifiers and regressors is used in the field of computational paralinguistics, partly owing to the diverse requirements of the variety of different tasks.

Fusion (optional). This stage exists if information from different input streams is to be fused at the 'late semantic' level (fusion of labels and scores) rather than at an early level such as the feature level (see, for example, Bocklet *et al.* 2010).

Encoding (optional). Once the final label has been assigned, the information needs to be represented in an optimal way for further processing in complex systems such as spoken language dialogue systems (De Melo and Paiva 2007). Here, standards should be followed to ensure reusability. Examples of such standards are: VoiceXML, Extensible MultiModal Annotation markup language (EMMA: Baggia *et al.* 2007), Emotion Markup Language (EmotionML: Schröder *et al.* 2007), or Multimodal Interaction Markup Language (MIML: Mao *et al.* 2008). Besides the target label, additional information such as confidence scores should also be encoded together with the label in order to allow for disambiguation strategies in subsequent processing steps.

Audio databases (training/adaptation phase). Audio databases contain the stored audio of and ground-truth labels for exemplary speech, to be used for model learning and system evaluation. In addition, a transcription of the spoken content, for example, may be given and/or the labelling of further target tasks.

Acoustic model (AM). The acoustic model contains a numerical representation of the learnt dependencies between the acoustic observations and the labels (classes, or continuous values in the case of regression).

Language model (LM). The language model contains statistical dependencies between units (words, syllables, etc.) of the input language.

In the following chapters, all the above steps except for fusion and encoding are explained in more detail: acoustic features and pre-processing (Chapter 8), linguistic features (Chapter 9), and feature reduction, modelling, and evaluation (Chapter 11). Then examples of practical application using these steps and methods are described in Chapter 12.

References

Baggia, P., Burnett, D. C., Carter, J., Dahl, D. A., McCobb, G., and Raggett, D. (2007). *EMMA: Extensible MultiModal Annotation markup language*. Manual. http://www.w3.org/TR/emma/.

Batliner, A., Seppi, D., Steidl, S., and Schuller, B. (2010). Segmenting into adequate units for automatic recognition of emotion-related episodes: a speech-based approach. *Advances in Human-Computer Interaction*. 15 pp.

Bocklet, T., Stemmer, G., Zeissler, V., and Nöth, E. (2010). Age and gender recognition based on multiple systems – early vs. late fusion. In *Proc. of Interspeech*, pp. 2830–2833, Makuhari, Japan.

De Melo, C. and Paiva, A. (2007). Expression of emotions in virtual humans using lights, shadows, composition and filters. In A. Paiva and R. Prada, and R. W. Picard (eds), *Affective Computing and Intelligent Interaction*, pp. 549–560. Springer, Berlin.

Eyben, F., Wöllmer, M., and Schuller, B. (2010). openSMILE – The Munich Versatile and Fast Open-Source Audio Feature Extractor. In *Proc. of the 9th ACM International Conference on Multimedia, MM*, pp. 1459–1462, Florence.

Jolliffe, I. T. (2002). *Principal Component Analysis*. Springer, Berlin.

Mao, X., Li, Z., and Bao, H. (2008). An extension of MPML with emotion recognition functions attached. In H. Prendinger, J. C. Lester, and M. Ishizuka (eds), *Proc. of the 8th International Conference on Intelligent Virtual Agents IVA*, volume 5208 LNAI, pp. 289–295, Berlin. Springer.

Pachet, F. and Roy, P. (2009). Analytical features: A knowledge-based approach to audio feature generation. *EURASIP Journal on Audio, Speech, and Music Processing*.

Pudil, P., Novovicova, J., and Kittler, J. (1994). Floating search methods in feature selection. *Pattern Recognition Letters*, **15**, 1119–1125.

Schmidt, M. N. and Olsson, R. K. (2006). Single-channel speech separation using sparse non-negative matrix factorization. In *Proc. of Interspeech*, pp. 2–5, Pittsburgh, Pennsylvania.

Schröder, M., Devillers, L., Karpouzis, K., Martin, J.-C., Pelachaud, C., Peter, C., Pirker, H., Schuller, B., Tao, J., and Wilson, I. (2007). What should a generic emotion markup language be able to represent? In A. Paiva, R. Prada, and R. W. Picard (eds), *Affective Computing and Intelligent Interaction*, pp. 440–451, Berlin. Springer.

Schuller, B., Batliner, A., Steidl, S., and Seppi, D. (2011). Recognising realistic emotions and affect in speech: State of the art and lessons learnt from the first challenge. *Speech Communication*, **53**(9/10), 1062–1087.

Schuller, B., Steidl, S., Batliner, A., Schiel, F., Krajewski, J., Weninger, F., and Eyben, F. (2013a). Medium-term speaker states – a review on intoxication, sleepiness and the first challenge. *Computer Speech and Language, Special Issue on Broadening the View on Speaker Analysis*.

Schuller, B., Steidl, S., Batliner, A., Burkhardt, F., Devillers, L., Müller, C., and Narayanan, S. (2013b). Paralinguistics in speech and language – state-of-the-art and the challenge. *Computer Speech and Language*, 27, 4–39.

Ververidis, D. and Kotropoulos, C. (2006). Fast sequential floating forward selection applied to emotional speech features estimated on DES and SUSAS data collection. In *Proc. of European Signal Processing Conf. (EUSIPCO 2006)*, Florence.

8

Acoustic Features

Just because your voice reaches halfway around the world doesn't mean you are wiser than when it reached only to the end of the bar.

(Edward R. Murrow)

There's no reason for increased volume. I am scanning your interrogatives quite satis-factorily.

(K.I.T.T., car in US TV series *Knightrider*)

This chapter describes the extraction of acoustic features from the speech signal. We will touch on digitalisation of the speech signal, pre-processing and enhancement, short time analysis for time-domain and frequency-domain low-level descriptor (LLD) extraction. The choice of LLDs is based on those most frequently found in the field of computational paralinguistics (Schuller *et al.* 2011, 2013a,b).

8.1 Digital Signal Representation

In order to process the speech signal digitally, the analogue signal $s_{\mathrm{ana}}(t)$ in continuous time t is represented by a sequence of equidistant impulses at the time instants $t = f(k\Delta t)$ (Parsons 1987). The area of these impulses is proportional to the analogue amplitude $s_{\mathrm{ana}}(k\Delta t)$. If the length of the sample impulse $a(t)$ is chosen very short and given that the area of the impulse is constantly unity, the sampling can be expressed by a discrete convolution of $s_{\mathrm{ana}}(t)$ and $a(t)$ in time steps Δt (Ruske 1993):

$$s_{\mathrm{ana},T}(t) = \sum_{k=-\infty}^{+\infty} \Delta t \cdot s_{\mathrm{ana}}(k\Delta t) \cdot a(t - k\Delta t). \tag{8.1}$$

Computational Paralinguistics: Emotion, Affect and Personality in Speech and Language Processing, First Edition.
Björn W. Schuller and Anton M. Batliner. © 2014 John Wiley & Sons, Ltd. Published 2014 by John Wiley & Sons, Ltd.

For ideally short sampling impulses $a(t)$ can be approximated by a Dirac impulse (Parsons 1987):

$$a(t) = \begin{cases} 0 & \text{for } |t| > \frac{\tau}{2} \\ \frac{1}{\tau} & \text{for } |t| \le \frac{\tau}{2}, \end{cases} \tag{8.2}$$

$$\lim_{\tau \to 0} a(t) = \delta(t).$$

In the discrete convolution one can exchange the function $s_{\text{ana}}(k\Delta t)$ with $s_{\text{ana}}(t)$ and change the order:

$$s_{\text{ana},T}(t) = s_{\text{ana}}(t) \cdot \sum_{k=-\infty}^{+\infty} \Delta t\, \delta(t - k\Delta t). \tag{8.3}$$

The process of sampling can therefore be represented as a product of the function $s_{\text{ana}}(t)$ with a sampling function. This sampling function is an infinite series of Dirac impulses periodic with the sampling period Δt (Parsons 1987).

For the choice of the sampling period Δt and the sampling frequency f_{sample}, the sampling theorem applies:

$$f_{\text{sample}} = \frac{1}{\Delta t} \ge 2B. \tag{8.4}$$

For further explanations the reader is referred to Oppenheim *et al.* (1996). At this point suffice it to say that, if the sampling frequency is chosen too low, aliasing artefacts will be audible when an analogue signal is reconstructed from the digital signal representation (i.e., when playing the signal on a loudspeaker).

If the sampling theorem holds, the perfect reconstruction of the original analogue signal is possible, if the periodic high-frequency spectral parts – which are introduced by the Dirac sampling – are cut off by an ideal Küpfmüller low-pass filter (Oppenheim *et al.* 1996). The corresponding convolution in the time domain can be interpreted as interpolation with infinite (in time) sinc functions $(\sin(x))/x$ which in their sum reconstruct the complete original signal:

$$s_{\text{ana}}(t) = \sum_{k=-\infty}^{+\infty} s_{\text{ana},T}(k\Delta t)\, \text{sinc}\left(\pi \frac{t - k\Delta t}{\Delta t}\right). \tag{8.5}$$

In practice, an ideal low-pass filter is not feasible. Thus, the sampling frequency has to be chosen higher than the actual bandwidth of the signal. For speech digitalisation typically 16 kHz is used because important frequencies in speech are normally below 7.5 kHz. For music, frequencies up to 20 kHz are audible and relevant, thus the typical sampling rate of HiFi audio compact discs (CDs) is 44.1 kHz.

Next to the time discretisation by sampling, the continuous analogue amplitude values need to be discretised to digital values (Wendemuth 2004). The word length w (in bits) of the binary number is limited (typically between 8 and 32 bits). The number Q of quantisation steps for w-bit encoding is:

$$Q = 2^w. \tag{8.6}$$

This limited number of steps results in a deviation between the original value and its quantised counterpart, which is called the quantisation error. This leads to quantisation noise in the reconstructed signal. For equally sized quantisation intervals (linear quantisation), this noise can be estimated in terms of the signal-to-noise ratio (SNR) r_q as:

$$r_q = 10 \log \frac{P_S}{P_N} \simeq 20 \, lg \, Q = 20 \, lg \, 2^w \quad \text{[dB]}, \tag{8.7}$$

where P_S is is the power of the analogue signal, and P_N is the power of the quantisation noise. For longer word lengths the following approximation holds:

$$r_q \simeq 6 \, \text{dB/bit}. \tag{8.8}$$

Better SNR values can be achieved by adapting the quantisation steps to the signal characteristics such as by the International Telecommunication Union's (ITU) A-law which is primarily used in Europe or the μ-law primarily used in Northern America and Japan in the telephony sector.

An example of a digitised speech waveform is shown in Figure 8.1 for the spoken phrase 'computational paralinguistics'. The acoustic parameters described in the following will be illustrated for this example.

8.2   Short Time Analysis

Speech signals are time varying, that is, they change over time (Deller *et al.* 1993). Thus, signal parameters are also time-varying. However, one can make the assumption that these parameters change at magnitudes slower than the signal samples themselves. The progress over time of the parameters can be considered as new signals sampled at a considerably lower frequency as compared to the original signal (O'Shaughnessy 1990). Their sampling frequency will henceforth be referred to as parameter sampling frequency, in contrast to the signal sampling frequency as was introduced above.

In short time analysis the signal is considered in a given short interval within which the speech signal is assumed to be stationary (Deller *et al.* 1993). To this end, a weighting of the signal in the time domain by a 'window' function $w(\tau)$ is carried out. The window emphasises the signal's values close to the time instant t and suppresses distant values. The faded signal part at time t can be described by a multiplication with the window function as:

$$s_{\text{ana}}(\tau)w(t - \tau). \tag{8.9}$$

In particular, two opposing constraints influence the choice of the window length T. Most importantly, the window needs to be long enough to allow the parameter of interest to be reliably determined. At the same time, however, it needs to be short enough to ensure that the

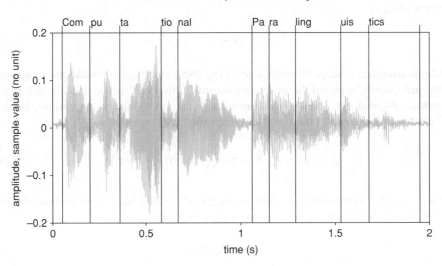

Figure 8.1 Plot of a digital speech waveform of the spoken phrase 'computational paralinguistics' spoken by two male speakers (the authors of this book; 'computational' by BS, 'paralinguistics' by AB). The signal amplitude on the y-axis is the value of the samples $s(t)$ quantised to 16-bit (range $-32\,767$ to $32\,767$) and then normalised to the range -1.0 to 1.0. It is dimensionless and has no SI unit here, as it is a digital sample value without a direct physical meaning such as voltage or power. All derived measures, such as signal energy, thus also do not have a unit

parameter does not change within the window, that is, the speech signal is 'quasi-stationary' within the window. As a result, a compromise has to be made which leads to an uncertainty by analogy with Heisenberg's uncertainty principle, the Heisenberg–Gabor limit, which says that a function cannot be both time- and band-limited:

$$\Delta\tau \cdot \Delta f \geq \frac{1}{4\pi}, \tag{8.10}$$

where $\Delta\tau$ and Δf are the uncertainty in time and frequency.

Furthermore, the sampling theorem also applies for the choice of the parameter sampling instants t. Common window lengths for speech analysis are 20–40 ms. However, the windows usually overlap by approximately 50% if the window function is a soft function and not a rectangular one. Typical speech parameter sampling frequencies are therefore around 100 Hz, that is, the window is typically shifted in steps of 10 ms. The parameter sampling frequency is often measured in frames per second because the windows of the audio signal are usually referred to as 'audio frames'. Usually, the audio signal values outside the window are set to zero. This is called the 'stationary' approach. On the other hand, the non-stationary approach assumes the signal outside of the window to be undefined. Alternatively, one can attempt to synchronise the window T with the speech signal's fundamental period T_0 in order to profit from the signal's inherent periodicity and reduce the distortions which arise from windowing to a minimum.

A crucial factor thus is the choice of the optimal windowing function. In fact, this choice depends on the type of parameter to be determined. For parameters in the time domain, rectangular windows often are sufficient. For a frequency transformation, window functions with a narrow and rectangular spectrum, but which also decay rapidly in the time domain (cf. the Heisenberg–Gabor limit), are needed. A compromise are comparably 'soft' window functions which rise and fall slowly in time and thus also in frequency. In addition, one wishes to avoid side maxima in both domains. Consider the rectangular window, for example. In the frequency domain its correspondent is the wavy sinc function. The Gaussian function – at the other extreme – has no side maxima in either domain. However, it is of infinite length in both domains. In general, the reduction of side maxima comes at the cost of a wider main maximum. Among common window functions are (represented for the interval $[-\frac{T}{2}, +\frac{T}{2}]$):

- The rectangular window. This window has the narrowest main maximum in the frequency domain, that is, the smallest bandwidth. However, it has large side maxima – the first with an amplitude of -16 dB:

$$w(\tau) = \begin{cases} 1 & \text{for } \tau = -\frac{T}{2}, \ldots, +\frac{T}{2} \\ 0 & \text{otherwise.} \end{cases} \tag{8.11}$$

- The Hamming window. This window is most commonly employed for parameters in the frequency domain. Its side maxima are the smallest at -42 dB, even at frequencies close to the main maximum:

$$w(\tau) = 0.54 + 0.46 \cos\left(2\pi \frac{\tau}{T}\right) \quad \text{and} \quad \tau = -\frac{T}{2}, \ldots, +\frac{T}{2}. \tag{8.12}$$

- The Hanning window. In the time domain this window can be represented either as a $\cos^2$ window or as a Hamming window with different constants. In comparison to the Hamming window, the time domain function is zero at the beginning and end of the period interval:

$$w(\tau) = \cos^2\left(\pi \frac{\tau}{T}\right) \quad \text{and} \quad \tau = -\frac{T}{2}, \ldots, +\frac{T}{2}. \tag{8.13}$$

The three window functions are shown in Figure 8.2 within the interval $[-\frac{T}{2}, +\frac{T}{2}]$; following the stationary approach, they are set to zero outside the interval.

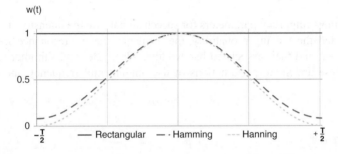

Figure 8.2 Frequently used window functions for speech analysis: Rectangular, Hamming, and Hanning

8.3   Acoustic Segmentation

For many applications a continuous audio stream can be directly analysed frame-by-frame. In other applications, however, chunking is required for 'supra-segmental' analysis (cf. Chapter 10). In the case of speech analysis, such chunks could be speech or non-speech segments, voiced or unvoiced segments, syllables, words, or larger syntactic entities.

Often speech segments in between 'silences' are to be analysed. The 'silences' might be filled with background noise, and the events of interest could be words or acoustic events, for example. The process of discriminating speech from silence and/or background noise is called voice activity or speech activity detection (VAD/SAD). The simplest method is to apply a threshold to the energy envelope of the speech signal. Usually, a hysteresis with two thresholds is used, to introduce dependence on the signal's past. If the first threshold is exceeded, a second, lower threshold may not be underrun during a given time frame in order to detect an onset of speech. If we assume the background noise changes less quickly than the speech signal of interest, we can use an adaptive algorithm to obtain the thresholds. First, we determine the histogram of the signal level. In the described case with background noise, this results in a distinct maximum at the average level of the background noise. This level plus a delta can then be used for speech onset determination. This histogram then needs to be updated on-line to allow adaptation to varying noise and speech levels.

More complex solutions are based on multi-dimensional feature information and machine learning. These approaches can be trained well to the signals of interest and therefore usually allow better results – but at the cost of greater effort. Both approaches, however, can be efficiently combined: only when a speech onset is expected owing to the signal energy thresholding is the classifier-based decision used to ensure that the speech onset belongs to the type of signal we are interested in.

8.4   Continuous Descriptors

In this section the most common speech parameters, also referred to as acoustic features, or acoustic low-level descriptors (LLDs), are described. In the description we assume the case of digitised audio, that is, the signal is denoted by $s(k)$ in the discrete time domain with the discrete time index k for the index of the current sample value. Further, in the short time analysis of the signal we require a second time variable: a discrete frame index n for the instant of measurement of parameters in contrast to the running time index k of the audio samples themselves.

Among the most important parameters for speech signals are the intensity, the fundamental frequency F_0 and the voicing probability, the formants, that is, resonance frequencies F_X of the vocal tract, and their associated bandwidths. Also, jitter and shimmer are sometimes of interest. These are micro-perturbations of the fundamental frequency and the intensity, respectively.

8.4.1   Intensity

Rather than modelling the actual perceived intensity which – according to psychoacoustic models – depends on the pitch, duration, and spectral shape of a stimulus (Zwicker and Fastl

1999), a basic LLD is frequently used: the physical energy E of the signal $s(k)$, defined as (Kießling 1997)

$$E = \sum_{k=-\infty}^{+\infty} s^2(k). \tag{8.14}$$

For a frame at time n the energy $E(n)$ is:

$$E(n) = \sum_{k=n-\frac{N}{2}}^{n+\frac{N}{2}+1} [s(k)w(n-k-1)]^2, \tag{8.15}$$

assuming a window function $w(k)$ which is non-zero for $k = n - \frac{N}{2}, \ldots, n + \frac{N}{2} + 1$ and typically rectangular. The same assumptions are also made for the following parameters. A numerically more constrained alternative to the energy E is the root means square (RMS) energy, which is defined as the square root of $E(n)$. It is plotted in Figure 8.3 (bottom) for our example utterance 'computational paralinguistics'.

Figure 8.3 also shows a logarithmic representation of the energy, which is closer to the way we perceive the loudness of a signal: a signal that has double the amplitude of another signal is perceived far less than twice as loud, that is, our hearing sensation is non-linear in this respect. A loudness measure even closer to human perception of loudness is the loudness shown in Figure 8.4. It loosely follows the concepts of the Zwicker loudness model introduced in (Zwicker and Fastl 1999), but makes several simplifying assumptions. Technically this loudness measure is implemented as the sum of the auditory spectral bands introduced in Section 8.4.7.

8.4.2   Zero Crossings

The number of zero crossings per time within a frame, that is, the zero crossing rate (ZCR: Deller *et al.* 1993), is defined as

$$\text{ZCR}(n) = \sum_{k=n-\frac{N}{2}}^{n+\frac{N}{2}-1} s_0(k), \quad \text{with } s_0(k) = \begin{cases} 0 & \text{if } \text{sgn}[s(k)] = \text{sgn}[s(k-1)] \\ 1 & \text{if } \text{sgn}[s(k)] \neq \text{sgn}[s(k-1)]. \end{cases} \tag{8.16}$$

The zero crossing rate (Figure 8.5) provides information about the frequency distribution (Furui 1996): for a pure sine oscillation the number of zero crossings is twice the sine's frequency. Since the general audio signal is usually a complex compound of different frequency components, the ZCR mainly indicates whether or not the signal contains strong low-frequency components – in this case the ZCR would be low. This is very useful for telling whether a speech signal is voiced or not. If it is voiced, it usually has a low ZCR.

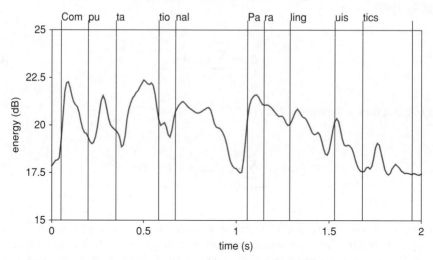

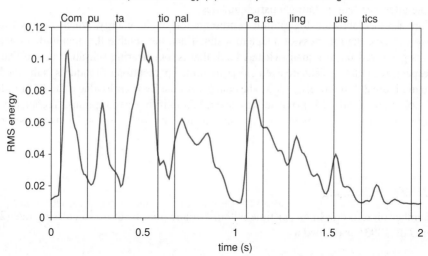

Figure 8.3 Logarithmic energy (top) and root mean square (RMS) energy (bottom) for the phrase 'computational paralinguistics'

8.4.3 Autocorrelation

Another useful basic descriptor is the autocorrelation function (ACF) $R(d)$, here the short-time ACF. For signals which are unrestricted in time it is defined as

$$R(d) = \sum_{k=-\infty}^{+\infty} s(k)s(k+d), \qquad (8.17)$$

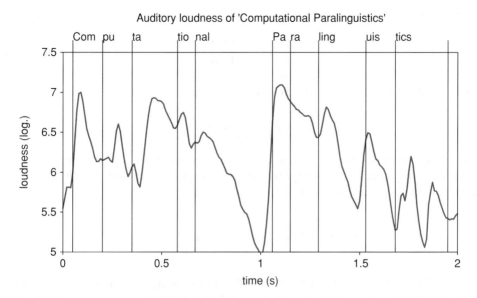

Figure 8.4 Auditory loudness computed from the perceptual linear prediction (PLP) auditory model for the phrase 'computational paralinguistics'. In contrast to the simple energy measures, the emphasised parts of the utterance and syllable nuclei are better correlated with maxima (especially for syllables 'tio' and 'nal')

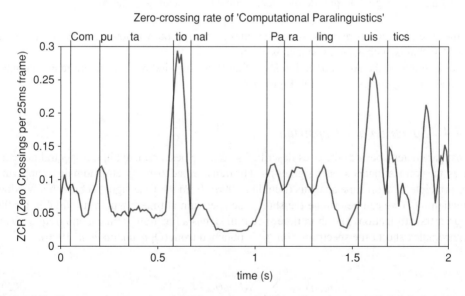

Figure 8.5 Zero crossing rate plotted for the spoken phrase 'computational paralinguistics'. A significantly higher zero crossing rate is visible for noise-like sibilants, /S/ in 'tio' and /s/ in 'uis' and 'tics' (SAMPA notation)

or the normalised short-time ACF as

$$r(d) = \frac{R(d)}{R(0)}, \tag{8.18}$$

where d is the delay parameter. The ACF is axisymmetric:

$$R(-d) = R(d). \tag{8.19}$$

For time-varying signals, the short-time ACF can be defined in a non-stationary (the signal is weighted only once) or stationary (the signal is weighted twice) way. For the commonly used stationary approach, at time n,

$$R(n, d) = \sum_{k=-\infty}^{+\infty} s(k)w(n - k - 1)s(k + d)w(n - k - 1 - d). \tag{8.20}$$

This definition is in accordance with the ACF for infinite signals. The finite limits result from setting values outside the window equal to zero. Some important characteristics of the ACF are (O'Shaughnessy 1990):

- $R(0)$ is a global maximum which is identical to the energy of the analysed signal;
- the ACF of a periodic signal is periodic itself;
- scaling of the signal amplitude by a factor a results in scaling of the ACF by a^2.

In the case of a (quasi-)periodic signal structure, a shift of the window has little influence on the ACF, that is, a phasing invariance is given.

An example of the ACF applied to short time analysis windows for the utterance 'computational paralinguistics' is given in Figure 8.6.

8.4.4 Spectrum and Cepstrum

Most parameters, besides the ones described so far, are best extracted in the spectral domain. Thus, the non-stationary speech signals must be transformed to the spectral domain by applying the principle of short time analysis (Schuller 2006): from the time signal $s(k)$ with a window function $w(k)$ we can determine the short time spectrum at time k with n as variable for the Fourier transformation. The short time spectrum is thus a function of time n and frequency m. The complex short time spectrum $S(m, n)$ is obtained from (Oppenheim *et al.* 1996):

$$S(m, n) = \sum_{k=n-\frac{N}{2}}^{n+\frac{N}{2}-1} s(k)w(n - k - 1)e^{\frac{-j2\pi mk}{N}}. \tag{8.21}$$

An example visualisation of a short time spectrum (a so-called spectrogram) of our example phrase 'computational paralinguistics' is shown in Figure 8.7.

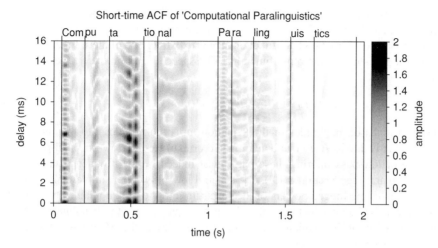

Figure 8.6 Short-time window autocorrelation function for the utterance 'computational paralinguistics'. Periodic patterns are clearly visible for voiced parts such as vowels. Low energy (white) of the ACF can be seen for the silences, closure sounds, and noise-like sibilants

To improve readability in the following, we switch back to an analogue frequency description with f as the continuous frequency.

According to the simplified linear source filter model of speech production, a speech signal can – in the frequency domain – be seen as a multiplication (convolution in the time domain) of the excitation function $E(f)$, the excitation transfer function $G(f)$, the transfer function of the vocal tract $H(f)$, and a transfer function $R(f)$ of the radiation into the space outside the human body scaled by an amplitude factor A (Deller *et al.* 1993; Fant 1973). Most prosodic

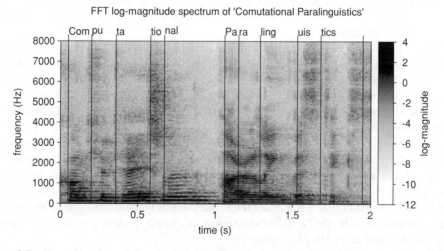

Figure 8.7 Short-time discrete Fourier transform (DFT) spectrogram of spoken phrase 'computational paralinguistics'. Voiced sounds visible through periodic patterns and formants through energy concentrations at higher frequencies (especially for 'para' and 'ling')

information (intonation, melody, etc.) is contained in the excitation information, while all other linguistic and paralinguistic information (segments/phones/phonemes, speaker traits, etc.) is contained in the vocal tract transfer function. Thus, to extract one of these functions, we must eliminate the influence of the other functions. The speech signal in the frequency domain is given by:

$$S(f) = E(f) \cdot G(f) \cdot H(f) \cdot R(f) \cdot A. \tag{8.22}$$

Taking logarithms on both sides of this equation turns the product into a summation. The signal part that is due to $E(f)$ can be eliminated or extracted by low- or band-pass filtering. In the case of low-pass filtering these parts must indeed be low-frequency in order not to cut away formants (see Section 8.4.8). The spectral low-pass can be implemented by a back-transformation of the logged powers of the spectrum into the time domain. This leads to the *cepstrum*, with the independent variable d being the 'quefrency' (O'Shaughnessy 1990). The variable d is a unit of time which corresponds to the delay in the ACF; this is the reason for the choice of the same identifier.

In the cepstrum the additive overlay of the individual components of the linear source filter model remains:

$$
\begin{aligned}
x(d) &= IDFT[\log |S(f)|^2] \\
&= IDFT[\log |E(f)|^2 + \log |G(f)|^2 + \log |H(f)|^2 + \log |R(f)|^2 + \log |A|^2] \quad (8.23) \\
&= e(d) + g(d) + h(d) + r(d) + A,
\end{aligned}
$$

where $(I)DFT$ is the (inverse) discrete Fourier transformation, and $e(d)$, $g(d)$, $h(d)$, and $r(d)$ are the time-domain equivalents of their capitalised counterparts in the frequency domain. The cepstrum is real-valued if it is computed from the amplitude or power spectrum because both are axisymmetrical (Deller *et al.* 1993). The desired low-pass can be obtained by cutting the cepstrum after the fundamental period T_0. Variants of the cepstrum use other back-transformations such as the discrete cosine transformation (DCT) or principal component analysis for decorrelation.

Figure 8.8 shows an example of a cepstrum. The upper parts (above 4 ms on the y-axis) are related to the excitation function and show the first peak of the periodic Dirac pulse excitation function at T_0 for the vowels and a noise-like excitation (without dominant peaks) for the consonants and pauses.

If we instead map the power spectrum onto mel scale bands by using triangular overlapping filters equidistant on the mel scale and then take the logarithms of the powers in each mel frequency band, we obtain the widely used so-called Mel-frequency cepstral coefficients (MFCCs: Davis and Mermelstein 1980). The mel scale takes human hearing perception into account, where lower frequencies are resolved better by human hearing than higher ones (Zwicker and Fastl 1999). An example plot of the Mel-band spectrogram and the actual MFCC can be found in Figure 8.9. The MFCC (bottom plot) is evidently more noise-like, due to the decorrelation with the DCT. The conversion of a linear frequency scale with frequency f in hertz to a mel scale $Mel(f)$ is given by:

$$Mel(f) = 2595 \cdot \log \left(1 + \frac{f}{700} \right). \tag{8.24}$$

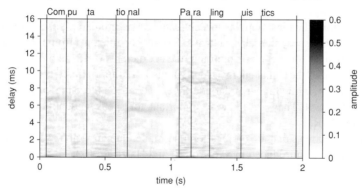

Figure 8.8 Short-time DFT-based cepstrum of the spoken phrase 'computational paralinguistics'. Peaks at 6 ms delay (left-hand side) and 8 ms delay (right-hand side) are clearly visible and indicate the first impulse of the periodic Dirac impulse sequence as excitation signal. Peaks and patterns close to 0 ms delay resemble the vocal tract impulse response

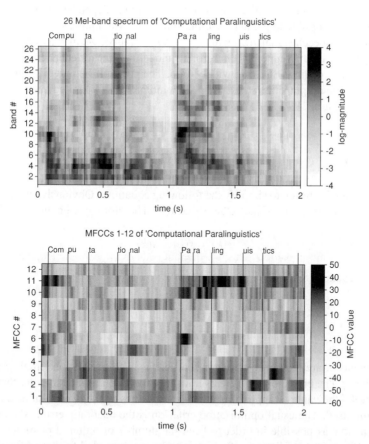

Figure 8.9 Mel-band spectrogram (top) and 12 MFCCs (bottom) for the spoken phrase 'computational paralinguistics'. The MFCC plot looks more noise-like, due to the decorrelation of the individual coefficients; in the mel-band spectrum the structure which is visible in the DFT spectrogram can still be seen, especially the vocal tract resonances, while the fine-grained structure from F_0 is almost eliminated

MFCCs are among the most popular speech features for automatic speech recognition and a large number of other speech processing tasks. Usually, coefficients from 0 up to 16 or higher are used. In particular, for speech recognition, the coefficients 0–12 are most frequently used. Coefficient 0 describes the signal energy. Coefficients 1–12 (approximately) primarily describe the phonetic content, while higher-order coefficients describe more the vocal tract, and thus, speaker characteristics. They are likewise important for speaker identification systems, but less relevant for ASR.

There exist alternatives that are more tailored to model human factors, such as human factor cepstral coefficients as introduced in Skowronski and Harris (2002).

8.4.5 Linear Prediction

A simple model for the production of speech is based on the assumption that voiced sounds – in particular, vowels – can be well modelled by the vocal tract resonance frequencies, the *formants* (Deller *et al.* 1993). From this we can conclude that subsequent samples of a speech signal are not independent but rather correlated, that is, linear dependencies exist among them. Thus, it should be possible to predict a sample value $s(k)$ with knowledge of its predecessors.

Given a digital speech signal $s(k)$, with $k = -\infty, \ldots, +\infty$, we may assume that the long-term average (the direct component) is zero. To model the linear dependencies, the methods of linear predictive coding (LPC) are widely used. The principle behind this is a linear system which describes an output value $s(k)$ as a linear combination of a limited number of preceding values $s(k - i)$ (Furui 1996):

$$\hat{s}(k) = -\sum_{i=1}^{p} a_i s(k - i). \tag{8.25}$$

The negative sign is chosen to simplify the following equations. Obviously, one cannot expect an error-free estimation $\hat{s}(k)$ of the actual value $s(k)$. Therefore, one obtains an error $e(k)$,

$$e(k) = s(k) - \hat{s}(k). \tag{8.26}$$

Substituing equation (8.25):

$$s(k) = -\sum_{i=1}^{p} a_i s(k - i) + e(k). \tag{8.27}$$

The coefficients a_i are the linear predictor coefficients. The summation length p is the order of the predictor. The predictor coefficients now have to be determined in order that – within a given interval – the values k approximate the actual values of $s(k)$ well, that is, the prediction error is minimised. The usual optimisation criterion is the quadratic error. Also, the order p should be as low as possible in order to keep the number of required predictor coefficients low (Furui 1996). The predictor coefficients need to be computed for each frame, due to the non-stationarity of speech signals (short time analysis, Section 8.2). It can be seen that the predictor polynomial represents a digital filter of the order p which can be used to produce either the speech signal $s(k)$ or the error signal $e(k)$, with $e(k)$ or $s(k)$ respectively as input signals. The coefficients a_i fully describe the linear system. If one uses the speech signal as input to the predictor, the system represents a digital transversal filter and the error signal is obtained as

$$e(k) = s(k) + \sum_{i=1}^{p} a_i s(k-i). \qquad (8.28)$$

In the following, we will use the z-transformation for the mathematical derivation. The (two-sided) z-transformation is given as:

$$S(z) = \sum_{k=-\infty}^{+\infty} s(k) z^{-k}. \qquad (8.29)$$

With the z-transformations $E(z)$ and $S(z)$ of the signals $e(k)$ and $s(k)$, and applying the rule of the z-transformation that $s(k-i)$ corresponds to $S(z)z^{-i}$ in the z-domain, we have

$$E(z) = S(z) \left(1 + \sum_{i=1}^{p} a_i z^{-i} \right), \qquad (8.30)$$

and for the transfer function $H(z)$,

$$H(z) = \frac{E(z)}{S(z)} = 1 + \sum_{i=1}^{p} a_i z^{-i}. \qquad (8.31)$$

In the inverse case, when the system is excited by the error signal and produces the speech signal, the filter is a mere recursive filter and the transfer function is the reciprocal. This is a very simple model for speech production, where the vocal tract is seen as linear filter which is excited by a regular series of pulses created by the vocal chords (for voiced sounds). The excitation pulses are not linearly predictable by the above method with a low number of predictor coefficients and thus produce prediction errors. For unvoiced sounds, the excitation is given by white noise. The vocal tract transfer function has only poles, that is, the system represents an all-pole model (Deller *et al.* 1993). These poles can be computed directly from the predictor coefficients a_i. These have to be determined for a given order p such that the deviation between the estimated signal and the real signal is minimal.

The quadratic error α within the interval of analysis (for the moment from $k = -\infty$ to $+\infty$, later within the open window region) is

$$\alpha = \sum_{k} e(k)^2 = \sum_{k} \left[\sum_{i=0}^{p} a_i s(k-i) \right]^2. \qquad (8.32)$$

Note the inclusion for simplicity of a coefficient a_0 which is constant and equal to 1. In order to find the minimum of this error function with respect to the coefficients, we differentiate partially with respect to each predictor coefficient and set the derived error equal to zero:

$$\frac{d\alpha}{da_i} = \sum_k \left[2s(k-i) \sum_{j=0}^{p} a_j s(k-j) \right] \overset{!}{=} 0. \tag{8.33}$$

After exchanging the order of the sums we get

$$\sum_{j=0}^{p} a_j \underbrace{\sum_k s(k-i)s(k-j)}_{r_{i,j}} \overset{!}{=} 0. \tag{8.34}$$

We can now substitute the so-called correlation coefficients $r_{i,j}$ as shown above. This results in a system of linear equations which allows us to apply linear algebra methods to solve for the p predictor coefficients a_j, which will not be detailed here:

$$\sum_{j=0}^{p} a_j r_{i,j} = 0, \quad \text{for } i = 1, \ldots, p. \tag{8.35}$$

It is, however, noteworthy that the predictor error α_p within an interval of analysis decreases monotonously with increasing predictor order p as the estimation of the signal improves:

$$\alpha_p \leq \alpha_{p-1}. \tag{8.36}$$

Linear prediction is also relevant in the frequency domain where in fact it is closely related to the ACF. The minimisation of the prediction error in the time domain – according to Parseval's theorem – results in a corresponding minimisation in the frequency domain. It can thus be shown that the filter yields – in the spectral domain – the smoothed envelope of the original fine-grained spectrum (O'Shaughnessy 1990). At the same time the digital filter transforms the error signal into a white spectrum. This means that its corresponding time signal is either a series of delta pulses – such as the pulse train excitation in case of voiced sounds – or white noise – as in the case of unvoiced excitation (Fant 1973).

Now let us first determine the LPC spectrum of the inverse filter denoted by a subscript 'inv'. Because it is a mere transversal filter (see above), its impulse response is identical to the LPC coefficients a_i (extended by $a_0 = 1$ at time 0):

$$h_{\text{inv}}(k) \equiv 1, a_1, a_2, \ldots, a_p. \tag{8.37}$$

The discrete complex spectrum is obtained directly by applying the discrete Fourier transform:

$$H_{\text{inv}}(m) \equiv DFT(h_{\text{inv}}(k)), \quad \text{with } m = m\Delta f, \tag{8.38}$$

$$\Delta f = \frac{1}{N\Delta t} = \frac{f_{\text{sample}}}{N}, \quad N = p + 1. \tag{8.39}$$

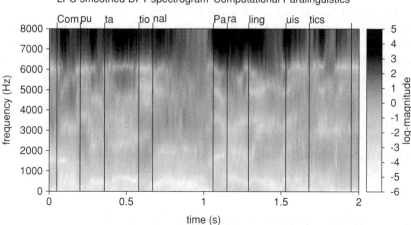

Figure 8.10 Short-time LPC spectrum (approximating the spectral envelope) for the utterance 'computational paralinguistics'. White lines show minima and black areas show smoothed maxima resembling the spectral envelope. The black areas (maxima) are correlated with the location of the formants and the maxima in the spectrogram. The fine-grained periodic structure caused by the periodic excitation signal is gone in the LPC spectrum

And with the DFT,

$$H_{\text{inv}}(m) = \sum_{k=0}^{p} h_{\text{inv}}(k)e^{-j2\pi mk/N}. \tag{8.40}$$

The DFT thus has $\left[\frac{p+1}{2}\right] + 1$ real values and $\frac{p+1}{2}$ imaginary values. We now compute the squares of the absolute values of the complex spectrum and obtain the power spectrum with $\left[\frac{p+1}{2} + 1\right]$ values. Figure 8.10 shows an example of short-time LPC spectra for our example utterance. The smooth nature of this spectrum can clearly be seen. In the example 256 bins are used (i.e., the same resolution as for the fast Fourier transform (FFT) spectrogram in Figure 8.7).

In case of the recursive all-pole model (denoted by the subscript 'rec') we have

$$H_{\text{rec}}(z) = \frac{1}{H_{\text{inv}}(z)} \tag{8.41}$$

and

$$H_{\text{rec}}(m) = \frac{1}{H_{\text{inv}}(m)}. \tag{8.42}$$

Taking logarithms,

$$\log[H_{\text{rec}}(m)] = -\log[H_{\text{inv}}(m)]. \tag{8.43}$$

On the logarithmic scale we thus only have to change the sign in order to obtain the spectrum of the recursive filter from the spectrum of the inverse filter.

As the LPC filter can only have poles, it can model formants of vowels, but not nasal sound or anti-formants, which would be zeros in the transfer function. Unlike the short time spectra computed via plain DFT, the LPC spectra are very smooth and do not show the waviness of the fundamental frequency. This comes, however, at the cost that for noise-like patterns like fricative sounds, LPC modelling is not well suited due to the spectrum still being approximated by p poles.

As speech spectra fall by approximately 6 dB per octave, the efficiency of the LPC analysis can be improved by *a priori* emphasis of higher frequencies with a first-order 'pre-emphasis' filter with the high-pass transfer function (O'Shaughnessy 1990)

$$H_{\mathrm{pre}}(z) = 1 - \mu z^{-1}, \tag{8.44}$$

where the pre-emphasis coefficient μ is usually chosen around 0.9.

In order to model the vocal tract transfer function $H(z)$ adequately, all important formants need to be captured by the model. For each formant, a pole pair is required according to Fant (1973). This fact results in a minimum predictor order p_{min} equal to twice the number of the formants. In practice, the actual predictor order is higher than the theoretical minimum by two or three to ensure proper capture of all formants in non-ideal conditions.

As stated above, the error α in the analysis interval falls monotonically with increasing predictor order p. For low p the error falls rapidly at first until all formants are captured by the model. At and after approximately $p = 16$, it remains almost the same. Another significant decrease takes place once the fine-grained structure of the spectrum caused by the fundamental frequency and its harmonics is captured by the linear model (typically if p exceeds the length of the fundamental period). However, an error always remains due to non-linearities and time-varying aspects.

8.4.6 Line Spectral Pairs

Line spectral pairs (LSPs) or frequencies (LSFs) are sometimes used for channel transmission of LPCs owing to their reduced sensitivity to quantisation noise, stability, and their ability to be interpolated. The basic principle of LSPs is the decomposition of the linear predictor (LP) polynomial for $H(z)$ as given in equation (8.31) (Kabal and Ramachandran 1986) into

$$P(z) = H(z) + z^{-(p+1)} H(z^{-1}) \tag{8.45}$$

and

$$Q(z) = H(z) - z^{-(p+1)} H(z^{-1}), \tag{8.46}$$

where $P(z)$ and $Q(z)$ correspond to transfer functions of the vocal tract with the glottis closed and opened, respectively. These two functions have roots only on the unit circle, unlike $H(z)$, which can have roots anywhere in the z-plain. Thus, $P(z)$ and $Q(z)$ are palindromic and antipalindromic polynomials, respectively. For the determination of the LSPs, we evaluate

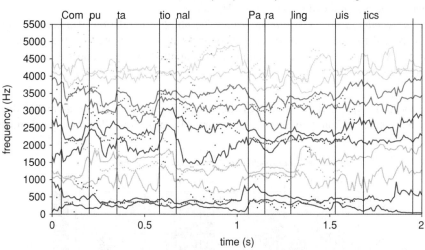

Figure 8.11 Line spectral pairs (LSPs) 1–5 (solid lines) and formant candidates 1–5 (dots) for 'computational paralinguistics'. Each pair of line spectral frequencies and the corresponding formant are plotted in the same shade of grey. The line spectral pairs – in theory – should enclose the corresponding formant as upper and lower bounds for voiced sounds. As can be seen in the plot, this is true in most cases

$P(e^{j\omega})$ and $Q(e^{j\omega})$ in a grid search for $\omega = 0, \ldots, \pi$, that is, we need to solve for the roots of the two polynomials of order $p + 1$. These roots are all complex symmetrical pairs $\pm\omega$ – hence the name LSPs (Furui 1996). Two roots are located at 0 and p; $p/2$ further roots need to be determined for $P(z)$ and $Q(z)$. The final result is p roots, that is, the same number as there are LPC coefficients. Figure 8.11 shows LSPs for our example utterance.

8.4.7 Perceptual Linear Prediction

While the LP coefficients are well suited to describing the phonetic content by good approximation of high-energy regions and filtering of the more speaker-specific fine-grained harmonic structure of the speech spectrum due to the source, they do violate principles of human hearing. Perceptual linear prediction (PLP) thus extends LP by psychophysics of the human hearing in order to base computations on an auditory spectrum estimate. The principles incorporated in the PLP procedure are:

- *Critical band spectral resolution.* Due to the linear model, LP coefficients treat all frequencies equally, whereas in the human auditory system the spectral resolution is roughly linear up to 800 or 1000 Hz and decreases non-linearly thereafter. PLP overcomes this by remapping the frequency axis according to the Bark scale and integrating the energy in the critical bands for a a critical band spectrum approximation.
- *Equal-loudness hearing curve.* To simulate the increased human sensitivity to the mid-frequency range of the audible spectrum at a conversational speech level, in PLP the critical

band spectrum is multiplied by an equal loudness curve which attenuates frequency ranges which are below or above the 400–1200 Hz range.

- *Intensity–loudness power law of hearing*: The non-linear relation of a sound's physical intensity and its subjectively perceived loudness is approximated by the power law of hearing. A cube-root amplitude compression is applied to the loudness-equalised critical band spectrum.

The spectrum thus derived shows less detail and is characterised by a smaller dynamic range (see Figure 8.12 and compare with Figure 8.9, top). This allows for good modelling

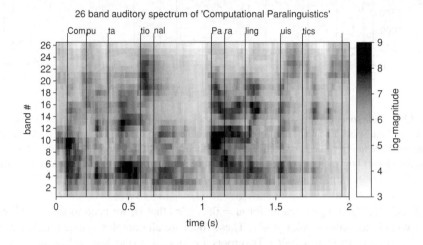

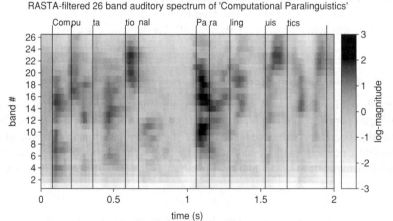

Figure 8.12   Auditory spectrogram (top) as used for perceptual linear prediction and RASTA (RelAtive SpecTrA) filtered auditory spectrogram (bottom) shown for the spoken phrase 'computational paralinguistics'. The auditory spectrum is very similar to the mel-band spectrum, except for the scaling of the magnitudes and the visible attenuation of the lower (below 4) and higher bands (above 24). The RASTA filtered spectrum shows more smoothness along the time axis: high-frequent and near-stationary parts are eliminated, while units occurring at the rhythm of speech are preserved

by a low-order all-pole model. After estimation of the auditory-like spectrum, the spectrum is converted to ACF values r. Then, these autocorrelations serve as inputs to standard LPC analysis, resulting in PLP coefficients (Hermansky 1990). By a standard recursion, these coefficients can be converted to cepstral coefficients, yielding PLP-CC.

Interestingly, PLP features achieve results similar to LP features at a lower predictor order, which is an indication that they encode the speech information in a more optimal way. This reduces the complexity of the following processing, especially the parameters needed in a learning algorithm.

An extension of PLP are RASTA (RelAtive SpecTrA) PLP coefficients (Hermansky *et al.* 1992). These aim at improved noise robustness in cases where there is a mismatch between recording conditions for training and testing data. In the RASTA method, a band-pass filter is applied to each band envelope of the critical band spectrum estimate. The filter emphasises frame-to-frame envelope changes between 1 Hz and 10 Hz with the following transfer function:

$$H(z) = 0.1 \cdot \frac{2 + z^{-1} - z^{-3} - 2z^{-4}}{z^{-4} \cdot (1 - 0.98z^{-1})}. \tag{8.47}$$

The idea behind this is that the vocal tract positions change at different speed in conversational speech usually higher than channel effects do, or background noise does. Moreover, human hearing seems to be less sensitive to slowly varying stimuli (Hermansky 1990).

In detail, the processing steps of RASTA-PLP are: DFT, logarithm, frame-to-frame band-pass filtering, equal loudness curve, power law of hearing, inverse logarithm, inverse DFT, solving linear equations for LPCs, and cepstral recursion if cepstral parameters are required.

8.4.8 Formants

Formants are resonance frequencies of the vocal tract transfer function. They vary according to the spoken content (i.e., the phonemes). In particular, the lower resonance frequencies of the vocal tract, that is, formants F_1 and F_2, are well correlated with the phonetic content. They allow for mapping of vowels to regions in the F_1, F_2 plane. In several languages (e.g., Dutch) F_3 also plays an important role for the spoken content, whereas the higher formants describe speaker characteristics.

For vowels and non-nasal consonants the transfer function of the vocal tract $H(z)$ can be approximated as an all-pole transfer function (Section 8.4.5). This corresponds to a mere recursive digital filter, as is implemented by linear prediction. The poles of $H(z)$ are considered the formants of the speech signal. When speaking about formant extraction, we usually aim to describe – in order of relevance – the centre frequency, the bandwidth, and the amplitude. Formants are mostly computed from LPCs, but also via some methods based on short time spectra. In those spectra the formants are seen as dominant maxima, for example, in the spectral envelope or – in ideal cases – even directly from the time-domain speech signal (Rigoll 1986). There are, however, a number of difficulties when using a spectral representation as a starting point for formant extraction. Most dominantly, single spikes might exist which exceed the vocal tract's resonance frequencies in amplitude and/or overlap with them – for example, from the fundamental frequency or from external noise. Next, the resonance frequencies or formants can be too close to each other, leading to their 'melting' due to limited spectral resolution. These fundamental problems can only be eased by LPC analysis.

In the following we consider formant analysis by linear prediction (Broad and Clermont 1989). The purely recursive linear prediction filter creates a smooth envelope of the short time spectra. Spectral maxima are modelled well, whereas areas of low spectral energy are not. In the linear model, speech production is modelled by the chain of speech generation (see Section 8.4.4), starting with the excitation $E(z)$ (periodic or noise), excitation spectrum $G(z)$, vocal tract $H(z)$ and radiation $R(z)$ (Fant 1973; Parsons 1987). However, we model the poles of the spectral function $S(z)$ of the speech signal. That is, in the transfer function $H_{LP}(z)$ of the prediction filter we do not know which of these components the poles originate from and $H_{LP}(z)$ cannot thus be directly assumed as the optimal estimation of $H(z)$. Rather, we have to determine which of the poles of $H_{LP}(z)$ are due to formants (McCandless 1974). The poles of the filter polynomial can first be determined by suitable numerical solver algorithms such as the Newton–Raphson method. This algorithm is initiated by an estimate of the first pole and then calculates the polynomial value and its derivative. Next, iteratively improved estimates are calculated. The iteration terminates once the delta of subsequent solutions is smaller than a predefined threshold. The polynomial can then be divided by this pole and the algorithm starts again on the now reduced polynomial until all poles are computed. At the end, a re-iteration for each pole with the full polynomial helps to overcome limited numerical precision of the first estimation round. Because the vocal tract position and thus the poles change comparably slowly over time we can speed up this process by using the poles from the previous speech frame as initialisation. For our phrase 'computational paralinguistics' the formant extraction result for F_1–F_3 is shown in Figure 8.13.

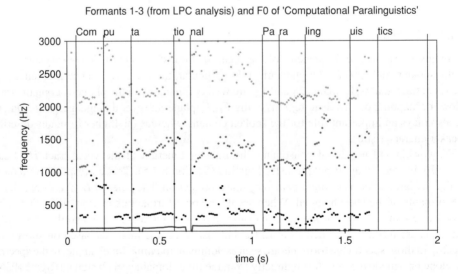

Figure 8.13 Formant candidates for $F_1 - F_3$ (dots) and fundamental frequency F_0 (solid line) shown for the spoken phrase 'computational paralinguistics'; formant candidates obtained via factorisation of LPC polynomial, no smoothing or post-processing is applied to formant candidates

As mentioned earlier, a method to determine formants which is not based on LPC is by smoothing short time spectra obtained through a DFT. The idea is to obtain a smoothed spectral envelope (just as in the case of LP), which is freed from the waviness caused by the fundamental frequency. Due to the harmonics of the fundamental frequency, the spectrum shows equidistant maxima at a distance of F_0. Obviously, these maxima can easily be confused with formants if the spectrum is not smoothed. Smoothing of the spectrum can be obtained by convolving it with a smoothing function. However, this method is not very precise.

If formant analysis is based on the spectral appearance, a peak-picking algorithm is needed to decide on the 'right' maxima among all possible extreme values of the spectral envelope. This holds both for spectral smoothing and linear prediction spectra. Typically, peak candidates are first found for each speech frame, then their evolution over time is taken into consideration in a second stage of smoothing.

Altogether, formant tracking has not been perfectly solved hitherto (Gläser *et al.* 2010). Among the main challenges are unfavourable signal conditions, in particular insufficient spectral resolution in the case of neighbouring formants of similar amplitude and peaks from other sources (not formants). Moreover, in a strict sense formants are only defined for vowels radiated via the mouth. The usage of the nasal cavity changes the real transfer function of the vocal tract significantly, as new nasal formants are added and formants might be compensated by so-called anti-formants, that is, zeros in the transfer function (Deller *et al.* 1993). Such a deletion of formants can also occur due to zeros in the excitation spectrum $G(z)$. In addition, depending on the speaker and the phoneme, often the amplitude of the formants F_3 and above is too low in comparison to surrounding noise. This is mostly the case for dark vowels. It makes an accurate detection of these higher formants problematic. Finally, there is no ground truth for formant trajectories – only gold standards – when formant extraction algorithms are tested on natural, spontaneous speech. Only small data sets with manual expert formant labels exist. One such set is the MSR-UCLA VTR database, for example. This set is a partition of the TIMIT corpus (see Section 6.2.9), manually labelled by expert phoneticians (Deng *et al.* 2006). An alternative, yet less realistic, approach to validity measurement is the use of synthesised speech. Exact formant positions are known from synthesiser parameters (Fulop 2010). Due to these problems, formant tracking is still an ongoing area of research and new approaches are still introduced, such as biologically inspired algorithms based on gammatone filter banks, (e.g., Gläser *et al.* 2010).

The tracking of anti-formants, on the other hand, is seldom of interest. Only a couple of references are given at this point to help the interested reader. The autoregressive moving average (ARMA) method is presented in Miyanaga *et al.* (1986), where a filter with an autoregressive part to handle the poles and a moving average part to handle the zeros is used. A more common method is to use the reciprocal or logarithmic vocal tract transfer function and then apply the same methods as for the poles (Steiglitz 1977).

8.4.9   *Fundamental Frequency and Voicing Probability*

The fundamental frequency F_0 or the fundamental period length T_0 plays a key role among speech parameters. Human perception is far more sensitive to changes in the fundamental frequency than to changes in other speech signal parameters (Zwicker and Fastl 1999). Therefore, a precise detection of F_0 is crucial and has a significant influence on the performance of paralinguistic speech analysis algorithms, as shown in the authors' work on emotion recognition in speech (Batliner *et al.* 2007), for example. F_0 detection might seem an easy task at first, because we only have to determine the period length of a quasi-periodic signal (Hess 1983). However, several factors make it more challenging than this. In fact, it is one of the most difficult tasks in speech signal analysis. As we know, speech production is a non-stationary process. The position of the vocal tract during articulation may change rapidly, leading to significant changes of the shape of the time-domain signal. This may occur also from one fundamental period to the next (Kießling 1997). Furthermore, the multiplicity of articulator positions of the human vocal tract used, in combination with the multiplicity of human voices, results in a huge variety of possible time-domain structures of the speech signal. Then, narrow-band lower-order formants could be confused with the fundamental frequency. In particular, the first formant of female voices can overlap with F_0 because it is typically found around 200–1400 Hz. Also, the excitation signal of the human voice itself is not always regular and does not always have a regular periodicity. This effect is especially strong for pathological voices, but holds also for all healthy voices. The voice can further switch into the 'strohbass' register, which is characterised by a very low frequent and irregular excitation as low as 25 Hz (laryngealisations, see Section 4.2.3). The range of the fundamental frequency across a large number of speakers (including children) spans almost four octaves (50–800 Hz). Finally, the transmission channel causes distortions and band limitations. This is especially problematic in the case of (narrow-band) telephone speech (300–3400 Hz), where the actual fundamental frequency (typically 50–250 Hz) is significantly suppressed. This case is known as a 'missing fundamental'. Our hearing system creates the sensation of a 'virtual pitch' based on the periodic structure of the higher harmonics of F_0 (Zwicker and Fastl 1999).

These issues have led to a large number of pitch detection algorithms (PDAs) for various conditions, none of which works to full satisfaction in all conditions (Heckmann *et al.* 2010). Some algorithms aim at determination of the fundamental period T_0, while others aim at detection of F_0. Due to the relation

$$F_0 = \frac{1}{T_0}, \tag{8.48}$$

the two are interchangeable and will not be differentiated in what follows. In algorithms where T_0 is to be determined, it is considered as a momentary value, that is, the time from the beginning of one period to the beginning of the following one. If F_0 is to be determined, the average period length of average frequency over a short time analysis window is considered. If the speech signal was strictly periodic, both definitions would lead to the same result.

Each PDA can be sub-divided into three steps: *pre-processing* with the aim of data reduction; the actual *extraction*; and *post-processing*, usually with the aim of smoothing the overall pitch trajectory, for example, with the Viterbi algorithm (see Section 11.2.2) (Hess 1983).

Independent of these steps, all PDAs can be grouped into two families (O'Shaughnessy 1990): those operating in the short-time domain (mostly applying a DFT), where two or three consecutive fundamental periods are typically observed within one frame; and those operating in the continuous-time domain, processing the input sample by sample and tracking each single pitch period.

We deal first with the short-time domain PDAs. The most obvious way is to determine the fundamental frequency directly from the short-time spectra by finding the lowest spectral maximum. However, this is not robust enough (e.g., in the case of a missing fundamental frequency as outlined above). Better results are obtained by analysing the sub-harmonic structure of the power spectrum (Hermes 1988). F_0 results as the largest common divisor of the frequencies of all harmonics. An efficient approach is to compress the power spectrum affinely along the frequency axis in the ratios 1:2, 1:3, etc., and then add the compressed spectrum to the original spectrum. By a coherent contribution of all higher-order harmonics of F_0, the peak at F_0 is emphasised. An example of the F_0 contour obtained with this algorithm is shown in Figure 8.14. This method is known as sub-harmonic summation (SHS). A related

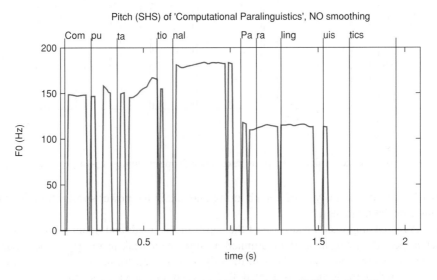

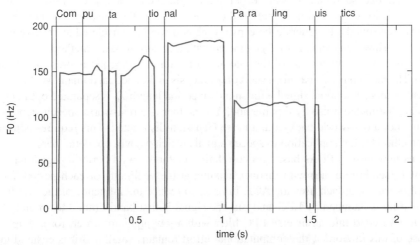

Figure 8.14 F_0 extracted with the sub-harmonic summation (SHS) method: without post-processing (top) and with Viterbi smoothing as post-processing (bottom). A higher pitch is observable for the first speaker (first word) compared to the second speaker (second word). The smoothed result shows longer continuous voiced segments (not always correct, for example, at the 'p' in 'com-pu' there should be a short unvoiced pause) due to costs for changes between voiced and unvoiced.)

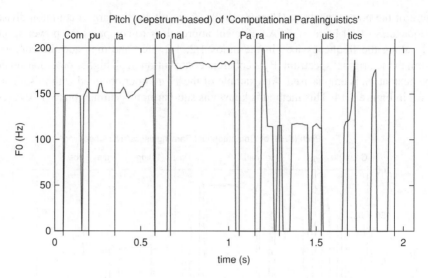

Figure 8.15 F_0 extracted with the cepstrum based method; no smoothing or post-processing is applied

approach is the direct analysis of neighbouring maxima in the power spectrum to determine the fundamental frequency.

An alternative approach makes use of the cepstrum (Ahmadi and Spanias 1999): the fundamental period T_0 can then be determined as significant peak at the right end along the quefrency axis (Section 8.4.4). Close to the origin of the cepstrum, the vocal tract impulse response including the formants is located. In the case of unvoiced sounds the excitation function is noise-like, such that no peaks occur at the right end of the cepstrum and the energy in this region is lower. By a simple threshold decision one can thus distinguish between voiced and unvoiced sounds. In general, the cepstral method can be considered as relatively robust. Figure 8.15 shows an example of F_0 extracted with a cepstrum-based method.

A method which makes use of the maximum likelihood principle is presented in Botros (1999). The method is also based on short time analysis in the spectral domain. Within a limited segment in time, a periodic signal with unknown period length T_0 is separated optimally from Gaussian-distributed noise by this method. Yet, neither is a real-world speech signal ideally periodic, nor a real-world background noise Gaussian-distributed. This requires adjustments of the method for the application to speech signals which we will not detail here.

Let us now turn to PDAs based on correlation methods. As a periodic signal has a periodic ACF (see Figure 8.6) with distinct maxima at the beginning of each period, the most straightforward approach uses the ACF. In order to reduce the influence of the first formant, the spectrum is flattened by LPC analysis: the signal is first band-limited to around 800 Hz. Next, it is inserted into an inverse LPC filter with a low predictor order, for example, of 4. Because of this low order, the computational effort remains small and it is ensured that the fundamental frequency is well preserved in the residual signal, whereas the first formant is eliminated. This is known as the simplified inverse filtering technique (SIFT: Markel 1972). Using the LPC residual signal, however, gives best results only in the case of non-disturbed speech with sufficient presence of high frequencies. For low, dark vowels the error signal has a rather low amplitude and will be dominated by noise, which is not attenuated by the LPC filter.

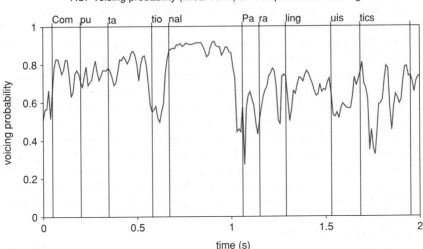

Figure 8.16 Voicing probability computed from the first distinct peak of the autocorrelation function in relation to the total signal energy

In the ACF based PDAs, F_0 is then determined by the first significant peak after the inherent one at the origin of the ACF (Boersma 2001). Wrong period values need to be eliminated or interpolated at this stage and potential changes in the period need to be foreseen. The ACF method can further be used to determine the harmonicity of the speech signal, that is, the harmonics-to-noise ratio (HNR): the ACF's first peak at the origin reflects the overall signal's energy. The HNR is then obtained by relating the peak at the origin to the next distinct peak – an example of this linear HNR can be seen in Figure 8.16. If this peak is considerably lower, we have clear evidence of a non-periodic signal. A logarithmic HNR can be calculated by

$$\mathrm{HNR}(n) = 10 \cdot \log \frac{\mathrm{ACF}(T_0)}{\mathrm{ACF}(0) - \mathrm{ACF}(T_0)}, \tag{8.49}$$

where T_0 is the fundamental period.

In summary, PDAs based on short time analysis are typically robust against noise, bandwidth limitation at the lower frequency end, and phase distortion. They do not, however, permit a period-by-period determination of T_0 which is needed if we want to measure the pitch period aligned micro-perturbations of F_0 and energy (jitter and shimmer; see Section 8.4.10).

Therefore we now consider PDAs which operate in the time domain. These can be characterised by the amount of effort put into pre-processing. The two extremes here are that either there is no pre-processing done, and the pitch period extraction stage operates on the original signal, or the pre-processing filters out everything except for the fundamental oscillation.

PDAs in the time domain analyse the speech signal period by period and determine the periods' boundaries (Hess 1983). This makes them more susceptible to local deviations caused by noise, for example, and thus less reliable than the majority of the short-time PDAs. In the case of highly non-periodic excitation signals, however, they usually provide better results.

We discuss PDAs operating directly on the time domain signal. The fundamental period is the response of the vocal tract to a single excitation pulse. The vocal tract is a lossy and passive linear system, thus, the impulse response is a sum of exponentially dampened oscillations. We can therefore expect maxima and minima at the beginning of each period to be more significant than towards the end. This allows us to determine T_0 with a peak search. There are, however, several problems which prevent this approach from working in practice. One of these is that F_1 is dampened only weakly whereas the signal envelope changes comparably faster, and in the case of a phase-distorted signal the formants may appear as if excited at different moments in time. This makes time-domain analysis rather difficult. However, these PDAs are very fast because only comparisons and decisions are needed in an implementation. The overall processing is as follows. The influence of higher-order formants is eased by low-pass filtering and LPC analysis (see the correlation-based methods discussed earlier). Next, all maxima and minima are found with a peak picking algorithm. Those which are not significant are eliminated until a single, most dominant extreme value remains per period. Post-processing, which takes the temporal context (e.g., previous periods) into account, can correct or eliminate obviously error-prone candidates.

The disadvantages of these approaches are that they do not work if the actual fundamental is not present in the signal (virtual pitch), and that it is difficult to implement the low-pass filtering because the range of possible F_0 values might not be known beforehand, and thus could potentially be very large and include the first and possibly second formant.

8.4.10   Jitter and Shimmer

Jitter and shimmer are considered as micro-prosodic descriptors as opposed to the prosodic descriptors intensity and pitch dealt with so far. Like the HNR, they describe the quality of the excitation signal and thus the quality of the voice. They belong to the group of voice quality features. Figure 8.17 shows an example plot.

Jitter is the deviation of the length of the fundamental period from one period to the next. This information is particularly helpful in speaker age or voice pathology determination, for example, as with increasing age or certain pathological conditions, the regularity of the periodic excitation decreases. Also the heart rate can have an influence on jitter (Orlikoff and Baken 1989). We can distinguish between the period-to-period (or local) jitter J_{pp}, which is the deviation from one period to the next and is given by

$$J_{pp} = T_0(n) - T_0(n - 1),   \tag{8.50}$$

and the period or cycle jitter J_c of the deviation of the current fundamental period from the 'ideal' fundamental period $\overline{T}_0$ which is obtained by averaging all pitch periods in the analysis interval,

$$J_c = T_0(n) - \overline{T}_0.   \tag{8.51}$$

Jitter is known to be particularly high at the beginning and end of a sustained vowel sound.

In a similar way, shimmer is the variation of the amplitude from one period to the next. It is typically measured on a logarithmic scale in decibels. According to Haji *et al.* (1986), a healthy person's shimmer is between 0.05 and 0.22 dB.

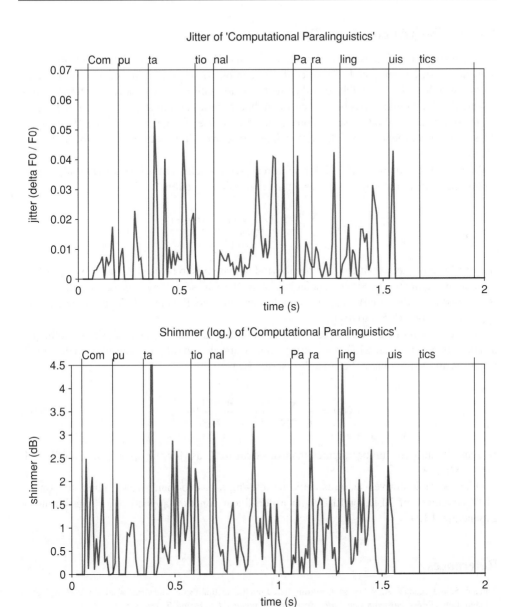

Figure 8.17 Jitter (top) and shimmer (bottom) for the utterance 'computational paralinguistics' in voiced regions (both are zero for unvoiced regions). Jitter in particular is higher at the onset (beginning) of vowels, where the periodic excitation signal is not yet stable, or the analysis window (50 ms for F_0 extraction) contains voiced and unvoiced parts – where the unvoiced parts cause high jitter and shimmer values

8.4.11   Derived Low-Level Descriptors

From all the above LLDs, derived features – what Pachet and Roy (2009) call *analytical features* – can be computed. These could be combinations of the above, for example, the energy weighted by the pitch, or other features normalised by the pitch or energy. Little research in this direction has been done so far, but it seems a promising avenue for the future. More established methods in speech processing are the use of delta regression coefficients and the smoothing of the low-level descriptor time series. We will now briefly describe both these methods.

According to Young *et al.* (2006) delta regression coefficients (first differential) can be computed from a time series $(x(t))$ using the regression equation

$$d(t) = \frac{\sum_{i=1}^{W} i \cdot (x(t+i) - x(t-i))}{2 \sum_{i=1}^{W} i^2}, \tag{8.52}$$

where W is half the size of the symmetric window which is to be used for computation of the regression coefficients (a common default is $W = 2$). In the same way, acceleration coefficients (second-order differential) can be computed as delta coefficients of delta coefficients by applying equation (8.52) to $d(t)$.

In order to minimise artefacts caused by windowing and short time analysis, smoothing is usually applied to each LLD time series $x(t)$. A simple moving average filter of length W frames is applied to obtain the smoothed series $\hat{x}(t)$:

$$\hat{x}(t) = \frac{1}{W} \sum_{i=0}^{W-1} x\left(t - \frac{W-1}{2} + i\right). \tag{8.53}$$

W must be an odd number greater than or equal to 3, as, for the special case of $W = 1$, $\hat{x}(t) = x(t)$.

Other possible derived features could be low-/high-pass or band-pass filtered versions of LLD, envelopes of LLD, or LLD with a non-linear function applied (e.g., logarithmic or exponential LLD).

References

Ahmadi, S. and Spanias, A. S. (1999). Cepstrum-based pitch detection using a new statistical V/UV classification algorithm. *IEEE Transactions on Audio, Speech and Language Processing*, **7**, 333–338.

Batliner, A., Steidl, S., Schuller, B., Seppi, D., Vogt, T., Devillers, L., Vidrascu, L., Amir, N., Kessous, L., and Aharonson, V. (2007). The impact of F0 extraction errors on the classification of prominence and emotion. In *Proc. of ICPhS*, pp. 2201–2204, Saarbrücken.

Boersma, P. (2001). Praat, a system for doing phonetics by computer. *Glot International*, **5**, 341–345.

Botros, N. (1999). Speech-pitch detection using maximum likelihood algorithm. In *Proc. of the First Joint BMES/EMBS Conference*, volume 2.

Broad, D. J. and Clermont, F. (1989). Formant estimation by linear transformation of the LPC cepstrum. *Journal of the Acoustical Society of America*, **86**, 2013–2017.

Davis, S. B. and Mermelstein, P. (1980). Comparison of parametric representations for monosyllabic word recognition in continuously spoken sentences. *IEEE Transactions on Acoustic, Speech and Signal Processing*, **28**, 357–366.

Deller, J., Proakis, J., and Hansen, J. (1993). *Discrete-Time Processing of Speech Signals*. Macmillan, New York.

Deng, L., Cui, X., Pruvenok, R., Huang, J., Momen, S., Chen, Y., and Alwan, A. (2006). A database of vocal tract resonance trajectories for research in speech processing. In *Proc. of ICASSP*, pp. 369–372, Toulouse, France.

Fant, G. (1973). *Speech Sounds and Features*. MIT Press, Cambridge, MA.

Fulop, S. A. (2010). Accuracy of formant measurement for synthesized vowels using the reassigned spectrogram and comparison with linear prediction. *Journal of the Acoustical Society of America*, **127**, 2114–2117.

Furui, S. (1996). *Digital Speech Processing: Synthesis, and Recognition*. Signal Processing and Communications. Marcel Dekker, New York, 2nd edition.

Gläser, C., Heckmann, M., Joublin, F., and Goerick, C. (2010). Combining auditory preprocessing and Bayesian estimation for robust formant tracking. *IEEE Transactions on Audio, Speech and Language Processing*, **18**, 224–236.

Haji, T., Horiguchi, S., Baer, T., and Gould, W. (1986). Frequency and amplitude perturbation analysis of electroglottograph during sustained phonation. *Journal of the Acoustical Society of America*, **80**, 58–62.

Heckmann, M., Joublin, F., and Nakadai, K. (2010). Pitch extraction in human-robot interaction. In *Proc. of IEEE/RSJ International Conference on Intelligent Robots and Systems (IROS)*, Taipei, Taiwan.

Hermansky, H. (1990). Perceptual linear predictive (plp) analysis for speech. *Journal of the Acoustical Society of America*, **87**, 1738–1752.

Hermansky, H., Morgan, N., Bayya, A., and Kohn, P. (1992). RASTA-PLP speech analysis technique. In *Proc. of ICASSP*, volume 1, pp. 121–124, San Francisco.

Hermes, D. J. (1988). Measurement of pitch by subharmonic summation. *Journal of the Acoustical Society of America*, **83**, 257–264.

Hess, W. (1983). *Pitch Determination of Speech Signals*. Springer, Berlin.

Kabal, P. and Ramachandran, R. P. (1986). The computation of line spectral frequencies using Chebyshev polynomials. *IEEE Transactions on Acoustics, Speech, & Signal Processing*, **34**, 1419–1426.

Kießling, A. (1997). *Extraktion und Klassifikation prosodischer Merkmale in der automatischen Sprachverarbeitung*. Berichte aus der Informatik. Shaker, Aachen.

Markel, J. (1972). The SIFT algorithm for fundamental frequency estimation. *IEEE Transactions on Audio and Electroacoustics*, **20**, 367–377.

McCandless, S. (1974). An algorithm for automatic formant extraction using linear prediction spectra. *IEEE Transactions on Acoustics, Speech, & Signal Processing*, **22**, 134–141.

Miyanaga, Y., Miki, N., and Nagai, N. (1986). Adaptive identification of a time-varying ARMA speech model. *IEEE Transactions on Acoustics, Speech, & Signal Processing*, **34**, 423–433.

Oppenheim, A. V., Willsky, A. S., and Hamid, S. (1996). *Signals and Systems*. Prentice Hall, Upper Saddle River, NJ, 2nd edition.

Orlikoff, R.-F. and Baken, R. (1989). The effect of the heartbeat on vocal fundamental frequency perturbation. *Journal of Sport and Health Research*, **32**(3), 576–582.

O'Shaughnessy, D. (1990). *Speech Communication*. Adison-Wesley, Reading, MA, 2nd edition.

Pachet, F. and Roy, P. (2009). Analytical features: A knowledge-based approach to audio feature generation. *EURASIP Journal on Audio, Speech, and Music Processing*.

Parsons, T. (1987). *Voice and Speech Processing*. McGraw-Hill, New York.

Rigoll, G. (1986). A new algorithm for estimation of formant trajectories directly from the speech signal based on an extended Kalman-filter. In *Proc. of ICASSP*, volume 11, pp. 1229–1232, Tokyo, Japan.

Ruske, G. (1993). *Automatische Spracherkennung. Methoden der Klassifikation und Merkmalsextraktion*. Oldenbourg, Munich, 2nd edition.

Schuller, B. (2006). *Automatische Emotionserkennung aus sprachlicher und manueller Interaktion*. Doctoral thesis, Technische Universität München, Munich.

Schuller, B., Batliner, A., Steidl, S., and Seppi, D. (2011). Recognising realistic emotions and affect in speech: State of the art and lessons learnt from the first challenge. *Speech Communication*, **53**, 1062–1087.

Schuller, B., Steidl, S., Batliner, A., Schiel, F., Krajewski, J., Weninger, F., and Eyben, F. (2013a). Medium-term speaker states – a review on intoxication, sleepiness and the first challenge. *Computer Speech and Language, Special Issue on Broadening the View on Speaker Analysis*.

Schuller, B., Steidl, S., Batliner, A., Burkhardt, F., Devillers, L., Müller, C., and Narayanan, S. (2013b). Paralinguistics in speech and language – state-of-the-art and the challenge. *Computer Speech and Language*, **27**, 4–39.

Skowronski, M. D. and Harris, J. G. (2002). Human factor cepstral coefficients. *Journal of the Acoustic Society of America*, **112**, 2279–2279.

Steiglitz, K. (1977). On the simultaneous estimation of poles and zeros in speech analysis. *IEEE Transactions on Acoustics, Speech, & Signal Processing*, **25**, 229–234.

Wendemuth, A. (2004). *Grundlagen der stochastischen Sprachverarbeitung*. Oldenbourg, Munich and Vienna.

Young, S., Evermann, G., Gales, M., Hain, T., Kershaw, D., Liu, X., Moore, G., Odell, J., Ollason, D., Povey, D., Valtchev, V., and Woodland, P. (2006). *The HTK Book*. Cambridge University Engineering Department. For HTK version 3.4.

Zwicker, E. and Fastl, H. (1999). *Psychoacoustics – Facts and Models*. Springer, Berlin.

9

Linguistic Features

Language is the source of misunderstandings.

(Antoine de Saint-Exupéry)

Language most shews a man: Speak, that I may see thee.

(Ben Jonson)

Let us now take a look at linguistic features – again, based on the most frequently encountered types in the field of computational paralinguistics (Cambria *et al.* 2013; Schuller *et al.* 2011, 2013a,b). This will touch upon in-domain data-driven methods as well as domain-independent knowledge-based approaches. A multiplicity of methods exist when dealing with linguistic analysis, some of which including deeper linguistic analysis. Thus, we can only present a subset of the predominant approaches.

9.1 Textual Descriptors

So far, we have dealt with acoustic parameters that capture the 'tone of voice'. However, methods that take into account the linguistic content have repeatedly proven their value for computational paralinguistics tasks, such as emotion recognition (Devillers *et al.* 2003; Polzin and Waibel 2000; Schuller *et al.* 2004b, 2005). These can be used in isolation (e.g., Schuller 2012) or in combination with acoustic analysis methods (e.g., Devillers and Vidrascu 2006).

These methods are based on converting unstructured text into a machine-readable representation, such as feature vectors. The dimensions of these feature vectors are usually given by a 'vocabulary' that is determined during the training phase of the recognition algorithm.

The feature vectors are often extracted from text, or human transcriptions of speech. Such 'ground truth' transcriptions by humans are not necessarily perfect; for example, human error rates of 4% are reported on transcription of spontaneous speech in Lippmann (1997). However, transcription by humans can be used as a 'canonical' reference allowing, for instance, for comparison of results of semantic analysis, eliminating the effect of varying speech recognition results as a confounding factor.

Still, in many real-world tasks such as spoken document retrieval, the transcription has to be determined by means of automatic speech recognition (ASR). Especially in the case of

Computational Paralinguistics: Emotion, Affect and Personality in Speech and Language Processing, First Edition.
Björn W. Schuller and Anton M. Batliner. © 2014 John Wiley & Sons, Ltd. Published 2014 by John Wiley & Sons, Ltd.

spontaneous speech, devising the exact transcription can be a highly challenging problem (Athanaselis *et al.* 2005; Mesaros and Virtanen 2009; Steidl *et al.* 2010; Wöllmer *et al.* 2009). It has to be noted, however, that perfect speech recognition accuracy does not seem to be a necessary precondition for robust recognition of paralinguistic information (Metze *et al.* 2011; Seppi *et al.* 2008). Small substitution errors (such as misrecognition of suffixes) are often eliminated by reducing words to their stems (see below) before linguistic feature extraction; other substitutions, insertions and deletions are not critical unless they change the 'tone' of the content with respect to the paralinguistic trait being analysed, such as replacing a word connoted with positive affect by a similar sounding one with negative connotation.

This section presents different approaches that were mostly introduced for the processing of text comprised of words; yet, they can be transferred to any domain dealing with sequences of symbols. Particularly in spoken language analysis, non-linguistic vocalisations such as sighs and yawns (Russell *et al.* 2003), laughs (Campbell *et al.* 2005; Truong and Leeuwen 2005), cries (Pal *et al.* 2006), and coughs (Matos *et al.* 2006) can also be considered to be 'words' (Batliner *et al.* 2006; Schuller *et al.* 2006). Generalising this to multimedia analysis, behavioural events can be modelled as words, as in Eyben *et al.* (2011). In the following – for the sake of simplification – we will speak of 'words' consisting of 'characters' representing the basic string units of analysis, and use 'speech' and 'text' in the sense of 'audio with symbolic content' and 'symbolic content'.

9.2   Preprocessing

When written text is analysed, some pre-processing usually has to be done. Delimiter characters such as spaces or punctuation are commonly used for segmentation, also called tokenisation, of chunks of text into smaller units of analysis, such as words. Furthermore, capitalisation is mostly normalised (e.g., by converting everything to lower case), in order to avoid using different descriptors for the same word in different capitalisations. This can be extended so as to allow certain word replacement rules, for example, to cover for varieties such as British English, American English, and Australian English, or even to consider words as identical below a specified edit distance (e.g., Schuller *et al.* 2004a).

9.3   Reduction

It is commonly found that only a fraction of the words contained in a text actually convey relevant information about the paralinguistic task to be analysed (e.g., affect-related words). While this can be addressed in the feature space through feature selection (e.g., by information gain) or reduction schemes (see Section 11.1), information reduction at the symbolic level is often considered more meaningful. Two techniques are usually applied to this end: stopping and stemming.

9.3.1   Stopping

Stopping is based on a set of rules deciding which words are irrelevant and should hence be removed from the text. These rules can be defined *a priori* by experts, such as the exclusion of function words (e.g., 'the'), or can be derived from data. Such data-driven stopping can be achieved by removing words that occur very frequently (because they are assumed to be uncorrelated with the target task), and also by excluding very rare words, such as words occurring less often than a predefined minimum word frequency. The latter is done since for such rare words, it is hard to obtain reliable statistics on the relation with the target task, as would be required by data-based machine learning methods.

9.3.2   Stemming

Stemming is a reduction method applied in natural language processing. It is used to map different morphological forms of the same word to a single symbol, the 'stem'. For natural language, this means, for example, merging different flexions, such as 'written', 'writing', 'writes', or members of the same word family, such as 'writer' in this example, to a stem symbol, 'writ'. A number of algorithmic approaches exist to automatically compute these mappings, among which Porter's stemmer (Porter 1980) is probably the most popular. This algorithm was designed for English, although modifications exist for multiple languages, such as Dutch (Kraaij and Pohlmann 1994). It models each word as a string of the form $[C](VC)^m[V]$, where the symbols C and V denote sequences of one or more consecutive consonants or vowels, respectively, and the superscript operator m indicates m-fold repetition of its string argument. m is also called the *measure* of the word. The algorithm itself consists of five steps, each applying a set of replacement rules to the word. Each of these rules can be associated with certain pre- and post-conditions restricting their application. Examples of these rules include the removal of plural and participle endings, as well as the replacement of suffixes (e.g., ATION → ATE; IVENESS → IVE). An example of a post-conditioned rule would be '$(m > 0)$ LY → ε' (ε denotes the empty string), which would reduce 'cheaply' to 'cheap' but leave 'reply' unchanged, because 'rep' would have a measure of 0 while 'cheap' has a measure of 1. Should more than one rule match in a step, only the rule with the longest matching suffix is applied. Still, due to the simple string replacement approach, Porter's stemming algorithm can lead to information loss at the semantic level, for example, when 'relativity' (as used in a text about physics) and 'relatives' are both stemmed to 'relativ'. Besides information reduction, an advantage of stemming is that words may be assigned to the correct stem even if some small speech recognition errors are present (e.g., substitution of 'writer' by 'writing').

9.3.3   Tagging

Tagging can be thought of as a generalisation of stemming to arbitrary symbol sequences (not just natural language), mapping each word to an equivalence class ('tag'). An example of tagging in natural language processing is part-of-speech (POS) tagging, also known as grammatical tagging or word-category disambiguation, which leads to very compact representations. Equivalence classes, in this case, comprise 'open' word classes such as adjectives adverbs, nouns, verbs excluding auxiliary verbs, and interjections (Batliner *et al.* 1999). Additionally, 'closed' word classes are defined which contain defined sets of auxiliary verbs, clitics, coverbs, conjunctions, determiners (articles, quantifiers, demonstrative adjectives, and possessive adjectives), particles, measure words, adpositions (prepositions, postpositions, and circumpositions), preverbs, pronouns, contractions, and cardinal numbers. Sometimes only auxiliary verbs and particles are tagged (Batliner *et al.* 2006). In the case of POS tagging, simple word-based string processing is usually not considered sufficient because of ambiguities; hence, in order to also use the context in which a word appears, techniques such as dynamic programming or hidden Markov models are applied for automatic POS tagging. An alternative is to use higher semantic concepts as tags, such as generally positive or negative terms (Batliner *et al.* 2006).

9.4 Modelling

9.4.1 Vector Space Modelling

After text pre-processing, linguistic descriptors are extracted, most commonly in the form of feature vectors with real-valued components (vector space modelling). The dimensions of these vectors (the vocabulary) usually correspond to either words ('bag-of-words'), sequences of words ('bag-of-N-grams'), or sequences of characters ('bag-of-character-N-grams').

Bag-of-Words

Bag-of-words (BoW) can be though of as the simplest method of converting a sequence of symbols into a real-valued feature vector, by counting the occurrences of symbols in the string to be analysed. Hence, each feature represents the frequency of a specific symbol. It is assumed that recognition is based on 'sufficiently' long sequences of symbols, such as paragraphs of texts, or even whole weblog entries. Mathematically, every such sequence is denoted by $S = (w_1, \ldots, w_S)$, where $S = |S|$ is the sequence length, and $w_i, i = 1, \ldots, S$, are the words in that sequence, possibly pre-processed by stemming or tagging as introduced above.

Of all the words w_i in the sequence, the BoW method considers all those that are contained in a predefined vocabulary, that is, a finite set of words $V = \{w_1, \ldots, w_V\}$, with $V = |V|$ being the size of this vocabulary. Typically, one chooses a vocabulary based on a training set $\mathcal{L}$ from the task to be analysed. In the simplest case, it is constituted of all words that occur at least once in $\mathcal{L}$.

Once a vocabulary is defined, each sequence S_j can be mapped to a BoW feature vector $\underline{x}_j = (x_{i,j})_i^T$ of dimension V. The value of $x_{i,j}$ can be simply defined as the number $f_{i,j}$ of occurrences of the word w_i in the sequence S_j. As a simplification, for example for naïve Bayes classifiers, binary word occurrence features $x_{i,j} \in \{0, 1\}$ can be used, indicating the presence (1) or absence (0) of a word i in S_j. Alternatively, the logarithm of the number of occurrences can be used, leading to 'term frequency' (TF) features:

$$\text{TF}_{i,j} = \log\left(c + f_{i,j}\right). \tag{9.1}$$

Here, c is a trivial constant that ensures that the above expression is well defined for $f_{i,j} = 0$. In most cases, c is set to 1.

An alternative transformation of $f_{i,j}$ takes into account the fact that frequently occurring words bear little information as to the task to be analysed. Instead of completely removing these words from the text, as done by data-based stopping (see Section 9.3.1), $f_{i,j}$ is downscaled by the 'inverse document frequency' (IDF), which is defined as $\log(|\mathcal{L}|/L_i)$, where L_i is the number of training sequences containing word w_i. The term 'document frequency' stems from the text information retrieval domain. This results in the feature

$$\text{IDF}_{i,j} = f_{i,j} \cdot \log \left(\frac{|\mathcal{L}|}{L_i} \right). \tag{9.2}$$

Naturally, both the TF and IDF transformations can be combined, which results in the so-called TFIDF (term frequency and inverse document frequency) approach:

$$\text{TFIDF}_{i,j} = \log \left(1 + f_{i,j} \right) \cdot \log \left(\frac{|\mathcal{L}|}{L_i} \right). \tag{9.3}$$

The above transformations do not take into account the fact that feature vectors might stem from sequences of varying length, which might result in vastly different representations of word sequences that are actually similar regarding the relative frequencies of words. Besides applying transformations to the individual components of the feature vectors $\underline{x}_j$, 'global' normalisation of all $\underline{x}_j$ can be used. Often, vectors are normalised such that all vectors have the same norm. The norm is usually chosen to be the average norm of the vectors corresponding to the training set $\mathcal{L}$. This results in the following transformation:

$$\underline{x}_j^{\text{norm}} = \frac{\frac{1}{|\mathcal{L}|} \sum_{k=1}^{L} |\underline{x}_k|}{|\underline{x}_j|} \cdot \underline{x}_j. \tag{9.4}$$

The norm $| \cdot |$ can be chosen as the L_1 norm, resulting in a normalisation by the sequence lengths, or as the Euclidean norm, etc.

A disadvantage of the BoW method is the modelling of isolated words without their 'left' and 'right' neighbouring context in a string. Thus, BoW ignores word positions or word dependencies. N-grams partly overcome this. A simple extension thus combines these BoW and N-grams.

Bag-of-N-Grams

Bag-of-N-grams (BoNG) modelling is a straightforward extension of BoW modelling that takes into account neighbouring relations between symbols. Instead of allowing only single symbols in the vocabulary, additionally 'N-grams' of words are considered. The term N-gram stems from language modelling in ASR; in our context, it simply denotes a fixed length sequence of symbols, where N is the length. Typically, bigrams (2-grams) or trigrams (3-grams) are used in addition to single words ('unigrams'). N-grams of different order N can be combined arbitrarily – one usually defines the minimum N-gram length g_{min} and the maximum N-gram length g_{max} of the feature space.

In complete analogy to the BoW approach, a vocabulary of N-grams is determined on the training set, and frequencies $f_{i,j}$ of N-gram i in sequence $\mathcal{S}_j$ are computed – either as 'raw' frequencies, or using any of the transformations listed above (TF, IDF, etc.) Note that the N-grams are usually determined after performing stopping and stemming – this usually increases the number of observations per N-gram and helps model the context between meaningful words. Even with stopping and stemming at word level, however, N-gram modelling leads to a combinatorial explosion – note that the size of the feature space grows exponentially with increasing N (V^N). To avoid this, N-grams are usually only considered if they occur with a certain minimum frequency. The feature representation resulting from combination of N-grams of different order and subsequent discarding of rare N-grams bears some similarity to the common 'back-off' language models in ASR that allow the combining of likelihoods of lower-order N-grams instead of explicitly modeling all possible N-grams. A disadvantage of BoNG modelling is its sensitivity to speech recognition errors, as for correct determination of the feature value several consecutive words have to be recognised accurately.

Bag-of-Character-N-Grams

The N-gram principle underlying the BoNG approach can also be applied at the character level, rather than the word level. This results in the bag-of-character-N-grams (BoCNG) approach. The vocabulary thus consists of fixed-length sequences (N-grams) of characters, instead of N-grams of words, and it is determined by analogy with BoNG – with tokens corresponding to characters instead of words. In particular, it is common to combine N-grams with different lengths, ranging from a minimum string length of c_{min} characters to a maximum string length of c_{max} characters, and to discard occurrences of rare N-grams in order to prevent combinatorial explosion of the feature space. Similarly to the use of word-internal or cross-word phoneme context in ASR, N-grams can be extracted just from whole words (such that 'word boundary' would contain the 3-gram 'o-r-d' but not 'd-b-o'), or across word boundaries (such that 'word boundary' would also contain 'd-b-o'). Stopping can be applied at the word level before determining the BoCNG space, while stemming at word level is usually left out since the BoCNG approach can be thought of as implicit stemming: words from the same family share certain character substrings, resulting in similar BoCNG representations.

An advantage of BoCNG over BoNG is its ability to handle unseen compound words, if these consist of substrings contained in the feature space. This may be relevant for 'open-vocabulary' languages such as German, which allow the formation of long compound words. For instance, the word 'spracherkennung' (speech recognition) could be represented by the features 's-p-r-a-c-h' and 'e-r-k-e-n-n' in a BoCNG space with $c_{max} \geq 6$, while it would be out-of-vocabulary in a BoW space comprising the (stemmed) feature dimensions 'sprach' and 'erkenn'.

The BoCNG principle can in turn be generalised to word sub-units other than characters. For example, to merge phonetically similar vocabulary entries, N-grams of phonemes as determined by an ASR engine can be used, which has proven to be advantageous (Iurgel 2007). Many other variants of N-gram vector space modelling can be conceived, such as N-grams of syllables, applying the same core principles (reduction, vocabulary selection, frequency transformation) as outlined above.

9.4.2   On-line Knowledge

All of the approaches presented so far are data-driven in the sense that a human-labelled set of training text is required for the particular task to analyse. In contrast, so-called open-domain methods rely on on-line knowledge sources that are publicly available and provide linguistic and common-sense knowledge about words, concepts and phrases, for example, in the form of lexical and semantic relations. These relations are represented in machine-readable form, for example, as semantic networks where words or concepts are represented as nodes in a graph and relations are represented by named links (Jurafsky and Martin 2000). A more specific type of on-line knowledge source is made up of annotated dictionaries, where properties of terms are stored as tags, without modelling relations between terms. Then, instead of relying on training data to compute statistics on the relation of word occurrence and the paralinguistic task being analysed, as done in the data-driven methods presented above, expert rules specific for the task are combined with general-purpose syntactic and semantic knowledge.

This chapter concludes by outlining an example algorithm to derive the sentiment of the author from written reviews of movies, purely based on on-line knowledge sources (Schuller and Knaup 2011; Schuller et al. 2009). Before presenting the algorithm in more detail, three exemplary open-domain linguistic information sources are introduced, which are also used by the algorithm.

ConceptNet

ConceptNet (Havasi et al. 2007) is part of the Commonsense Computing Initiative aiming at machine understanding of text written by humans. Hence, it contains a large set of semantic networks of concepts in a machine-readable format, focusing on common-sense knowledge extracted from public sources such as Wikipedia. Furthermore, the 'wisdom of the crowd' is included by crowd-sourcing of non-specialist humans, through an interface that allows for editing by users, and includes appropriate measures to avoid false claims and detect mistakes to a certain extent (Havasi et al. 2007). Concepts consist of one or more words, such as 'sleep' or 'watch a movie'. ConceptNet does not allow for retrieving concepts by syntactic category information, and hence does not feature word sense disambiguation; however, sufficiently specific concepts can usually be formulated to avoid ambiguities. Concepts are normalised by removal of punctuation, stopping, stemming, and ordering of stems in alphabetic order, so that minor syntactic variations are merged, and word order is ignored (Havasi et al. 2007). Figure 9.1 shows the histogram of the size of the concepts in ConceptNet (for version 3). As can be seen, the lion's share of concepts in this knowledge source consists of multiple words.

Concepts are interlinked by relations with intuitive meaning, such as 'IsA', 'UsedFor', or 'PartOf'. Thus, it is possible to represent knowledge in the form of predicates. Each predicate consists of two concepts and a relation, for example, 'UsedFor(bed, sleeping)' ('A bed is used for sleeping').

Figure 9.2 shows an example storage of predicates in ConceptNet from the domain of movies. Here, 'movie' is connected to 'actor' by the 'PartOf' relation, and also to 'fun' by a 'HasProperty' relation; thus, it is obvious that a concept can be part of many relations. Note that relations are unidirectional, since inversion of a predicate is not always meaningful. Negated predicates are also supported. Each predicate stored in ConceptNet has a confidence score which can be increased/decreased by users in a crowd-sourcing fashion (Havasi et al. 2007). The current (as of 2013) version 5 of ConceptNet covers multiple languages, including English, French, Spanish, German, Japanese, Chinese, Hindi, and Arabic.

General Inquirer

As an example of a 'dictionary' type of knowledge source, let us briefly introduce the General Inquirer database (Stone et al. 1966). In this database, terms are mapped to tags, some of

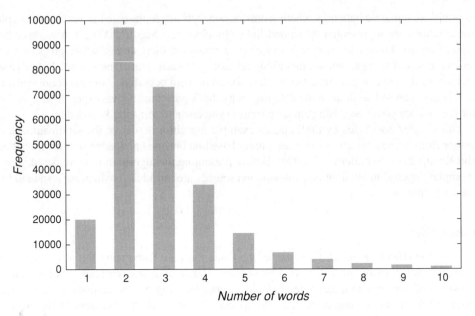

Figure 9.1 Histogram based on ConceptNet 3: frequency of concept occurrence plotted against size of concepts in words (Schuller and Knaup 2011)

which correspond to paralinguistic dimensions; for example, for sentiment analysis from text, the tags 'Positiv' and 'Negativ' (*sic*) are of particular interest. For this specific domain, 1915 terms, such as 'adore', 'master' and 'intriguing', are assigned the 'Positiv' tag, while the 'Negativ' tag comprises terms such as 'accident', 'lack', and 'boring'. For rudimentary word sense disambiguation, terms are partially annotated with POS information, definitions, and frequencies of occurrence.

WordNet

Finally, as an example of a source of lexical information, let us present the WordNet database. Based on current psycholinguistic and computational theories of human lexical memory (Fellbaum 1998), it groups words into equivalence classes by synonymy – these equivalence classes are accordingly called *synsets* – and connects synsets by lexical or semantic relations.

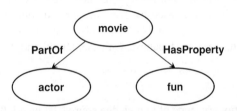

Figure 9.2 An example of a concept and relations within the concept in ConceptNet

This is the main difference from ConceptNet which features relations expressing common-sense knowledge. Examples of WordNet relations include *hyponymy* (a word is a specific instance of a more general word), *meronymy* (a word is a constituent part of another word), or *antonymy* (a word is the opposite of another word). Despite the conceptual difference from ConceptNet, some of these relations are found in similar form in the latter; for example, meronymy arguably corresponds to the 'PartOf' relation in ConceptNet. In contrast to ConceptNet, entries in WordNet are strictly separated into syntactic categories, including nouns, verbs, adjectives, and adverbs.

Example Algorithm

Based on the three knowledge sources introduced so far, an effective algorithm for linguistic analysis can be derived. While the algorithm was designed with the target domain of movie review valence estimation (sentiment analysis) in mind (Schuller and Knaup 2011; Schuller *et al.* 2009), some of the core principles can arguably be applied to other paralinguistic domains of interest, such as gender or personality analysis.

The idea of the algorithm is to find the verbs and nouns (as filtered out by POS tagging) which are 'closest' to words related to the domain of interest (such as affect) as determined by General Inquirer tags. WordNet then serves to replace words unknown to General Inquirer by synonyms, and ConceptNet is used to 'filter out' expressions not relating to the domain of interest. The algorithm is represented in Figure 9.3 as a flow-chart. Let us now detail the individual processing steps.

First, the text is split into sequences S of words or similar entities. The sequences S are then analysed by a syntactic parser for POS tagging. The POS classes (openNLP notation in parentheses) include adjective (JJ), adverb (RB), determiner (DT), verb (VB), and noun (NN). For example, the sequence 'a carefully designed plot' will be tagged as 'a/DT carefully/RB designed/VB plot/NN'. Since we are not interested in complete coverage of the syntax of longer sequences, but rather simple expressions that are similar to the predicates contained in the knowledge sources introduced above, the parser performs rather simple chunking into shorter units such as noun phrases (NP), verb phrases (VP), or prepositional phrases (PP). In particular, this kind of parsing produces a 'flat' structure which is very well suited for the subsequent processing steps. Words of the DT class and punctuation are removed in the pre-processing step.

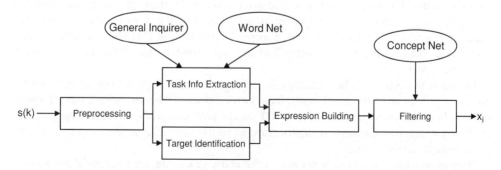

Figure 9.3 Open-domain linguistic analysis based on on-line knowledge sources

The linguistic content of each sentence is converted to one or more *ternary expressions* (T-expressions) of the form ⟨*target, verb, source*⟩. The three parts of the T-expression are required to occur in the same sentence. T-expressions originate in research on automatic question answering (Katz 1997), and were subsequently ported to paralinguistic text analysis, particularly sentiment classification in product reviews (Yi *et al.* 2003).

The 'target' of a T-expression refers to a feature term, for example, a movie in the case of sentiment analysis in movie critics. What is considered a 'feature term' has to be defined by expert rules, or a knowledge source such as ConceptNet can be used. In particular, terms can be filtered by using ConceptNet's 'PartOf' or 'HasProperty' relations. For example, in movie review valence estimation, the predicate 'PartOf(actor, movie)' suggests that if the subject of the sequence is an actor, a T-expression related to the movie domain can be built. Additionally, databases of named entities can be employed; in the example, to filter movie titles and actor names. Furthermore, in this step, personal pronouns (e.g., 'it') can be resolved based on other expressions built from the sequence.

Next, sources need to be identified for each target, that is, words conveying the actual information of interest such as affect, gender, or personality. Sources are identified by syntactic as well as semantic restrictions. Syntactic restrictions ensure that the source is being directed to the target in question, and involve breaking down the sequence into units of statements, and associating a source with the target if and only if both occur in the same section without a *border indication* separating them. These border indications are subordinating conjunctions, prepositional phrases, coordinating conjunctions, and punctuation such as commas or colons. Semantic restrictions relate to finding sources connected to the task at hand. Here, this is exemplified by the concept of valence.

To find the source candidates, first all verbs are retrieved from the section of the target, and General Inquirer is used to determine their value v. A word w_i is assigned a value $v(w_i) = 1$ if it has the General Inquirer tag *Positiv*, and a value $v(w_i) = -1$ if it is tagged as *Negativ*. Should a word not exist in General Inquirer, WordNet synsets are used to look up its synonyms for potential matches. After this process, some of the the words in the sequence are tagged with valence. For all verbs, the *siblings* – the direct neighbours – are first scanned for adverbs. If the adverb has a valence tag, a T-expression of the form ⟨*target, verb, adverb*⟩ is generated. Thus, for example, 'a/DT carefully/RB designed/VB plot/NN', can result in a T-expression such as ⟨*plot, designed, carefully*⟩ – although 'design' does not have a valence value, and 'careful' has positive valence. If no valence-related adverbs are found, but the verb carries a valence value (e.g., 'love'), the verb is considered to be the source. Note that multiple expressions can exist for a given target if there are multiple source words contained in the phrase being analysed. For example, since two adverbs are contained in 'a/DT carefully/RB designed/VB, superbly/RB executed/VB plot/NN' ('carefully' and 'superbly'), and both of these have valence values, this phrase results in two corresponding T-expressions: ⟨*plot, designed, carefully*⟩ and ⟨*plot, executed, superbly*⟩.

It remains to define a fall-back strategy for the case where the sequence considered does not contain a verb. In this case, for example, *binary expressions* of the form ⟨*target, adjective*⟩ can be built. In the movie review example, 'an/DT excellent/JJ setting/NN' could be represented as the binary expression ⟨*setting, excellent*⟩, since no verb exists in this sequence, yet 'excellent' has a positive valence value.

Once expressions have been generated by the above procedure, the strength of the relation of source and target has to be determined. This strength can be measured as the distance of source and target in the sequence. One makes the assumption that a source is mostly directed at the 'closest' target. A simple method is to only keep the expression with the shortest distance between the target word and its target for further processing. Alternatively, a maximum distance can be enforced (Morinaga *et al.* 2002; Turney and Littman 2003); though it has been shown (Zhang and Ye 2008) that this can degrade performance when applied to sentiment analysis. In the following, a more generic approach will be presented that computes a score function s for each expression containing target t_i and source w_i. The design of the score function is inspired by Ding *et al.* (2008). It is based on the multiplicative inverse of the distance of source and target, but with an additional constant factor c and an exponent e for 'fine-tuning':

$$s(w_i, t_i) = c \cdot v(w_i) \cdot \frac{1}{D(w_i, t_i)^e}. \tag{9.5}$$

Here, setting $c > 1$ boosts the score s in the case of small distance of w_i and t_i, while it has little effect for $D(w_i, t_i) \gg 1$. $v(\cdot) \in \{-1, 1\}$ denotes knowledge-based assignment of words to a binary label. In the case of sentiment analysis, it corresponds to the valence tag ('Negativ' $= -1$, 'Positiv' $= +1$) assigned by General Inquirer. That said, the function v could obviously denote any binary label for a computational paralinguistics task (e.g., personality). $D(w_i, t_i) \geq 1$ is the word distance between w_i and t_i. An exponential decay function with exponent e is used to further boost or lower the score based on the distance. For $e > 1$, the score decreases more rapidly for greater $D(w_i, t_i)$; $e < 1$, leads to a slower decrease of the score for larger distances.

The final output of a sequence – the accumulated score S of the N expressions it contains – is

$$S = \sum_{i=1}^{N} s(w_i, t_i). \tag{9.6}$$

A binary class (-1 or 1) can be chosen by taking the sign of S,

$$\text{sgn}(S) = \begin{cases} 1 & u \geq 0 \\ -1 & u < 0, \end{cases} \tag{9.7}$$

or S serves as a feature for classification or regression in combination with data-driven analysis.

In some cases, expression (9.6) cannot be evaluated because no target or source words are found in a sentence. In this case, a fall-back mechanism has to be employed, returning, for example, $S = 0$. Failure to extract target words can be caused by very short sequences, or colloquial language: colloquial terms are sparsely contained in general-purpose dictionaries.

References

Athanaselis, T., Bakamidis, S., Dologlu, I., Cowie, R., Douglas-Cowie, E., and Cox, C. (2005). ASR for emotional speech: Clarifying the issues and enhancing performance. *Neural Networks*, **18**, 437–444.

Batliner, A., Buckow, J., Huber, R., Warnke, V., Nöth, E., and Niemann, H. (1999). Prosodic feature evaluation: Brute force or well designed? In *Proc. of ICPhS*, pp. 2315–2318, San Francisco.

Batliner, A., Steidl, S., Schuller, B., Seppi, D., Laskowski, K., Vogt, T., Devillers, L., Vidrascu, L., Amir, N., Kessous, L., and Aharonson, V. (2006). Combining efforts for improving automatic classification of emotional user states. In *Proc. of IS-LTC 2006*, pp. 240–245, Ljubljana.

Cambria, E., Schuller, B., Xia, Y., and Havasi, C. (2013). New avenues in opinion mining and sentiment analysis. *IEEE Intelligent Systems Magazine, Special Issue on Concept-Level Opinion and Sentiment Analysis*, **28**, 15–21.

Campbell, N., Kashioka, H., and Ohara, R. (2005). No laughing matter. In *Proc. of Interspeech*, pp. 465–468, Lisbon.

Devillers, L. and Vidrascu, L. (2006). Real-life emotions detection with lexical and paralinguistic cues on human-human call center dialogs. In *Proc. of ICSLP*, pp. 801–804, Pittsburgh.

Devillers, L., Vasilescu, I., and Lamel, L. (2003). Emotion detection in task-oriented spoken dialogs. In *Proc. of ICME 2003, Multimedia Human-Machine Interface and Interaction*, pp. 549–552, Baltimore, Maryland.

Ding, X., Liu, B., and Yu, P. S. (2008). A holistic lexicon-based approach to opinion mining. In *Proc. of the International Conference on Web Search and Web Data Mining (WSDM '08)*, pp. 231–240, Palo Alto, California.

Eyben, F., Wöllmer, M., Valstar, M., Gunes, H., Schuller, B., and Pantic, M. (2011). String-based Audiovisual fusion of behavioural events for the assessment of dimensional affect. In *Proc. of the International Workshop on Emotion Synthesis, rePresentation, and Analysis in Continuous spacE, EmoSPACE 2011, held in conjunction with the 9th IEEE International Conference on Automatic Face & Gesture Recognition and Workshops, FG*, pp. 322–329, Santa Barbara, CA.

Fellbaum, C. (1998). *WordNet: An Electronic Lexical Database*. MIT Press, Cambridge, MA.

Havasi, C., Speer, R., and Alonso, J. (2007). ConceptNet 3: A flexible, multilingual semantic network for common sense knowledge. In *Recent Advances in Natural Language Processing*, Borovets, Bulgaria.

Iurgel, U. (2007). *Automatic Media Monitoring Using Stochastic Pattern Recognition Techniques*. Ph.D. thesis, Technische Universität München, Germany.

Jurafsky, D. and Martin, J. H. (2000). *Speech and Language Processing*. Prentice Hall, Upper Saddle River, NJ.

Katz, B. (1997). From sentence processing to information access on the World Wide Web. In *Proc. of the AAAI Spring Symposium on Natural Language Processing for the World Wide Web*, pp. 77–86, Stanford, CA.

Kraaij, W. and Pohlmann, R. (1994). Porter's stemming algorithm for Dutch. In L. Noordman and W. De Vroomen (eds), *Informatiewetenschap 1994: Wetenschappelijke bijdragen aan de derde STINFON Conferentie*, pp. 167–180, Tilburg, The Netherlands.

Lippmann, R. P. (1997). Speech recognition by machines and humans. *Speech Communication*, **22**(1), 1–16.

Matos, S., Birring, S., Pavord, I., and Evans, D. (2006). Detection of cough signals in continuous audio recordings using hidden Markov models. *IEEE Transactions on Biomedical Engineering*, **53**, 1078–1083.

Mesaros, A. and Virtanen, T. (2009). Automatic recognition of lyrics in singing. *EURASIP Journal on Audio, Speech, and Music Processing*, **2009**. 11 pp.

Metze, F., Batliner, A., Eyben, F., Polzehl, T., Schuller, B., and Steidl, S. (2011). Emotion recognition using imperfect speech recognition. In *Proc. of Interspeech*, pp. 478–481, Makuhari, Japan.

Morinaga, S., Yamanishi, K., Tateishi, K., and Fukushima, T. (2002). Mining product reputations on the Web. In *Proc. of the Eighth ACM SIGKDD International Conference on Knowledge Discovery and Data Mining (KDD '02)*, pp. 341–349, New York.

Pal, P., Iyer, A., and Yantorno, R. (2006). Emotion detection from infant facial expressions and cries. In *Proc. of ICASSP*, pp. 809–812, Toulouse.

Polzin, T. S. and Waibel, A. (2000). Emotion-sensitive human-computer interfaces. In *Proc. of the ISCA Workshop on Speech and Emotion*, pp. 201–206, Newcastle, Co. Down.

Porter, M. F. (1980). An algorithm for suffix stripping. *Program*, **14**, 130–137.

Russell, J., Bachorowski, J., and Fernandez-Dols, J. (2003). Facial and vocal expressions of emotion. *Annual Review of Psychology*, **54**, 329–349.

Schuller, B. (2012). Recognizing affect from linguistic information in 3D continuous space. *IEEE Transactions on Affective Computing*, **2**, 192–205.

Schuller, B. and Knaup, T. (2011). Learning and knowledge-based sentiment analysis in movie review key excerpts. In A. Esposito, A. M. Esposito, R. Martone, V. Müller, and G. Scarpetta (eds), *Toward Autonomous, Adaptive, and Context-Aware Multimodal Interfaces: Theoretical and Practical Issues: Third COST 2102 International Training School, Caserta, Italy, March 15-19, 2010, Revised Selected Papers*, volume 6456/2010 of *Lecture Notes on Computer Science (LNCS)*, pp. 448–472. Springer, Berlin.

Schuller, B., Müller, R., Rigoll, G., and Lang, M. (2004a). Applying Bayesian belief networks in approximate string matching for robust keyword-based retrieval. In *Proc. of the 5th IEEE International Conference on Multimedia and Expo, ICME*, pp. 1999–2002, Taipei, Taiwan.

Schuller, B., Rigoll, G., and Lang, M. (2004b). Speech emotion recognition combining acoustic features and linguistic information in a hybrid support vector machine-belief network architecture. In *Proc. of ICASSP*, pp. 577–580, Montreal, Canada.

Schuller, B., Müller, R., Lang, M., and Rigoll, G. (2005). Speaker independent emotion recognition by early fusion of acoustic and linguistic features within ensembles. In *Proc. of Interspeech*, pp. 805–809, Lisbon.

Schuller, B., Köhler, N., Müller, R., and Rigoll, G. (2006). Recognition of interest in human conversational speech. In *Proc. of Interspeech*, pp. 793–796, Pittsburgh, PA.

Schuller, B., Schenk, J., Rigoll, G., and Knaup, T. (2009). "The Godfather" vs. "Chaos": Comparing linguistic analysis based on online knowledge sources and bags-of-N-grams for movie review valence estimation. In *Proc. of the 10th International Conference on Document Analysis and Recognition, ICDAR*, pp. 858–862, Barcelona, Spain.

Schuller, B., Batliner, A., Steidl, S., and Seppi, D. (2011). Recognising realistic emotions and affect in speech: State of the art and lessons learnt from the first challenge. *Speech Communication*, **53**, 1062–1087.

Schuller, B., Steidl, S., Batliner, A., Schiel, F., Krajewski, J., Weninger, F., and Eyben, F. (2013a). Medium-term speaker states – a review on intoxication, sleepiness and the first challenge. *Computer Speech and Language, Special Issue on Broadening the View on Speaker Analysis*.

Schuller, B., Steidl, S., Batliner, A., Burkhardt, F., Devillers, L., Müller, C., and Narayanan, S. (2013b). Paralinguistics in speech and language – state-of-the-art and the challenge. *Computer Speech and Language*, **27**, 4–39.

Seppi, D., Gerosa, M., Schuller, B., Batliner, A., and Steidl, S. (2008). Detecting problems in spoken child-computer interaction. In *Proc. of the 1st Workshop on Child, Computer and Interaction, WOCCI, ICMI 2008 post-conference workshop)*, Chania, Greece.

Steidl, S., Batliner, A., Seppi, D., and Schuller, B. (2010). On the impact of children's emotional speech on acoustic and language models. *EURASIP Journal on Audio, Speech, and Music Processing, Special Issue on Atypical Speech*, **2010**.

Stone, P., Kirsh, J., and Cambridge Computer Associates (1966). *The General Inquirer: A Computer Approach to Content Analysis*. MIT Press, Cambridge, MA.

Truong, K. and Leeuwen, D. v. (2005). Automatic detection of laughter. In *Proc. of Interspeech*, pp. 485–488, Lisbon.

Turney, P. D. and Littman, M. L. (2003). Measuring praise and criticism: Inference of semantic orientation from association. *ACM Transactions on Information Systems*, **21**, 315–346.

Wöllmer, M., Eyben, F., Keshet, J., Graves, A., Schuller, B., and Rigoll, G. (2009). Robust discriminative keyword spotting for emotionally colored spontaneous speech using bidirectional LSTM networks. In *Proc. of ICASSP*, pp. 3949–3952, Taipei, Taiwan.

Yi, J., Nasukawa, T., Bunescu, R., and Niblack, W. (2003). Sentiment Analyzer: Extracting sentiments about a given topic using natural language processing techniques. In *Proc. of the IEEE International Conference on Data Mining (ICDM)*, pp. 427–434, Melbourne, FL.

Zhang, M. and Ye, X. (2008). A generation model to unify topic relevance and lexicon-based sentiment for opinion retrieval. In *Proc. of the 31st Annual International ACM SIGIR Conference on Research and Development in Information Retrieval (SIGIR '08)*, pp. 411–418, New York.

10

Supra-segmental Features

The whole is more than the sum of its parts.

(Aristotle)

You will win because you have enough brute force. But you will not convince.

(Miguel de Unamuno)

When we are interested in information which is 'hidden' in the way low-level features evolve over a given timespan, then we need to extract 'supra-segmental' features from the low-level frame-wise features. In fact, this is usually the way to represent paralinguistic feature information, as it provides a greater reduction of information that otherwise may depend too strongly on the phonetic content (Schuller *et al.* 2009). Such information could be affect that is expressed by a speaker within an utterance, mood expressed over a series of utterances, or speaker states such as alcohol intoxication or sleepiness, and traits such as age, gender, and speaker identification. The general principle of supra-segmental features is to obtain a single, fixed length feature vector which describes a sequence of low-level descriptors (LLDs) of possibly variable length (Ververidis and Kotropoulos 2006).

Common methods to obtain a supra-segmental feature vector are: mapping of the LLDs to a single vector by applying functionals to the LLD time series (Section 10.1), stacking of low-level feature frames optionally followed by a dimensionality reduction, for example, by principal component analysis (Section 10.3).

When we compute a supra-segmental feature vector, the first step is to define the segments or units of analysis. As discussed in Section 8.3, feasible units of segmentation for speech are either continuous segments of speech activity which are separated by silences or non-speech segments, or semantic segments such as words, phrases, or sentences which can be determined from ASR transcriptions. These segments will vary in length from one segment to the next. This fact is a considerable problem which we encounter and have to consider when we compute supra-segmental feature vectors. Not all methods can handle segments of variable length (e.g., stacking of feature frames).

An alternative for segmentation is to use segments of fixed length. This can be either a subdivision of the variable-length segments into fixed-length sub-segments, or a systematic 'blind' segmentation of the complete input into segments with a typical fixed length of a

Computational Paralinguistics: Emotion, Affect and Personality in Speech and Language Processing, First Edition.
Björn W. Schuller and Anton M. Batliner. © 2014 John Wiley & Sons, Ltd. Published 2014 by John Wiley & Sons, Ltd.

few seconds. As another dynamic length segmentation approach, segmentation based on the Bayesian information criterion (BIC) can be performed as suggested by Cettolo and Vescovi (2003). If the BIC measured between two adjacent sliding windows differs over a given threshold, a segment boundary between the two windows is assumed.

10.1 Functionals

Functionals are relations which map a (time) series of input values of arbitrary length to a single output value. Statistical descriptors such as mean and variance are examples of simple functionals. More advanced functionals are, for example, regression coefficients or the number of maxima found in the input series. They can be applied to all frame-level numeric feature vectors, both acoustic and linguistic. Moreover, functionals can be applied hierarchically, that is, another set of functionals can be applied to a series of supra-segmental feature vectors. We discuss this in Section 10.2.

To demonstrate the principle of mapping a time series of variable length to a feature vector of fixed length, we give the following example. Let us assume we apply $M = 4$ functionals (mean, variance, maximum and minimum value) to $N = 2$ LLD series (pitch, energy) both of length $L = 10$ frames. As we apply each functional to each LLD, we get $M \cdot N = 8$ supra-segmental features in a single vector which summarises all L input values.

The most common functionals used in computational paralinguistics are summarised by group in Table 10.1. The functional groups include:

Means Various types of mean values.

Moments Statistical nth-order moments.

Extremes Extreme values and range of input signal.

Percentiles Percentile values and percentile ranges. The nth percentile is the value below which $n\%$ of all values of the input are found. Standard percentiles are the first, second, and third quartiles, corresponding to the 25th, 50th (median), and 75th percentiles. A percentile range is the difference between two percentiles (larger percentile minus smaller percentile).

Regression Linear and quadratic (and higher) regression coefficients estimated by fitting a line or parabola (or equivalent function) to the input series by minimising the quadratic error between the line/parabola and the input series. Corresponding linear and quadratic regression errors are computed as absolute and quadratic differences between the input values and the estimated line and parabola.

Peaks Statistics describing the local extrema in the series and their distribution.

Segments Statistics describing the distribution of segments in the series. Segments can be identified in a number of ways ranging from high-level methods such as speech versus non-speech segments, to low-level methods such as delta-thresholding, for example, when the change in the input signal over N frames is higher than a given threshold α, a new segment boundary is detected.

Samples Copies of input series values at given time steps, usually relative to the full segment length, that is, at $n\%$ of the input.

Table 10.1   Common functionals in computational paralinguistics, ordered by type

Means

Arithmetic, quadratic, geometric mean
Mean of absolute values
Mean of all non-zero values

Statistical moments

Variance and standard deviation
Kurtosis and skewness
Centroid
Zero crossing rate and mean crossing rate

Extreme values

Maximum, minimum, range

Percentiles

Quartiles and inter-quartile ranges
Percentiles and various inter-percentile ranges

Peak statistics

Number of maxima/minima
Mean and standard deviation of maxima/minima
Mean distance and standard deviation of distance between maxima/minima
Peak mean to arithmetic mean ratio

Segment statistics

Number of segments
Mininum, mean, maximum segment length
Standard deviation of segment length

Samples

Copies of input values at fixed (relative) time steps, e.g., middle, end

Modulation Functionals which describe signal modulations: discrete cosine transformation (DCT) coefficients, modulation spectra, and linear predictive coding (LPC) coefficients, for example.

A standardised coding for feature names (including functionals) has been proposed by the CEICES initiative. The full coding scheme is printed in the Appendix (Section A.2).

10.2   Feature Brute-Forcing

The principle of applying functionals to time series can be expanded to an (infinite) multi-layer hierarchy. Functionals of LLDs can be computed over short (around 1 second) segments – resulting in first-level supra-segmental feature vectors. Then, functionals are again applied to a set of these supra-segmental feature vectors (e.g., over one sentence or speech segment).

This process can be repeated for even higher levels such as paragraphs or recording sessions. The set of functionals can be different for each level, and usually it is reduced more and more for higher levels. A large set of functionals is usually used at the first level. At higher levels, the mean and moments are usually the most promising functionals.

Now, by applying all available and commonly used functionals to all available LLDs (optionally in a hierarchical way) we can *brute-force* very large feature sets (Schuller *et al.* 2008). Not all combinations of low-level features and functionals make sense in general and some might not be relevant for all tasks; however, one can let machine learning methods deal with selection of relevant features for every task.

This method stands in contrast to designing a small set of features which are based on expert knowledge. For every paralinguistic task, expert studies must be carried out to find relevant acoustic and linguistic features which can then be implemented. The advantage of this method is that if the experts find 'good' features, and the features can be robustly extracted automatically without manual correction, such methods are computationally efficient due to the rather small number of features. However, experts have not identified perfect features for paralinguistics tasks, and the ones they have defined are not robustly extractable. As a small, blunt example we would mention gender recognition: given the fact that the pitch of female voices is twice as high as that of male voices, we can formulate a simple expert rule for speaker gender detection based on pitch as a single feature. However, pitch detection algorithms might return double or half of the actual pitch value (e.g., if the actual fundamental frequency is missing the algorithm might pick the first harmonic; see Section 8.4.9). In this case the gender detection result is likely to be incorrect.

The 'brute-force' approach, on the other hand, generates a large number of features which are not designed specifically for the task at hand. However, most of the features are robust to extract, and machine learning algorithms will select feature sets which are highly correlated to the task of interest. If a feature is unsuitable due to extraction errors (like pitch in the above example), it might not be well correlated with the class and discarded by the feature selection algorithm (see Section 11.1).

Figure 10.1 gives a full overview of the principle of hierarchical feature extraction.

10.3 Feature Stacking

The advantage of using functionals for supra-segmental modelling is that they can map variable-length input sequences to a fixed-size vector. However, information is lost in this process, depending on the type(s) of functionals used. For example, the arithmetic mean discards information about the variations over time, while the standard deviation describes these variations but discards information about the average value. This reduction of information is often desirable, especially when the segments are very long.

For some tasks, such as those dealing with identifying short speech events (e.g., non-linguistic vocalisations) in continuous streams, an alternative supra-segmental modelling approach might be better suited: feature frame stacking. The concept is very simple: all L frames (N low-level features) of an L-frame sequence are concatenated into a supervector. This results in an $N \cdot L$-dimensional feature vector for each segment.

A precondition for this method is that all segments must have the same length, so that the dimensionality of the supra-segmental feature vector remains constant. Also, the method is unsuitable for segments longer than approximately $L = 100$ (typically 1 second), because the resulting feature vector will be too large.

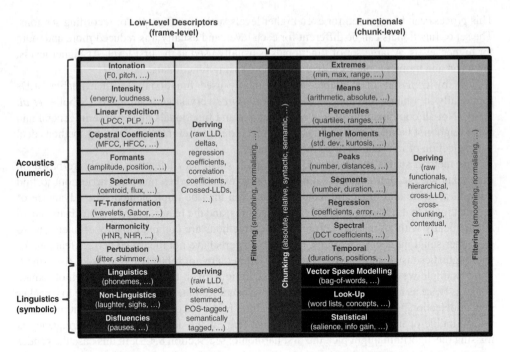

Figure 10.1 Systematic hierarchical speech feature brute-forcing. A horizontal division is shown by LLDs and functionals – the functionals are not calculated if frame-level processing is carried out. A vertical division into acoustic or signal-type features and linguistic or symbolic features is also shown

If feature stacking is to be applied to time series of variable length or long time series, a heuristic frame selection method could be applied, where only every nth frame is selected, or frames at given relative positions (cf. the 'samples' functional).

Moreover, in order to reduce the dimensionality of the supervectors (in the case of long segments), data reduction methods such as principal component analysis or linear discriminant analysis can be applied. As a side effect the resulting features in the transformed (and optionally reduced) feature space are also decorrelated, which can be beneficial for some machine learning algorithms which assume independence among the features. The features in the original space are highly correlated because values of the same feature at consecutive time instants are usually highly correlated for speech signals (see linear predictive coding, Section 8.4.5).

References

Cettolo, M. and Vescovi, M. (2003). Efficient audio segmentation algorithms based on the BIC. In *Proc. of ICASSP*, pp. 537–540, Hong Kong.

Schuller, B., Wimmer, M., Mösenlechner, L., Kern, C., Arsić, D., and Rigoll, G. (2008). Brute-forcing hierarchical functionals for paralinguistics: A waste of feature space? In *Proc. of ICASSP*, pp. 4501–4504, Las Vegas, NV.

Schuller, B., Steidl, S., and Batliner, A. (2009). The Interspeech 2009 Emotion Challenge. In *Proc. of Interspeech*, pp. 312–315, Brighton, UK.

Ververidis, D. and Kotropoulos, C. (2006). Emotional speech recognition: Resources, features, and methods. *Speech Communication*, **48**(9), 1162–1181.

11

Machine-Based Modelling

Much learning does not teach understanding.

(Heraclitus)

In this chapter, we deal with the actual machine-based modelling once a feature representation has been found. We start with the feature relevance analysis, leading on to the actual machine learning. We confine ourselves to the approaches encountered most often in the field of computational paralinguistics today (Schuller *et al.* 2011, 2013a,b). Beyond the machine learning algorithms presented and the variations thereof, there are almost infinitely many others. However, those chosen also present a good balance of basic methods, including static modelling of single feature vectors such as after extraction of supra-segmental features, as well as dynamic approaches that are suitable for modelling on a frame-by-frame level. The first type – static modelling – includes decision trees, support vector machines, and neural networks, together with the recently highly successful approach of memory modelling. As for the second type – dynamic modelling – hidden Markov models have been selected as a representative exemplary approach. This choice is motivated by the fact that they are the quasi-standard in many speech processing tasks. We include classification and regression to allow for discrete class-based and continuous modelling.

Finally, we deal with testing protocols, discussing partitioning and balancing of data, performance measures, and also result interpretation. The latter aspect is related to the 'magical numbers' discussion that was given above in Section 3.10, this time, however, going into detail for the calculation of such measures.

11.1 Feature Relevance Analysis

The feature extraction methods described so far generate large feature spaces. This is true for both the 'brute-forcing' of acoustic features by more or less exhaustive application of statistical functionals to contours of LLDs, and the vector space modelling of linguistic content. Large feature spaces are problematic both in classifier training and evaluation, as an increase in the number of features is usually accompanied by both a growth in the complexity of optimisation algorithms for training, and a growth in the size of the resulting model. Furthermore, increasing

Computational Paralinguistics: Emotion, Affect and Personality in Speech and Language Processing, First Edition.
Björn W. Schuller and Anton M. Batliner. © 2014 John Wiley & Sons, Ltd. Published 2014 by John Wiley & Sons, Ltd.

the number of model parameters comes with the danger of 'over-fitting': more features usually help to better fit a model to the characteristics of the training data, but this is prone to low generalisation to unseen test data. We have already given a hint on feature selection techniques tailored to linguistic features, such as stopping and stemming. This chapter now provides a unified view on this issue.

The most straightforward approach to feature selection is to evaluate each feature's merits by computing statistical and information-theoretic measures such as the correlation coefficient (CC) of the feature with (continuous-valued or ordinal) class labels (see equation (11.99)), or the information gain ratio (IGR) with respect to nominal class labels (see equation (11.6)). These can be used to obtain a ranking of features, keeping only a specified number of 'top' features, or those with CC or IGR above a minimum value.

This procedure, however, has two main disadvantages. First, especially in the context of feature brute-forcing, such selection methods are usually not sufficient, since for each information bearing feature there might be multiple features of similar nature which would then be selected as well, resulting in a highly correlated feature space after selection. To combine feature selection with feature space decorrelation, correlation-based feature selection (CFS) can be employed (Witten and Frank 2005). This is based on defining the following filter function M for the feature subset S with $k = |S|$ features:

$$M(S) = \frac{k \cdot CC_{cf}}{\sqrt{k + k(k - 1)CC_{ff}}},$$ (11.1)

In the above, CC_{cf} denotes the mean CC of features in S with the class label, and CC_{ff} is the average CC of features in S with each other. The underlying idea is that 'good' subsets of features have high predictive power regarding the class label, yielding a high value in the numerator of equation (11.1), and a low degree of redundancy among the features, yielding a small value in the denominator.

However, CFS does not solve the second disadvantage of statistical feature selection, which is that these statistics might not be related to the requirements of the classifier; for instance, support vector machines (SVMs) are robust to zero-information features, as will be explained below (cf. Section 11.2.1). Thus, information-theoretic measures or simple statistics of feature subsets are not always indicative of the expected classifier performance using these subsets. These considerations lead to the introduction of 'wrapper-based' evaluation, which directly uses the performance of the intended classifier as an evaluation measure (Schuller 2006; Witten and Frank 2005). In particular, this means that for every feature subset considered, a classifier has to be trained and evaluated, which can easily become computationally expensive. While this can be partly alleviated by tuning classifier (hyper)parameters, such as the SVM complexity constant (see Section 11.2.1), for fast training and evaluation, this comes at the risk of introducing a bias, as it is not clear how the feature subsets considered would behave in the final system using classifier parameters tuned for optimal system performance.

To provide a truly optimal feature set, the above criteria (wrapper-based evaluation, CFS, etc.) would have to be evaluated on all possible feature combinations. It is easy to see that there are $\binom{N}{|S|}$ possible subsets of size $|S|$ of a feature space of size N; thus, exhaustive evaluation of all possible feature combinations has exponential complexity in N, rendering it effectively infeasible for brute-forced feature spaces of several hundred or thousand features.

Thus, efficient search algorithms become mandatory. Probably the most straightforward algorithm is 'greedy hill climbing', This procedure starts from an initial ranking of the features (e.g., by CC), and iteratively selects the best of the remaining features, either by the rank of the feature itself, or by evaluation of the subset resulting from adding the feature to the current set (e.g., by CFS). This procedure, however, is prone to ending up in local minima ('nesting effect'). Thus it is usually combined with a 'backtracking' option that allows for removing features and replacing them by other candidates further down the list of relevant features, which might be more helpful in combination. Combining greedy hill climbing with backtracking yields the so-called 'floating search', and if greedy selection and backtracking are applied iteratively, the resulting algorithm is known as sequential floating forward search (SFFS).

In the case of large brute-forced feature sets, SFFS usually yields rather small subsets of the original feature set. If one merely wants to remove unnecessary features, one can 'invert' the above procedure ('backward' instead of 'forward' search) – one thus starts with the full feature set and iteratively removes features, for example, those with low IGR. However, if the initial set is large, this variant is usually much more expensive than forward search.

A number of further measures and search functions exist, and one can also add further combinations or alterations of features throughout search, usually by random injection or genetic algorithms to limit the search space (Pachet and Roy 2009; Schuller *et al.* 2005, 2006a,b).

All the above methods are purely data-based. Often, they yield a feature set that is both sub-optimal (due to the approximative nature of search algorithms) and hard to interpret, as it is usually a mixture of features from different groups. For example, a selected feature might have a counterpart of similar nature that is easier to interpret, which was not selected because it has slightly lower correlation with the target variable. Especially CFS and similar methods often yield 'grab bag' types of feature sets where the value of individual features is unclear, because they are more focused on decorrelation rather than relevance analysis. At this point, expert knowledge can be used to replace 'blind' statistical selection of feature subsets by semantic criteria, such as evaluating only feature subsets corresponding to functionals of a single low-level descriptor (LLD), or only a single functional type of several LLDs. While such feature selection is often inferior to a combination of SFFS and wrapper-based selection in terms of resulting system performance, it has the potential to deliver deeper insights into the nature of the features. For example, it could turn out that the median of the LLDs is better suited than the arithmetic mean.

In contrast to the feature selection methods described above that are closely related to feature relevance, feature reduction methods aim to compress the information contained in the full feature space. An example of a feature reduction method is principal component analysis (PCA) (see Kroschel *et al.* 2011), which computes an orthogonal transformation matrix such that the dimensions of the resulting feature space are uncorrelated and ordered by decreasing variance – then, usually only the first few dimensions are kept. This basically assumes that any kind of variance in the features has to be preserved and is thus susceptible to noisy or zero-information features. Linear discriminant analysis (LDA) (again see Kroschel *et al.* 2011) can be viewed as an extension of PCA that also considers class labels in computation of the transformation matrix, in order to spread class centres in the feature space. Both PCA and

LDA, however, can easily lead to feature representations that are very hard to interpret, since every dimension of the resulting feature space is a linear combination of potentially large numbers of input features with positive and negative coefficients.

11.2 Machine Learning

Based on the extraction of features as outlined in the previous chapter, a system has to be designed to map these features to the target variable of interest, such as the age or emotion of a speaker. One can generally distinguish between rule-based schemes and machine learning techniques. The former are based on expert knowledge as touched upon in Chapter 3; the latter – and these make up the vast majority of today's approaches – assume only the coarse structure of the classification or regression scheme to be given, for example, determining the target class using a sequence of threshold decisions (decision trees, see below). The exact parameters of the scheme are determined by statistical optimisation – usually, minimising the error that would occur when classifying a set of labelled training data.

One can very roughly subdivide machine learning based schemes into static and dynamic classification. The choice between these two is closely connected to the choice of frame-wise or supra-segmental features: static classifiers (or regressors) map a feature vector of fixed size to a target label, while dynamic classifiers handle feature sequences of varying length, allowing for time warping.

11.2.1 Static Classification

'Static classification' refers to the process of assigning a discrete class label to an unknown feature vector of fixed dimensionality. As examples of static classifiers, two classifiers which are frequently used in computational paralinguistics, are briefly described in the following: decision Trees and support vector machines.

Decision Trees

Let us begin with decision trees (DTs: Kroschel *et al.* 2011; Quinlan 1993). The structure of these is very similar to a rule-based classification scheme; however, instead of using expert knowledge, the sequence of rules to apply is determined automatically from training data. Thus, an advantage of DTs is that they produce human-readable sets of rules, making classification transparent and intuitive to understand (assuming that one starts from interpretable features, see above).

The type of DTs considered in this book perform comparisons with constants in order to decide to which next comparison to branch, until a class label decision can be reached. Mathematically, trees are typically considered as a sub-class of graphs. Starting from a generic undirected graph, the conditions for a tree are that the graph is acyclic and connected, that is, each node needs to be reachable by a path from each other node. The classification process can be modelled as a path in this graph. As a graph, a DT is defined by a set of nodes V and a set $E \subseteq V \times V$ of edges e, such that $e = (v_1, v_2) \in E$ represents a connection from node $v_1 \in V$ to node $v_2 \in V$. As a tree, the condition $|E| = |V| - 1$ holds.

A path of length P through the tree is a sequence $v_1, \ldots, v_P$, $v_k \in V$ with $(v_k, v_{k+1}) \in E$, $k = 1, \ldots, P - 1$. Each path starts at the 'root' node r, which is the one and only node in the

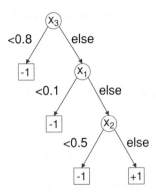

Figure 11.1   An example of a decision tree: a binary classification problem is solved in a feature space of dimension 3. Circles are root and inner nodes, rectangles are leaves (with class labels)

graph that does not have incoming edges, that is, E contains no element of the form (v, r), $v \in V$. Given a feature space of dimension N, a partial function

$$a : V \to \{1, \ldots, N\}$$

is defined, mapping all nodes that have outgoing edges (inner nodes) to features. This function is usually injective, that is, each feature only occurs at most once in the tree.

The edges traversed in a path correspond to branch decisions based on the values of these features. Each edge is assigned a feature interval. To determine the class label of a pattern vector $\underline{x} = (x_1, \ldots, x_N)^T$, one follows a path through the tree that satisfies the following criterion: starting at the root node, the nodes v in the path are interconnected by the edge for which $x_{a(v)}$ is within this edge's interval. The number of outgoing edges of a node depends on quantisation of the features into J_n intervals per feature n, resulting in a finite number of outgoing edges. In particular, the root node and each inner node v have $J_{a(v)}$ outgoing edges. The quantisation intervals are determined in the learning process (see below). An important special case is the binary decision tree: here, each inner node has two outgoing edges with intervals of the form $]-\infty, \xi]$ and $]\xi, +\infty[$ corresponding to a threshold decision at each node. The intervals are determined using simple 'binning' as in histogram calculation (Witten and Frank 2005); an alternative is to use sigmoid functions for the decisions (Landwehr *et al.* 2005).

Furthermore, each path ends at a 'leaf'. The 'leaves' are the nodes b without an outgoing edge, that is, for which in E there exists no (b, v) with $v \in V$. Based on a mapping from leaves to class labels, the class of the leaf in which the path ends is determined. An example of a binary DT is shown in Figure 11.1.

It remains to outline an optimisation criterion for determining the parameters of the tree. Generally, one aims to maximise the information gain for each node in the traversed path, with respect to correct classification using the 'remaining' features at each node (features that have not been used in the previous decisions). To this end, the Shannon entropy $H(Y)$ of the distribution of the class probabilities $(Y_1, \ldots, Y_M)$ can be employed:

$$H(Y_1, \ldots, Y_M) = -\sum_{i=1}^{M} Y_i \log_2(Y_i). \tag{11.2}$$

One now considers the amount of information needed to assign an instance to a class $i \in \{1, \ldots, M\}$. This information is determined on the training set $\mathcal{L}$ of pattern vectors $\underline{x}$ with known class attribution y according to

$$H(\mathcal{L}) = - \sum_{i=1}^{M} \hat{Y}_i \log_2(\hat{Y}_i), \quad \hat{Y}_i = \frac{|\mathcal{L}_i|}{|\mathcal{L}|}, \tag{11.3}$$

where $\mathcal{L}_i$ is the set of elements in $\mathcal{L}$ with class attribution i.

To determine an optimal tree structure, one uses an 'information gain' (IG) approach. One considers the value of each individual feature n with respect to the class assignment. This is measured as the remaining information needed for classification before and after observing the value of feature n. For each n the set $\mathcal{L}$ is divided into the subsets $\mathcal{L}_{n,j}$, $j = 1, \ldots, J_n$, on basis of the different values of n, such that $\mathcal{L}_{n,j}$ only contains those vectors where feature n is in the interval j. The remaining average information $H(\mathcal{L}|n)$ needed after observation of the feature n for the class assignment results as the weighted average of the information $H(\mathcal{L}_{n,j})$, as required to classify an element of the subset $\mathcal{L}_{n,j}$:

$$H(\mathcal{L}|n) = \sum_{j=1}^{J_n} \frac{|\mathcal{L}_{n,j}|}{|\mathcal{L}|} H(\mathcal{L}_{n,j}). \tag{11.4}$$

Based on this, the IG can be defined formally as the reduction in the entropy, that is, the information needed for the assignment, by addition of the feature n:

$$IG(\mathcal{L}, n) = H(\mathcal{L}) - H(\mathcal{L}|n). \tag{11.5}$$

In particular, if all elements in $\mathcal{L}$, whose features n have the same value, belong to the same class – this is the case, in particular, if a feature has a different value for each element in $\mathcal{L}$ – then $H(\mathcal{L}|n) = 0$, and one obtains a maximal $IG(\mathcal{L}, n)$. Thus, the above definition can be problematic since it tends to favour features with a large number of different values J_n, which can lead to over-fitting. One can compensate for this by introducing the information gain ratio (IGR)

$$IGR(\mathcal{L}, n) = \frac{IG(\mathcal{L}, n)}{H\left(\dfrac{|\mathcal{L}_{n,1}|}{|\mathcal{L}|}, \ldots, \dfrac{|\mathcal{L}_{n,J_n}|}{|\mathcal{L}|} \right)}. \tag{11.6}$$

The term in the denominator is called the *split information* and is computed according to equation (11.2). This is the information one obtains by splitting the set $\mathcal{L}$ according to the values of the feature n.

It remains to outline an effective algorithm to determine the optimal sequence of features in the DT. Similarly to the feature selection issue outlined above, an exhaustive enumeration of possible trees is generally not feasible due to the exponential complexity in N. Thus, one mostly relies on greedy algorithms, where at every step a feature is selected by a local optimisation criterion. A global optimum is thus not guaranteed. A popular instance of such a greedy algorithm for the training of DTs is the iterative dichotomiser 3 (ID3) algorithm (Quinlan 1983). ID3 is based on dynamic programming, determining sub-trees for subsets of features and concatenating them to yield a DT for the overall feature set. For a given set of features $\mathcal{M} \subseteq \{1, \ldots, N\}$ and training set $\mathcal{L}$, the algorithm proceeds as follows:

1. Termination 1: If all elements in $\mathcal{L}$ belong to class i, return a leaf labelled i.
2. Termination 2: If $\mathcal{M}$ is empty, return a leaf labelled by the most frequent class in $\mathcal{L}$.
3. Else proceed recursively: Search for the feature n' with the highest IG(R), that is,

$$n' = \arg \max_{n \in \mathcal{M}} \text{IG}(\mathcal{L}, n).$$

For all $j = 1, \ldots, J_{n'}$ construct a DT by ID3 on the feature set $\mathcal{M} - \{n'\}$ and the training set $\mathcal{L}_{n',j}$. Return a tree with the root labelled by the feature n' whose edges lead to the constructed DTs.

From the above, it is easy to see that ID3 always terminates, as in every recursive call the remaining set of features decreases, eventually leading to one of the 'termination' branches of the algorithm being executed. The recursive step is further visualised in Figure 11.2.

Extensions of ID3 are the C4.5 and J48 variants that introduce pruning of sub-trees (Quinlan 1987, 1993) for increased efficiency. During pruning, a whole sub-tree can be replaced by a leaf if the error probability is not significantly increased by this substitution. Note that this reduces the number of features, that is, an inherent feature selection by IG(R) is given.

DTs usually are not used with large feature sets as introduced above. This is because they are prone to over-fitting: assuming that the number of training vectors is small in comparison to the number of features, it is easy to perfectly model the training set by a DT. Typically, large feature sets are thus randomly sub-sampled, and for each subset a DT is built. In classification, the decisions of all subspace DTs are fused, for example, by majority voting, yielding the so-called random subspace method for building decision forests (Ho 1998). Random sub-sampling can also be used for selecting training instances, resulting in random forests (Quinlan 1996). Decision forests and random forests are known as competitive classifiers for paralinguistic tasks, delivering performance similar to the classifiers introduced below.

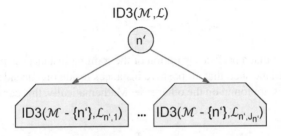

Figure 11.2   ID3 algorithm: a recursive call for the feature n'. The feature maximises the IG(R) for the classification of $\mathcal{L}$ within $\mathcal{M}$ (Kroschel *et al.* 2011)

Support Vector Machines

Probably the most frequently used classifiers in computational paralinguistics are *support vector machines* (SVMs). This popularity can be attributed to a few convenient properties – mainly their ability to handle large feature spaces (as those generated by feature brute-forcing), sparse features (features that are 'almost always' zero, such as frequently encountered in vector space modelling of linguistic features – see Chapter 9), and their robustness to over-fitting. These properties were pointed out in (Joachims 1998) in the context of classification by linguistic features, and they have since also been exploited for acoustic classification. Furthermore, SVMs can be easily extended to continuous class labels by considering support vector regression as introduced in Cortes and Vapnik (1995) – details are given in Section 11.2.3.

The core idea of SVMs is built around binary linear classifiers, optimised to provide the best possible separation between classes in the feature space – in contrast to other classifiers with linear decision boundaries, such as nearest neighbour classifiers. This optimisation criterion leads to classification based on so-called 'support vectors' lying in between the centres of gravity of the classes and defining the decision boundary. Support vectors are chosen by solving a quadratic optimisation problem, for which efficient algorithms are available. As a result, classification is based on a small subset of the learning instances, reducing the danger of over-fitting. Finally, to solve non-linear decision tasks, the 'kernel trick' is applied to map into a higher-dimensional decision space but retaining the low complexity of the support vector principle.

In general, SVMs are thus capable of discriminating between two classes, that is, solving binary decision problems. We first focus on this task. Multi-class SVMs can be designed based on a combination of binary SVMs, through diverse strategies, for example, by training SVMs for each pair of classes and summing the 'votes' for each class during recognition, or by forming a binary decision tree (cf. above) with threshold decisions replaced by SVM classification.

Starting from this broad picture, let us now flesh out the underlying mathematical principles. SVMs are trained based on a set of learning instances $\mathcal{L}$, $L = |\mathcal{L}|$, where each of the instances $\underline{x}_l \in \mathcal{L}$, $l = 1, \ldots, L$, is given a class label $y_l \in \{-1, +1\}$. The labels -1 and $+1$ are used in order to simplify the mathematics. We will refer to the patterns $\underline{x}_l$ with $y_l = +1$ and $y_l = -1$ as 'positive' and 'negative' instances, respectively. Thus we can write $\mathcal{L}$ as

$$\mathcal{L} = \{(\underline{x}_l, y_l) \mid l = 1, \ldots, L\}, \quad \text{where } y_l \in \{+1, -1\}. \tag{11.7}$$

Like any linear classifier, SVMs are defined by a hyperplane $H(\underline{w}, b)$ consisting of the normal vector $\underline{w}$ and the scalar bias b,

$$H(\underline{w}, b) = \{\underline{x} \mid \underline{w}^T \underline{x} + b = 0\}. \tag{11.8}$$

Let us assume first that a perfect separation of the training instances is possible, that is, that there exists a hyperplane such that all positive instances reside on one side of the plane while all negative instances are found on the other side. Mathematically, this can be expressed as the side conditions

$$y_l = +1 \Rightarrow \underline{w}^T \underline{x}_l + b > \varepsilon,$$
$$y_l = -1 \Rightarrow \underline{w}^T \underline{x}_l + b < -\varepsilon \tag{11.9}$$

which have to be fulfilled for all training instances. Assuming that such a hyperplane exists, a normalisation of the side conditions can be realised by appropriate scaling of $\underline{w}$ and b (Cristianini and Shawe-Taylor 2000) such that

$$
y_l = +1 \Rightarrow \underline{w}^T \underline{x}_l + b \geq +1,
$$
$$
y_l = -1 \Rightarrow \underline{w}^T \underline{x}_l + b \leq -1 \tag{11.10}
$$

holds for all training instances $\underline{x}_l, l = 1, \ldots, L$. Thus, we can define the margin of separation $\mu_{\mathcal{L}}$ as the minimum of the magnitude of the distances of all points $\underline{x}_1 \ldots \underline{x}_l$ in $\mathcal{L}$ to H:

$$
\mu_{\mathcal{L}}(\underline{w}, b) = \min_{l=1,\ldots,L} |D(\underline{x}_l)| \tag{11.11}
$$

where $D(\underline{x})$ is the signed distance of a point $\underline{x}$ to the hyperplane H,

$$
D(\underline{x}) = \frac{\underline{w}^T \underline{x} + b}{||\underline{w}||}. \tag{11.12}
$$

SVMs are based on the principle of 'maximum margin', that is, maximum discrimination between the two classes is reached by maximising the margin of separation. Mathematically, we look for the hyperplane $H^* = H(\underline{w}^*, b^*)$ with maximal value $\mu_{\mathcal{L}}^*(\underline{w}^*, b^*)$ to separate the training set $\mathcal{L}$. The instances $\underline{x}_l^{sv} \in \mathcal{L}$ which are closest to the hyperplane H^* are called *support vectors* of H^* with respect to $\mathcal{L}$. From (11.10) and (11.12) it follows that their distance $D^*(\underline{x}_l^{sv})$ from the hyperplane H^* is

$$
D^*(\underline{x}_l^{sv}) = \frac{\pm 1}{||\underline{w}||}. \tag{11.13}
$$

As a consequence, positive and negative instances are spread apart by at least a 'corridor width' of $2/||\underline{w}||$. The borders of this corridor are made up of the support vectors. This is illustrated in Figure 11.3.

Instead of maximising the width of the corridor, $2/||\underline{w}||$, one can minimise the expression $\frac{1}{2}\underline{w}^T \underline{w}$. This results in a convex minimisation problem, which has a unique solution $\underline{w}^*$. The minimisation has linear side conditions resulting from the separation condition (11.10):

$$
y_l(\underline{w}^T \underline{x}_l + b) - 1 \geq 0, \quad \text{with } l = 1, \ldots, L. \tag{11.14}
$$

This is a classical boundary value problem which can be solved using Lagrange multipliers, as explained by Cortes and Vapnik (1995) in detail.

The derivation so far has been based on the assumption that training instances can be separated flawlessly by a hyperplane in the feature space. Generally, however, this is not the case. We can model this in the current approach by allowing vectors to be placed on the 'wrong side', by introducing so-called 'slack' variables $\xi_l \geq 0, l = 1, \ldots, L$, giving the deviation from the 'ideal' boundary condition:

$$
y_l = +1 \Rightarrow \underline{w}^T \underline{x}_l + b \geq +1 - \xi_l,
$$
$$
y_l = -1 \Rightarrow \underline{w}^T \underline{x}_l + b \leq -1 + \xi_l. \tag{11.15}
$$

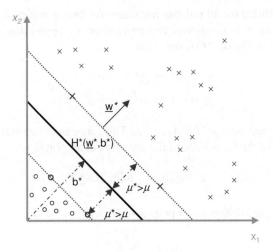

Figure 11.3 Example (optimal) hyperplane $H^*(\underline{w}, b)$ (thick solid line) in a feature space of dimension 2. The maximum margin of separation μ^* is shown as the two dotted parallel lines. "x" and "o" symbols represent instances of the two classes to be separated

Obviously, one wants as little deviation from the ideal boundary conditions as possible – thus, the expression

$$\frac{1}{2}\underline{w}^T\underline{w} + C \cdot \sum_{l=1}^{L} \xi_l \tag{11.16}$$

needs to be minimised, where C is an error weighting factor that needs to be determined – an example of a classifier 'hyperparameter' that is left to engineering, usually by considering a separate validation set (cf. Section 11.3.1). It can be shown that the above optimisation problem (the 'primal problem') is equivalent to the 'dual problem' of maximising

$$\sum_{l=1}^{L} a_l - \frac{1}{2}\sum_{k=1}^{L}\sum_{l=1}^{L} a_k a_l y_k y_l (\underline{x}_k^T \underline{x}_l), \tag{11.17}$$

with the side conditions

$$0 \le a_l \le C, \quad l = 1, \ldots, L, \tag{11.18}$$

$$\sum_{l=1}^{L} a_l y_l = 0. \tag{11.19}$$

In this model, the normal vector and bias of the hyperplane are given by

$$\underline{w} = \sum_{l=1}^{L} a_l y_l \underline{x}_l, \tag{11.20}$$

$$b = y_{l^*}(1 - \xi_{l^*}) - \underline{x}_{l^*}^T \underline{w}_{l^*}, \tag{11.21}$$

where l^* denotes the index of the training vector $\underline{x}_l$ with the largest coefficient a_l. This representation has an interesting interpretation. The normal vector $\underline{w}$ is now a weighted sum of training instances with the coefficients $a_l \leq C, l = 1, \ldots, L$; these weighting coefficients replace the slack variables ξ_l in the optimisation problem. The support vectors are the training instances $\underline{x}_l$ that satisfy $a_l > 0$. This is also called *support vector expansion* (Smola and Schölkopf 2004).

For the optimisation of the dual problem (11.17), L^2 terms of the form $\underline{x}_k^T \underline{x}_l$ have to be computed, resulting in quadratic complexity in terms of the number of training instances. However, there exist efficient dynamic programming algorithms to solve the dual problem, breaking it down into smaller optimisation problems that can be solved analytically and subsequently combining the results. One such algorithm is sequential minimal optimisation, which is introduced in detail by (Platt 1998).

Classification by linear SVMs as introduced above exactly corresponds to a simple linear classifier and is given by

$$d_{\underline{w},b}(\underline{x}) = \text{sgn}(\underline{w}^T \underline{x} + b), \tag{11.22}$$

where sgn is the sign function (9.7). It is easy to see that the decision rule (11.22) performs an implicit weighting of features by relevance. The vector $\underline{w}$ can be interpreted as a weight vector assigning each feature i to a weight w_i in the decision rule. A large absolute value of weight w_i means that the decision is heavily influenced by the value of feature i. Thus, SVMs are very robust to 'nuisance' dimensions carrying zero information with respect to the target class – geometrically, such dimensions will not affect the maximum margin hyperplane, causing the corresponding entries in $\underline{w}$ to vanish.

Linear SVMs as introduced so far are only able to solve pattern recognition problems where the instances belonging to the (two) different classes can be separated with a certain acceptable error by a hyperplane in the space $\underline{X}$. For classes that can only be separated non-linearly, one can apply non-linear transformations. Figure 11.4 depicts an example of a two-class problem in one-dimensional space which cannot be solved flawlessly by any kind of linear boundary. However, by mapping the vectors into a two-dimensional space by means of a quadratic vector-valued function $\mathbf{\Phi}$, the instances can be separated without error.

In general, such transformations $\mathbf{\Phi}$ are of the form

$$\mathbf{\Phi} : \underline{X} \to \underline{X}', \quad \dim(\underline{X}') > \dim(\underline{X}). \tag{11.23}$$

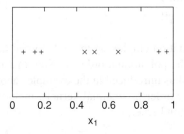
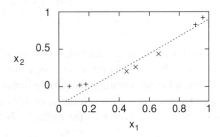

Figure 11.4 Example of projection into a higher-dimensional space to solve a binary problem that cannot be solved linearly in one-dimensional space (left). By mapping $\mathbf{\Phi} : x_1 \mapsto (x_1, x_1^2)$, linear separation is possible without error in the new higher-dimensional space (right) (Kroschel *et al.* 2011)

Applying this transformation to the normal vector $\underline{w}$ (11.20), we have

$$\underline{w} = \sum_{l:a_l>0} a_l y_l \Phi(\underline{x_l}). \tag{11.24}$$

The decision function $d_{\underline{w},b}(\underline{x})$ results in:

$$d_{\underline{w},b}(\underline{x}) = \text{sgn}(\underline{w}^T \Phi(\underline{x}) + b). \tag{11.25}$$

If no further constraints are put on Φ, this drastically increases the complexity of SVM training and classification, since every dimension of the resulting feature vectors has to be computed explicitly. This is where the so-called 'kernel trick' (Schölkopf and Smola 2002) comes into play. Note that for classifier training in the transformed space, only scalar products of the form $\Phi(\underline{x})^T \Phi(\underline{x}')$ have to be computed:

$$\underline{w}^T \Phi(\underline{x}) = \sum_{l:a_l>0} a_l y_l \Phi(\underline{x_l})^T \Phi(\underline{x}), \tag{11.26}$$

Explicit computation of Φ is also not needed for evaluation of the decision function. Thus, only the result of the scalar product $\Phi(\underline{x})^T \Phi(\underline{x}')$ has to be given as a function, the so-called 'kernel function' $K^\Phi(\underline{x}, \underline{x}')$, which fulfils

$$K^\Phi(\underline{x}, \underline{x}') = \Phi(\underline{x})^T \Phi(\underline{x}'). \tag{11.27}$$

The kernel function additionally needs to be positive semi-definite, symmetric, and satisfy the Cauchy–Schwarz inequality. In widespread use is the polynomial kernel of order p,

$$K_p^\Phi(\underline{x}, \underline{x}') = (\underline{x}^T \underline{x}' + 1)^p. \tag{11.28}$$

In this example, it can be easily seen how drastically the application of the kernel function K^Φ instead of the transformation Φ reduces the required computation effort. To compute a polynomial of order p in the space $\underline{X}$, a total of

$$\binom{\dim(\underline{X}) + p}{p} \approx \frac{\dim(\underline{X})^p}{p!} \tag{11.29}$$

terms would need to be calculated, while the computation employing the polynomial kernel requires only $O(\dim(\underline{X}))$ operations, regardless of the polynomial order p. Note that using $p = 1$ ('linear kernel') results in a simple linear SVM as introduced in the examples above.

Frequently used non-linear kernels, besides higher-order polynomial kernels, include the Gaussian kernel, also known as the radial basis function kernel,

$$K_\sigma^\Phi(\underline{x}, \underline{x}') = \exp\left(\frac{||\underline{x} - \underline{x}'||^2}{2\sigma^2}\right), \tag{11.30}$$

where σ is the standard deviation of the Gaussian, and the sigmoid kernel,

$$K^{\Phi}_{k,\Theta}(\underline{x}, \underline{x}') = \tanh(k(\underline{x}^T \underline{x}') + \Theta),\tag{11.31}$$

with amplification k and offset Θ.

There exist many further kernels for special requirements, such as the Kullback–Leibler (KL) divergence kernel frequently used in Gaussian mixture model (GMM) SVM 'supervector' construction. In this case, feature dimensions correspond to GMM mixture weights, means and variances; instead of measuring their distance by the standard scalar product, the KL divergence of the corresponding Gaussian distributions is (approximately) computed.

For non-numeric input, that is, symbol sequences, a string subsequence kernel can be used (Lodhi *et al.* 2002). This directly maps textual information to a high-dimensional feature space without explicit calculation of features. The underlying model is based on counting the number of observations of substrings in a given string, similar to BoCNG (see Section 9.4.1), but allowing non-contiguous substrings: for example, 'ser' exists in 'serene' (as in BoCNG), but also in 'superb'. However, in the latter case, the non-contiguity is penalised by reducing the number of observations by a decay factor $\lambda \in [0, 1]$. The mapping of text to this substring space can be modelled as a transformation Φ, yet an implicit calculation is done by using a kernel function $K^{\Phi}(s, t)$ for strings s and t. It is implemented effectively through string similarity, which can be computed by dynamic programming (see Section 11.2.2) (Lodhi *et al.* 2002).

Apart from special cases such as the above where a kernel function is tailored to the problem, the optimal kernel function for a generic classification or regression problem is usually found empirically, such as by cross-validation (see Section 11.3.1) using SVMs with different kernel functions and kernel hyperparameters (polynomial order, etc.) Recently, so-called multi-kernels have been introduced to provide a data-based solution to the search for optimal kernel functions (Yang *et al.* 2011). The basic idea is to consider a weighted sum of kernels as a kernel function, optimising alternately the hyperplane parameters and the weights of the kernels. This leads to implicit selection of the kernel function best suited to the problem, similar to the implicit selection of features and training instances as outlined above.

In computational paralinguistics, tasks are often correlated with each other – for example, detection of basic emotions such as happiness can be regarded as a task similar to binary classification of low and high arousal. Both tasks can be 'taught' to a single SVM classifier, joining the training vectors for both tasks ($\mathcal{L}_s$ and $\mathcal{L}_t$) in a single training set $\mathcal{L}_{s\cup t}$ and employing a kernel function respecting the task similarity $S_{s,t}$ between tasks s and t – meaning that a positive instance in $\mathcal{L}_s$ should also be a positive instance for task $\mathcal{L}_t$. In the general case of multiple tasks, this leads to the introduction of a positive definite and symmetric task similarity matrix, $\mathbf{S} = (S_{t_1,t_2})$, that is employed in a multi-task kernel function. For example, the linear kernel is extended to

$$K^{\mathrm{MT}}(\underline{x}^{(t_1)}, \underline{u}^{(t_2)}) = S_{t_1,t_2} \underline{x}^{(t_1)T} \underline{u}^{(t_2)}.\tag{11.32}$$

where $\underline{x}$ and $\underline{u}$ belong to tasks t_1 and t_2, respectively. Non-linear multi-task kernels are explained in detail in (Evgeniou *et al.* 2005). Multi-task learning is especially promising in cases where only little training data is available for a given task while a large training base exists for similar tasks (Widmer and Rätsch 2012).

Neural Networks

This section gives a short introduction to artificial neural networks (ANNs) and describes the extension for classification and regression of time series with long-range context called long short-term memory (LSTM) networks and their bidirectional variants (BLSTM). Combining LSTM with recurrent neural networks (RNNs) yields (bidirectional) long short-term memory recurrent neural networks ((B)LSTM-RNNs).

ANNs are capable of learning arbitrary (also non-linear) functions (Niemann 2003), and belong to the most popular machine learning algorithms, since the first mathematical models introduced by McCulloch and Pitts (1943). This model still provides the basis for today's ANNs (e.g., Schuller 2006). ANNs are motivated by the physiological neural networks in the central nervous system of vertebrates. The main information processing unit is the neuron. Via its axon a neuron emits a certain activity of electrical impulses (Rigoll 1994). These impulses are propagated to the synapses of other neurons through a complex network of connections. The activity of a single neuron is based on the cumulative activation at its input. A higher neuron activity results in a higher impulse frequency. An ANN thus consists of neurons and their directed connections. It is fully described by the network topology (the layout of the neurons and the connections), multiplicative weights for all the connections, and the type of neurons (e.g., the transfer function employed in the neurons) as well as the encoding of the output. Figure 11.7 shows an example of such an ANN. The feature vector $\underline{x} = \{x_i\}$, $i = 1, \ldots, N$, is applied as input to the network at its N input neurons. The values of the feature vector are multiplied by the weights w_i, $i = 0, \ldots, N$, which can be written as a weight vector $\underline{w} = \{w_i\}$. The additional weight w_0 has a special meaning: it represents a permanent additive offset rather than being a multiplicative weight. After application of the weights, the weighted inputs are summed, resulting in a value u. This u is then processed by the neuron with its (usually) non-linear transfer function $T(u)$. The output v of this function is the output of the neuron, which is forwarded to the next layer of neurons. For $T(u)$, in most cases a steep decision function is preferred. A visualisation of a single neuron is given in Figure 11.5.

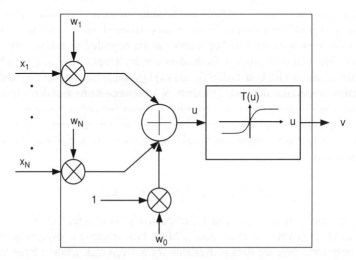

Figure 11.5   An example of an artificial neuron

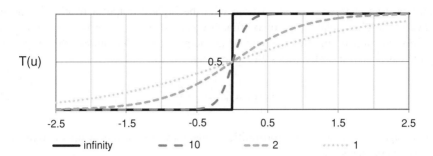

Figure 11.6 Sigmoid function for various values of the steepness parameter α. For $\alpha \to \infty$ a binary threshold decision function is approximated

A popular neuron transfer function is the sigmoid function

$$T(u) = \frac{1}{1 + e^{-\alpha u}},\qquad(11.33)$$

where α is the steepness parameter (see Figure 11.6). Other popular functions include the hyperbolic tangent function as a special case of the sigmoid function which has an extra additive offset, and the unit step function

$$T(u) = \begin{cases} 0 & \text{if } u < 0 \\ 1 & \text{if } u \geq 0. \end{cases}\qquad(11.34)$$

The sigmoid function is popular because it approximates an ideal threshold decision (cf. Figure 11.6) very closely (given a large α) while being differentiable. The differentiability is needed for training of the network with a gradient descent algorithm.

There are many different neuron topologies, that is, ways of arranging and connecting neurons. We now introduce the most important and most common of these.

Feed-Forward Neural Networks

A feed-forward neural network (FNN) has connections from the input neurons (top layer) to the output neurons (bottom layer) only. Upward connections, also called recurrent connections, that is, connections to the input layer, or connections that would form a loop in some way, are not allowed in this type of network. The best-known implementation of the FNN is the multilayer perceptron (MLP) (Deller *et al.* 1993): This has at least three layers: one input layer, one or more hidden layers, and one output layer. All connections feed forward from one layer to the next without backward connections – hence the name FNN. MLPs have no memory over time, that is, they treat all input patterns independently. The outputs $\hat{y}_j$, $j = 1, \ldots, M$, of the last layer can be written as a vector $\hat{\underline{y}}$. If the network is used as a regressor, usually only one output neuron is present and the output resembles the regressor value. A regressor for multiple (dependent) targets can thus be constructed if more than one output neuron is used.

For classification, the neuron activation values of the output layer have to be mapped to a discrete class representation. This is referred to as 'encoding'. Typically one output neuron per class is used. The class of the neuron with the highest output activation is chosen as classification result. Usually a 'softmax' function is used as a transfer function in the output layer. This function normalises the sum of all outputs to 1. This allows the outputs to be interpreted as posterior probabilities $P(j|\underline{x})$ of each class:

$$P(j|\underline{x}) = \hat{y}_j = \frac{e^u}{\sum_{j=1}^{M} e^u}. \tag{11.35}$$

The advantage of these posterior probabilities is that they provide a measure of confidence for the classification decision.

In the recognition phase, the net is computed layer by layer from the input layer to the output layer. For each layer the weighted sum of the inputs from the previous layer (or the input data, for the first layer) is computed for each neuron. The output of each neuron in the layer is computed via the neurons' transfer functions. With the softmax function at the output layer neurons and the given encoding, the class recognised is assigned with a maximum search. As an alternative, we could choose, for example, a binary encoding of the classes with the network's outputs, but this is not optimal in cases where classes with neighbouring indexes are not close to each other in the feature space.

Back-Propagation

Among the multiplicity of learning algorithms which are available for ANNs, the gradient descent based back-propagation algorithm (Rumelhart *et al.* 1987) is one of the most popular. Let $\underline{W} = \{\underline{w}_j\}$, $j = 1, \dots, J$, summarise the weight vectors $\underline{w}_j$ of a layer, where J is the number of neurons in this layer. To measure the progress of the network training, the mean square error (MSE) $E(\underline{x}, \underline{W})$ between the gold standard y and the network output $\hat{y} = f(\underline{x}, \underline{W})$ is used as the objective function:

$$E(\underline{x}, \underline{W}) = |y - \hat{y}|^2 \tag{11.36}$$

For simplicity, in the following we consider the case of a single output. An extension of the equations to multiple outputs is straightforward. Other objective functions can be used instead, among them the McClelland error or cross-entropy. Before the training of the weights with back-propagation, the weights must be initialised to non-zero values. This is typically done by a pseudo-random generator. Then, three steps follow for the back-propagation, repeated over a number of epochs (one epoch is one pass of all three steps) until a stopping criterion is reached. These three steps are:

1. Forward pass (computation of outputs, given a vector of inputs).
2. Computation of the objective function (e.g., the MSE as in equation (11.36).
3. Backward pass with weight adaptation by the corrective term

$$w_i \rightarrow w_i + \Delta w_i = w_i - \beta \cdot \frac{\delta E(\underline{x}, \underline{W})}{\delta w_i}, \tag{11.37}$$

where β is the update step size, which is to be determined empirically, and w_i is an individual weight in the network.

The iterative updating of the weights is typically stopped after a given maximum number of iterations or when the change in the output error (objective function result) from one iteration to the next is below a defined threshold (Schalkoff 1994). Experience is needed in order to be able to determine a 'good' set of learning parameters. However, automated approaches exist to learn these parameters. To avoid over-fitting to a given set of training data, a large enough number of training instances is required compared to the number of parameters (weights) in the network and the dimensionality of the feature vector.

An alternative to gradient descent based back-propagation is resilient propagation which incorporates the previous change of weights into the current change of weights (Riedmiller and Braun 1992).

By automatic learning of the weights in the input layer, ANNs are able to cope with redundant and irrelevant feature information. The learning process is further discriminative as the information over all classes is learnt at once (Rigoll 1994). Their high degree of parallelism is one of the main advantages for efficient implementation on modern multi-core architectures or graphics and general purpose processors with many hundreds of computation pipelines.

If the temporal context of a feature vector is relevant, that is, the vector is part of a time series, this context must be explicitly presented to a feed-forward network. This can be done, for example, by using a fixed-size sliding window which combines several feature vectors into a 'supervector', as in (Lacoste and Eck 2005).

Recurrent Neural Networks

Another technique for modelling past context with neural networks is to extend feed-forward networks with (cyclic) backward connections. The resulting network is then called a recurrent neural network (RNN). RNNs can theoretically map from the entire history of previous inputs to the current output. The recurrent connections build a kind of memory, which allows input values to persist in the hidden layer(s) over multiple time steps and thus influence the network output in the future. On sequences of finite length, RNNs can be trained by a modified back-propagation algorithm called 'back-propagation through time' (BPTT) (Werbos 1990). The network is first unfolded over time. Then the training follows the same steps as if training a FNN with back-propagation. However, in each epoch the outputs must be computed and processed in sequential order. Details are given in (Werbos 1990). If future context (i.e., data following the current frame) is also required to compute the outputs, one possible solution is to introduce a delay between the input values and the output targets. Figure 11.7 shows an example.

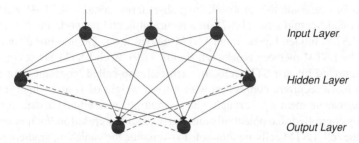

Figure 11.7 Example of a recurrent neural network, with three input-layer, two hidden-layer, and two output-layer neurons. The dashed line connections depict exemplary possible recurrent connections

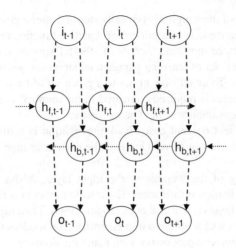

Figure 11.8 Bidirectional neural network: input i, output o, and two hidden layers processing input sequences forwards (h_f) and backwards (h_b) over time t (Dotted arrow connections)

A more elegant implementation of modelling of future temporal context is provided by a bidirectional recurrent neural network (BRNN). Two (sets of) independent hidden layers are used instead of one. Both are connected to the same input and output layers. The first hidden layer (set) processes the input sequence forwards and the second one backwards. The network therefore at any frame has access to the full input sequence in both forward and backward directions. There is no need to define the number of input frames which are stacked – and thus the amount of context that can be modelled. Figure 11.8 illustrates this principle. Bidirectional networks, however, must have the complete input sequence available before it can be processed.

Long Short-Term Memory

Although BRNNs have access to past and future information, the range of temporal context that can be modelled is limited to a few frames because of the 'vanishing gradient' problem described by Hochreiter *et al.* (2001): the influence of an input value decays (recurrent weight less than 1) or blows up exponentially (recurrent weight greater than or equal to 1) over time, as it cycles through the network with its recurrent connections and gets dominated by new input values. To overcome this problem, long short-term memory (LSTM) was introduced by Hochreiter and Schmidhuber (1997). In a recurrent neural network, the hidden layers are replaced by LSTM hidden layers. In an LSTM hidden layer, the non-linear units (neurons) are replaced by LSTM memory blocks (cf. Figure 11.9). Each LSTM block contains one or more self-connected linear LSTM cells. In each cell a so-called 'constant error carousel' is found which has a recurrent connection with a fixed weight of 1, and which implements a potentially permanent memory. Through 'gates' (multiplicative units) the data flow in the cell is dynamically controlled, and potentially the content could be erased and/or new content could be stored in the cell. LSTM cells are thus able to overcome the vanishing gradient problem and can learn the optimal amount of contextual information relevant for the learning task through

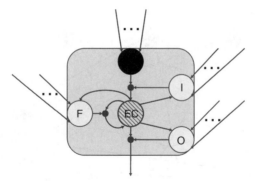

Figure 11.9 LSTM memory cell: the usual neuron (large black circle) on top is added by input (I), output (O), and forget (F) gates that collect activations controlling the cell by multiplicative units (small black circles). The actual memory is realised by a recurrent connection with the fixed weight 1.0 that maintains the internal state (error carousel: EC)

the weights at the inputs of the multiplicative gate units. Figure 11.10 highlights the vanishing gradient problem for RNNs and how it is overcome by LSTM (right).

An LSTM layer is composed of recurrently connected LSTM blocks, each of which contains one or more memory cells. Each memory cell has a constant error carousel and three multiplicative 'gate' units: the input, output, and forget gates. The gates perform functions analogous to read, write, and reset operations in computer memory cells. More specifically, the cell input is multiplied by the output activation of the input gate, the cell output by that of the output gate, and the previous cell memory state (from the error carousel) by the forget gate (cf. Figure 11.9). Usually, the same non-linear transfer function, denoted by T_g in what follows, is applied for all three gates. A popular choice is the hyperbolic tangent function. For the transfer function of the 'original' neuron (topmost unit in Figure 11.9) a sigmoid function, denoted

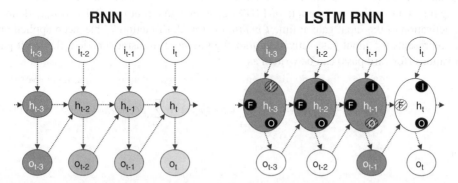

Figure 11.10 Vanishing gradient problem of a recurrent neural network (left) and of an LSTM recurrent neural network (right). The 'degree of memory' of past events is indicated by grey shading. i_t, h_t, o_t are input, hidden, and output layers, respectively, at time t. Recurrent connections are indicated by black arrows. LSTM: black circle and lined circle filled input, output, and forget gates (I, O, and F, respectively) indicate passive and active states

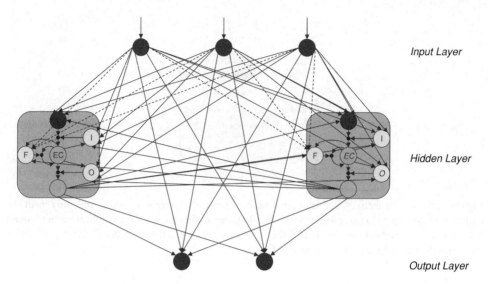

Figure 11.11 Example of enrichment of a recurrent neural network by LSTM cells

by T_i in what follows, is usually chosen. It is the actual input neuron of the LSTM cell, and it would be the only neuron in a standard FNN or RNN. Everything after the input neuron is the LSTM extension. The output transfer function of the LSTM cell after the error carousel is henceforth referred to as T_o. Popular choices for this function are sigmoid or softmax function. The weight of the recurrent connection in the error carousel is chosen as 1 in order to realise the permanent storage effect.

The overall effect is that the network is able to store and retrieve information over long periods of time. For example, as long as the input gate remains closed (i.e., its output activation is zero or close to zero), the internal state of the cell cannot be overwritten by the current input and can therefore be made available to the net later in the sequence when the output gate is open.

Figure 11.11 depicts an example of LSTM cells integrated into an RNN. If $\alpha_{\text{in},t}$ denotes the activation of the input gate at time t *before* the transfer function T_g has been applied and $\beta_{\text{in},t}$ represents the input gate activation *after* the (sigmoid) transfer function, the input gate activations (forward pass) can be written as

$$\alpha_{\text{in},t} = \sum_{i=1}^{I} w_{i,\text{in}} x_{i,t} + \sum_{h=1}^{H} w_{h,\text{in}} \beta_{h,t-1} + \sum_{c=1}^{C} w_{c,\text{in}} s_{c,t-1} \qquad (11.38)$$

and

$$\beta_{\text{in},t} = T_g(\alpha_{\text{in},t}), \qquad (11.39)$$

respectively. $w_{i,j}$ corresponds to the weight on the connection from node i to j. The indices i, h, and c enumerate the inputs $x_{i,t}$, the cell outputs from other blocks in the hidden layer, and the memory cells. I, H, and C are, respectively, the number of inputs, the number of cells in the hidden layer, and the number of LSTM cells in one block. Finally, $s_{c,t}$ corresponds to the *state* of a cell c at time t.

In a similar way, the forget gate ('for') activations before and after applying the transfer function T_g can be computed by

$$\alpha_{\text{for},t} = \sum_{i=1}^{I} w_{i,\text{for}} x_{i,t} + \sum_{h=1}^{H} w_{h,\text{for}} \beta_{h,t-1} + \sum_{c=1}^{C} w_{c,\text{for}} s_{c,t-1}, \tag{11.40}$$

$$\beta_{\text{for},t} = T_g(\alpha_{\text{for},t}). \tag{11.41}$$

The memory cell value $\alpha_{c,t}$ is a weighted sum of inputs at time t and hidden unit activations at time $t - 1$:

$$\alpha_{c,t} = \sum_{i=1}^{I} w_{i,c} x_{i,t} + \sum_{h=1}^{H} w_{h,c} \beta_{h,t-1}. \tag{11.42}$$

To determine the current state of a cell c, the previous state is scaled by the activation of the forget gate and the input $T_i(\alpha_{c,t})$ by the activation of the input gate:

$$s_{c,t} = \beta_{\text{for},t} s_{c,t-1} + \beta_{\text{in},t} T_i(\alpha_{c,t}). \tag{11.43}$$

The computation of the output gate ('out') activations follows the same method as the calculation of the input and forget gate activations. However, now the *current* state $s_{c,t}$, rather than the state from the previous time step, is considered:

$$\alpha_{\text{out},t} = \sum_{i=1}^{I} w_{i,\text{out}} x_{i,t} + \sum_{h=1}^{H} w_{h,\text{out}} \beta_{h,t-1} + \sum_{c=1}^{C} w_{c,\text{out}} s_{c,t}, \tag{11.44}$$

$$\beta_{\text{out},t} = T_g(\alpha_{\text{out},t}). \tag{11.45}$$

Finally, the LSTM cell output is determined as

$$\beta_{c,t} = \beta_{\text{out},t} T_o(s_{c,t}). \tag{11.46}$$

Note that the originally proposed version of the LSTM cell architecture contained only input and output gates. Forget gates were added later by Gers *et al.* (2000) in order to allow the memory cells to reset themselves whenever the network needs to *forget* past inputs completely.

Like RNNs, LSTM networks can be trained by the BPTT algorithm. They have demonstrated remarkable performance in a variety of pattern recognition tasks such as phoneme classification (Graves and Schmidhuber 2005), handwriting recognition (Graves 2008), keyword spotting (Wöllmer *et al.* 2011c), affective computing (Wöllmer *et al.* 2010), and driver distraction detection (Wöllmer *et al.* 2011b). LSTM networks can be either unidirectional or bidirectional. The bidirectional variant is known as BLSTM. Further details on LSTM can be found in Graves (2008).

11.2.2 Dynamic Classification: Hidden Markov Models

Speech and language are sequential, and an audio or text stream $\underline{X} = \{\underline{x}_1, \underline{x}_2, \ldots, \underline{x}_T\}$ accordingly yields a sequence of T feature vectors $\underline{x}_t$ at times t. So far, however, we have mostly dealt with classification of single feature vectors without the use of temporal context. One exception were the different types of RNNs and LSTM-RNNs discussed in Section 11.2.1. Yet, even these are not able to 'warp' data in time, for example, stretching or shortening of vowels in speech in comparison to a reference pattern or model.

The most frequently encountered algorithms for sequence classification are hidden Markov models (HMMs: Rabiner 1989), a simple form of dynamic Bayesian networks (DBNs). This is due to their dynamic modelling ability throughout different hierarchy levels and a well-formulated, generative stochastic framework. In ASR, for example, at a low level, phonemes are modelled; at a higher level, words are built from the phonemes. Each class i is modelled by a HMM which represents the probability $P(\underline{X}|i)$, where $\underline{X}$ is the 'observation' (the feature vector at a given time), which is generated by the HMM. Applying Bayes' rule allows $P(i|\underline{X})$ to be determined given that $P(\underline{X}|i)$ and the class priors $P(i)$ are known and the priors of the feature vector observations $P(\underline{X})$ are ignored.

A Markov model can be seen as a finite-state machine which can change its state at any step in time. In a HMM, at each time step t a feature vector $\underline{x}_t$ is generated depending on the current state s and the state-dependent emission probability $b_s(\underline{x})$. The likelihood of a transition from state j to state k is expressed by the state transition probability $a_{j,k}$ (O'Shaughnessy 1990). The probabilities $a_{0,j}$ are required to enter the model in a state j with a certain probability. In order to simplify calculations and algorithm implementation, a non-emitting initial state s_0 and a non-emitting final state s_F are usually defined (Kroschel et al. 2011). The structure of such a model is shown in Figure 11.12. In this example, the most commonly used type of HMM for speech processing is depicted – the left–right model. In this type, the state number cannot decrease over time, that is, only transitions from left to right are allowed. A specific variant is the 'linear' model, where no state can be skipped. Other topologies allow states to be skipped, such as the Bakis topology in which one state may be skipped. If any state can be reached from any other state with non-zero probability, that is, the states are fully connected, the model is referred to as 'ergodic'.

We speak of a 'hidden' Markov model, as the sequence of states remains unknown – only the observation sequence is known (Rabiner 1989) to an outside observer. Note that the 'Markov property' is valid, that is, that the conditional probability distribution of the hidden variable $s(t)$ at time step t, given the values of the hidden variable s at all times, depends only on

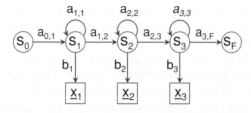

Figure 11.12 Example of a linear left–right HMM: three emitting states are shown. Squares, observations; circles, switching states; arrows, conditional dependencies

the hidden variable $s(t-1)$, that is, values of earlier steps in time have no influence (Jelinek 1997). Further, the observation $x(t)$ depends only on the value of the current state's hidden variable $s(t)$.[1]

The probability $P(\underline{X}|i)$ can be computed by summing over all possible state sequences:

$$P(\underline{X}|i) = \sum_{Seq} a_{s_0,s_1} \prod_{t=1}^{T} b_{s_t}(\underline{x}_t) a_{s_t,s_{t+1}}, \qquad (11.47)$$

where Seq stands for the set of all possible state sequences. For an efficient computation of this probability, the forward algorithm is applied. In principle, the forward algorithm is a more efficient recursive way to calculate the complete generation probability (see equation (11.27)) by use of 'forward variables' (see equation (11.53) below). Instead of summing over all possible state sequences, the size of the search space and the associated computational complexity can be reduced by the Viterbi algorithm. This algorithm only considers the instantaneously most probable state sequence, which results in a speed-up, however, at the cost of the global optimum (Jelinek 1997):

$$\hat{P}(\underline{X}|i) = \max_{Seq} \left\{ a_{s_0,s_1} \prod_{t=1}^{T} b_{s_t}(\underline{x}_t) a_{s_t,s_{t+1}} \right\}. \qquad (11.48)$$

In the recognition phase, the class i associated with the model that has the highest probability $P(\underline{X}|i)$ for a given input is chosen. To compute this probability for unknown inputs, the parameters $a_{j,k}$ and $b_s(\underline{x}_t)$ need to be known for each model. Just as for the static classifiers, these parameters are estimated in a training phase given a large set of training data. The most popular method for this purpose is the forward–backward algorithm (described below under 'Estimation').

In most speech application scenarios the state emission probabilities $b_s(\underline{x}_t)$ are modelled by Gaussian mixtures (GMs) – often loosely referred to as Gaussians. Such mixtures are linear superpositions of Gaussian functions. With the number of mixture components M and the 'mixture weight' of the mth component $c_{s,m}$, the emission probability density function (PDF) can be formulated as (Jelinek 1997)

$$b_s(\underline{x}_t) = \sum_{m=1}^{M} c_{s,m} \mathcal{N}(\underline{x}_t; \underline{\mu}_{s,m}, \underline{\Sigma}_{s,m}), \qquad (11.49)$$

where $\mathcal{N}(\cdot; \mu, \Sigma)$ is a multivariate Gaussian density with the mean vector $\underline{\mu}$ and the covariance matrix $\underline{\Sigma}$. Apart from such 'continuous' HMMs, 'discrete' HMMs are also used. Conditional probability tables for discrete observations $b_s(\underline{x}_t)$ are used in the latter case instead of GMs.

---

[1] Note that, for better readability, the time t is used in the subscript or argument in this section, following Rabiner (1989).

Estimation

The parameters of HMMs can be determined by the Baum–Welch algorithm (Baum *et al.* 1970), which is a special case of the expectation maximisation (EM) algorithm. If the maximum likelihood (ML) estimates of the means and covariances of the Gaussians for each state s are to be computed, we have to take into account that each observation vector $\underline{x}$ contributes to the state parameters. This is because the overall probability of an observation $\underline{x}$ is based on the summation of all possible state sequences. Thus, Baum–Welch estimation assigns each observation to each state in proportion to the probability of the observation of the respective feature vectors in the state. Denoting by $L_{s,t}$ the probability of the model being in state s at time step t, Baum–Welch estimation of the means and covariances of a single Gaussian PDF is obtained as follows:

$$\hat{\underline{\mu}}_s = \frac{\sum_{t=1}^{T} L_{s,t} \underline{x}_t}{\sum_{t=1}^{T} L_{s,t}}, \tag{11.50}$$

$$\hat{\underline{\Sigma}}_s = \frac{\sum_{t=1}^{T} L_{s,t} (\underline{x}_t - \underline{\mu}_s)(\underline{x}_t - \underline{\mu}_s)^T}{\sum_{t=1}^{T} L_{s,t}}, \tag{11.51}$$

where the hat symbol denotes estimated parameters.

The Gaussian PDFs are initialized and estimated with a single Gaussian mixture component at first. Then, 'up-mixing' to more Gaussian mixture components is achieved in a simple way by seeing the mixture components as sub-states. The mixture weights are now the state transition probabilities of these sub-states. These state transition probabilities are estimated by the relative frequencies

$$\hat{a}_{j,k} = \frac{A_{j,k}}{\sum_{s=1}^{S} A_{j,s}}, \tag{11.52}$$

where $A_{j,k}$ denotes the number of transitions from state j to state k, and S denotes the number of states of the HMM.

The forward-backward algorithm is applied for the computation of the probability $L_{s,t}$. The 'partial' forward probability $\alpha_s(t)$ for an HMM which represents the class i is defined as

$$\alpha_s(t) = P(\underline{x}_1, \ldots, \underline{x}_t, s_t = s | i). \tag{11.53}$$

This can be interpreted as the joint probability of having observed the first t feature vectors and being in state s at time step t. The recursion

$$\alpha_s(t) = \left[\sum_{j=1}^{S} \alpha_j(t-1) a_{j,s} \right] b_s(\underline{x}_t) \tag{11.54}$$

allows for an efficient computation of the forward probability, S being the number of emitting states.

The backward probability represents the joint probability of the observations from time step $t + 1$ to T:

$$\beta_s(t) = P(\underline{x}_{t+1}, \ldots, \underline{x}_T | s_t = s, i). \tag{11.55}$$

It can be computed by the recursion

$$\beta_j(t) = \sum_{s=1}^{S} a_{j,s} b_s(\underline{x}_{t+1}) \beta_s(t + 1). \tag{11.56}$$

To compute the probability of being in a state at a given time step, we have to multiply the forward and backward probabilities:

$$P(\underline{X}, s_t = s | i) = \alpha_s(t) \cdot \beta_s(t). \tag{11.57}$$

Thus L_{st} can be determined as

$$L_{st} = P(s_t = s | \underline{X}, i) = \frac{P(\underline{X}, s_t = s | i)}{p(\underline{X}|i)} = \frac{1}{p(\underline{X}|i)} \cdot \alpha_s(t) \cdot \beta_s(t). \tag{11.58}$$

If we make the assumption that the model has to reach the last state S when the last observation $\underline{x}_T$ is to be made, that is, the last observation has to be emitted in the last state, then the probability $P(\underline{X}|M_t)$ equals $\alpha_S(T)$. This means that Baum–Welch estimation can be executed as described above.

In the recognition phase, the Viterbi algorithm is usually applied to speed up the search over all possible paths. It is similar to computing the forward probability. However, the summation over all possible paths is replaced by a maximum search to allow for the following forward recursion:

$$\phi_s(t) = \max_j \{\phi_j(t - 1) a_{j,s}\} b_s(\underline{x}_t), \tag{11.59}$$

where $\phi_s(t)$ is the ML probability of observing the vectors $\underline{x}_1$ to $\underline{x}_t$ and being in state s at time step t of a given HMM which represents class i. Thus, the estimated ML probability $\hat{P}(\underline{X}|i)$ equals $\phi_S(T)$.

Hierarchical Decoding

HMMs are particularly suited for decoding of continuous speech, that is, both segmenting and recognising continuous streams of audio. In addition, their probabilistic formulation permits hierarchical analysis in order to unite knowledge at different levels. Let S be a sequence of speech units such as a spoken sentence. The sequence $\underline{X}$ of T feature vectors is from the phrase S (Ruske 1993). A classifier should provide an estimate $\hat{S}$ of the sequence of speech units which is as close as possible to the actual sequence S. According to Bayes' decision rule, the decision is optimal on average if the classifier chooses the class which – based on the current observation – has the highest probability. For the optimal decision it is thus required that

$$p(\hat{S}|\underline{X}) = \max_{S_j} p(S_j|\underline{X}),$$ (11.60)

where S_j are the possible observed sequences. We thus have to determine the probability for all possible sequences S_j. In practice it is scarcely possible to do this. Therefore, Bayes' law is applied and the equation is reformulated as follows:

$$p(S_j|\underline{X}) = p(\underline{X}|S_j)\frac{p(S_j)}{p(\underline{X})}.$$ (11.61)

As the probability $p(\underline{X})$ depends purely on the feature vector series $\underline{X}$ and therefore is independent of S_j, it can be neglected for the maximum search over all sequences S_j, yielding:

$$\underbrace{p(\underline{X}|S_j)}_{\text{AM}} \cdot \underbrace{p(S_j)}_{\text{LM}} \overset{!}{=} \max,$$ (11.62)

where the acoustic model (AM) and the language model (LM) represent the acoustic properties of the inputs and the semantics or syntax, respectively. In order to scale the influence of the LM, an exponential factor Λ – the so-called LM scale factor – can additionally be introduced. This leads to

$$p(\hat{S}|\underline{X}) = \max_{S_j} p(\underline{X}|S_j) \cdot p(S_j)^{\Lambda}.$$ (11.63)

Λ is usually determined in experiments or is learnt in a semi-supervised manner (White *et al.* 2009). It is typically in the range of 10 ± 5.

The sequence which maximises the above expression is output as the best estimate $\hat{S}$:

$$\hat{S} = \arg\max_{S\in\mathcal{U}} p(\underline{X}|S) \cdot p(S)^{\Lambda},$$ (11.64)

where $\mathcal{U}$ represents all possible/allowed sequences. Let us now assume that each sequence S_j is a sequence of speech units $a_1, a_2, a_3, \ldots, a_A$. In the following a single sequence S_j is highlighted for simplicity. For this sequence,

$$p(S_j) = p(a_1, a_2, \ldots, a_A).$$ (11.65)

If we further assume that the acoustic realisations of the speech units are independent of each other, the units can be modelled individually:

$$p(\underline{X}|S_j) = p(\underline{x}_1, \ldots, \underline{x}_i)p(\underline{x}_{i+1}, \ldots, \underline{x}_j)\cdots p(\underline{x}_{k+1}, \ldots, \underline{x}_A).$$ (11.66)

It is assumed that no pauses occur between these units and silences or pauses between words, for example, are treated as units of their own. Note that the unit boundaries $i, j, \ldots, k$

and unit number A are unknown at first and need to be determined by the classifier (decoding and classification, respectively).

In the same way each speech unit can contain a sequence of sub-units one level lower in the hierarchy – again assuming independence of the sub-units. Speech units could be phonemes, triphones, or syllables, for example. If the units are modelled by HMMs, the Viterbi algorithm can be applied on all three layers for the decoding (Ruske 1993): for the search of the state sequence within the HMMs, for the sequence of the individual phoneme HMMs, that is, $\hat{S}$, and for the sequence of the words.

At the phoneme transitions the LM is applied to model higher-level information such as word transition probabilities (Furui 1996). A language model can contain N-grams, for example, which model the conditional probability of a sequence of consecutive phonemes or words. The Viterbi path determines the best path through all layers and thus the best sequence of speech units. For an illustrative example, see the 'trellis' diagram in Figure 11.13 (Rabiner 1989).

If the 'vocabulary' size (i.e., the number of speech events) is very large, the Viterbi search can become computationally very demanding and slow.

Although at time t only a single column needs to be analysed in the trellis diagram (Figure 11.13), all emission probabilities in all states for all models need to be computed.

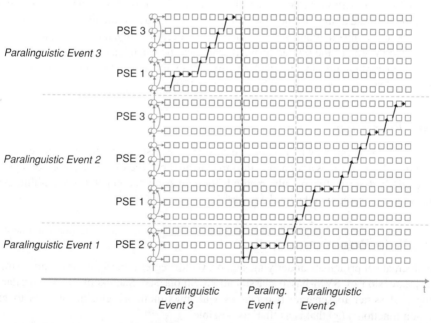

Figure 11.13 Viterbi search of the optimal state sequence to produce an observation series. The Trellis diagram shows speech units (e.g., words), referred to as 'paralinguistic events' (in the example, three different ones), and sub-units (e.g., phonemes) denoted as 'paralinguistic sub-events' (PSE, again, three different ones). A path is shown over time in upward arrows. Squares represent observations (feature vectors). HMMs (one per unit) are sketched in Bakis topology. After backtracking the sequence of paralinguistic events as Paralinguistic Event 3, then 1, then 2 is recognised as the best path, as is shown along the horizontal time axis

In the case of large vocabulary continuous speech recognition (LVCSR) this may easily require computation of more than 100 000 normal distributions every 10 ms (Ruske 1993). To speed up computations, we can make use of the fact that many paths in the trellis offer little or no hope of ever leading to the overall best path. These paths can be removed, leading to the 'beam search' method. The candidates with the lowest scores are 'pruned', keeping only the N highest scored paths. By accepting a sub-optimal solution (usually less than a 1% increase in error probability), a considerable speed-up and reduced memory consumption are achieved. The algorithm is implemented by smart list management in five consecutive steps (Lowerre 1976).

First, a list of all active states at time step t is set up. This list contains all points of the trellis diagram with a current probability that exceeds a given threshold.

Then, from this list of states all possible subsequent states in the next time step are computed that can be reached by the Viterbi path diagram. The algorithm works recursively as usual and the effect is the same as when applying the path diagrams in a backward direction. Next, the list of subsequent states is reduced by deleting states with path probabilities below the threshold, which is the actual pruning step. This threshold should be constantly adapted to the current step in time. Thus, the 'beam width' (number of paths considered per time step) is increased or reduced according to the validation of the concurring paths' ascent or decline. The beam width is decisive for the trade-off between higher accuracy (broadened width) and higher speed (narrowed width).

Optionally, at transitions between speech events (e.g., words), the value of the LM can be used and in any case a jump to the first state of the first model of the new speech event takes place. Further, the information for back-tracking is stored. Finally, the best sequence of units is obtained at the end of the pruned forward search by the usual backtracking.

In practical applications, this particularly efficient search algorithm can reach reductions of the number of states to be computed by a factor of 1000.

11.2.3   Regression

In contrast to classification, where a discrete class label is assigned to the input, in regression a continuous value or score is assigned as the target to an input vector or a sequence of input vectors. We briefly describe two popular methods for regression in computational paralinguistics: support vector regression and neural networks.

Support Vector Regression

The mathematical principles underlying support vector regression (SVR) are a fairly straightforward extension of SVM-based classification (Smola and Schölkopf 2004). Similarly to training a classifier to separate two classes with maximum margin, the goal is to find a regression function $f(\underline{x})$ that is as 'flat' as possible.

Let us first consider linear SVR, which defines a real-valued linear N-dimensional function

$$f(\underline{x}) = \underline{w}^T \underline{x} + b, \tag{11.67}$$

described by a vector $\underline{w}$ and a scalar b. Again, we assume a labelled set of training patterns $\mathcal{L}$, but the target labels y_l are now (conceptually) continuous ($y_l \in \mathbb{R}$) rather than binary

(-1 or $+1$). Analogously to the criterion for maximisation of the SVM classifier margin, flatness of $f(\underline{x})$ can then be achieved by minimising the dot product $\underline{w}^T\underline{w}$ under the conditions

$$
\begin{aligned}
y_l - \underline{w}^T\underline{x}_l - b &\le \varepsilon, \\
\underline{w}^T\underline{x}_l + b - y_l &\le \varepsilon,
\end{aligned}
\tag{11.68}
$$

similarly to the boundary conditions for SVM (11.9). Thus, the regression function is allowed to have a deviation of at most ε from the actual targets. Just as not all classification problems can be separated flawlessly by a linear boundary, in the general case not all targets y_l can be estimated using a linear function with only an error of at most ϵ. Thus, one again introduces non-negative slack variables ξ_l and ξ_l^* as for SVM training. The resulting boundary conditions for SVR are thus given by

$$
\begin{aligned}
y_l - \underline{w}^T\underline{x}_l - b &\le \epsilon + \xi_l, \\
\underline{w}^T\underline{x}_l + b - y_l &\le \epsilon + \xi_l^*.
\end{aligned}
\tag{11.69}
$$

Intuitively, this allows vectors to be mapped to values with an error greather than ε – but naturally, this is penalised by an error weighting term as in SVM training. This results in the 'primal problem', to minimise

$$
\frac{1}{2}\underline{w}^T\underline{w} + C\sum_{l=1}^{L}(\xi_l + \xi_l^*),
\tag{11.70}
$$

where C is the error weighting factor. As in the case of SVM, this problem can be transformed to an equivalent 'dual' optimisation problem, namely to maximise

$$
\sum_{l=1}^{L} y_l(a_l - a_l^*) - \varepsilon\sum_{l=1}^{L}(a_l + a_l^*) - \frac{1}{2}\sum_{k=1}^{L}\sum_{l=1}^{L}(a_k - a_k^*)(a_l - a_l^*)\underline{x}_k^T\underline{x}_l,
\tag{11.71}
$$

subject to the side conditions

$$
0 \le a_l, a_l^* \le C, \quad l = 1, \dots, L,
\tag{11.72}
$$

$$
\sum_{l=1}^{L}(a_l - a_l^*) = 0.
\tag{11.73}
$$

After determining the coefficients a_l and a_l^*, the vector $\underline{w}$ for the regression function is obtained as a weighted sum of training vectors,

$$
\underline{w} = \sum_{l=1}^{L}(a_l - a_l^*)\underline{x}_l.
\tag{11.74}
$$

Analogously to SVM, the vectors $\underline{x}_l \in \mathcal{L}$ for which $a_l - a_l^* \ne 0$ are the support vectors of the regression function (Smola and Schölkopf 2004). These correspond to the training vectors with an absolute deviation of exactly ε from the target value. Using the above, the linear regression function to determine the value for an unseen pattern vector $\underline{x}$ becomes

$$f(\underline{x}) = \sum_{l=1}^{L}(\alpha_l - \alpha_l^*)\underline{x}_l^T \underline{x} + b. \tag{11.75}$$

The bias b is given by

$$b = \max_l\{-\varepsilon + y_l - \underline{w}^T \underline{x}_l \mid a_l < C \text{ or } a_l^* > 0\} \tag{11.76}$$

Note the strict inequalities in the above equation.

The regression function as defined above can now be extended to non-linear regression by applying the same kernel trick as for SVM. The resulting non-linear regression function is obtained by simply substituting the standard dot product $\underline{x}_l^T \underline{x}$ by application of the kernel function K^{Φ}:

$$f(\underline{x}) = \sum_{l=1}^{L}(\alpha_l - \alpha_l^*)K^{\Phi}(\underline{x}_l, \underline{x}) + b. \tag{11.77}$$

Neural Networks

In Section 11.2.1 we introduced the concept of artificial neural networks in the light of classification. However, these networks are primarily regressors, as the output values of the neuron units are of continuous nature. For classification additional effort with a maximum search over N outputs and softmax activation function in the output layer is required.

In the case of regression, the output layer neurons often have linear transfer functions, to best cover the range of all possible output values without an inherent non-linear bias.

11.3 Testing Protocols

A crucial issue in computational paralinguistics is to select the classifier or regressor best suited to a given problem. This requires appropriate measures for performance assessment. In this section, the error in the classifier's predictions will be of most interest; while a number of further practical and conceptual aspects can be considered as performance criteria, such as computational complexity and on-line learning capability, these are usually properties known in advance.

11.3.1 Partitioning

So far, prediction errors have been considered as 'loss functions' for classifier training, for example, in the backpropagation algorithm for training of neural networks. The loss function is used to optimise the classifier parameters, such as the neuron weights or the support vectors of a SVM. Assessing only the training set error, however, is not sufficient, as any problem can be modelled with zero training set error by sufficiently complex models – trivially, by a 'look-up' function representing the pairs of training set instances and labels. Thus, the crucial issue is generalisation of the models to test cases not seen in training.

Ideally, systems such as emotion recognisers would be evaluated in real-life applications by users (see Schuller *et al.* 2009a). However, this is usually too costly for system development in a rapid prototyping fashion, such as choosing between different SVM kernel functions for a given recognition task. Hence, to evaluate systems in fair conditions resembling real-life applications, one usually relies on so-called 'held-out data': a portion of the available labelled data is not used in training but rather presented to the classifier after training, allowing assessment of the ability of the classifier/regressor to generalise to a certain degree. Holding out data also means not using these data in any 'tuning' of the steps in the chain of processing, including pre-processing, feature extraction (such as choosing the cut-off frequencies for Mel bands or the number of MFCCs), normalisation (such as determining the means and variances used in mean–variance normalisation), and selection of 'hyperparameters' for the learning algorithm (such as the error weighting factor in SVM optimisation, or the number of states and mixtures per state in an HMM). For the reasons given above, such 'tuning' should not be done on the training data, as this would cause a bias towards more complex models – yet tuning to a specific set of held-out test data is not optimal either, since it is not clear whether the adjustments made would generalise to other data. Thus, besides a training and testing partition, usually a disjoint development partition is needed for such tuning steps to simulate truly 'unseen' test data in the system evaluation, that is neither used for training nor for system tuning. In training of the final classifier, trained using the hyperparameters determined on the development set, this partition can be added to the training data.

The requirement of appropriate partitioning of labelled data adds a technical perspective to corpus engineering (Chapter 6). This is because a variety of criteria often need to be respected in evaluation. For example, it is often desired to have disjoint sets of subjects in the partitions of the database ('speaker' or 'subject independence'). If the latter criterion is not met, for example, an emotion classifier could learn how exactly the training subjects express their emotions, instead of generalising to a larger population. Furthermore, factors that influence the expression of the paralinguistic trait of interest (e.g., age or gender) should be similarly distributed in the partitions – for example, a system that has only seen adult speech in training cannot be expected to generalise to children's speech (Steidl *et al.* 2010). Finally, the distribution of the class labels should be similar among partitions, that is, resemble a uniform distribution. Balancing these factors with respect to the partitions is also known as 'stratification'. In order to prevent 'mis-tuning' of the system, in particular the data found in the development partition should be similar in nature to the test partition. Given large enough data sets, random splitting of the data often provides for stratification, but neglects other criteria such as speaker independence. In any case, it has to be ensured that the splitting can be reproduced, such as by delivering lists of training, development, and test instances along with the corpus. Only then can a corpus serve for comparative evaluations of systems.

Concerning the size of the partitions, a compromise has to be found between having enough training data and ensuring significance of results by providing enough instances to develop and test on. A 40–30–30 'percentage split' of the data is often used for training, development, and testing. In the case of learning schemes where model evaluation takes considerable time, such as for dynamic classification using HMMs, smaller test partitions are used. For smaller data sets, one often prefers *cross-validation* over a percentage split scheme. The idea is that by exchanging the 'roles' of the partitions (those used for training, development, and testing) in a cyclic fashion, a maximal amount of data can be used both for training and testing, without ever evaluating a model on data seen during training. To this end, one splits the overall data set into J sets of equal size (J-fold cross-validation), respecting stratification – in this case, one speaks of stratified cross-validation – and possibly other criteria as discussed above. In each cycle $j = 1, \ldots, J$, the data partition with index j is used for testing while training on the remaining ones, possibly holding out another partition, for example, $(j + 1) \bmod J$, for development. It is easy to see that by this scheme, after J cycles each partition has been used for testing once, and at the same time the maximum amount of training data was provided in each cycle. To evaluate the classification performance, one usually provides the mean of the evaluation measure (see below) reached for each cycle, and additionally the standard deviation to estimate the reliability of the evaluation. For $J = 3$, one obtains a scheme similar to a standard percentage split with swapping roles. Other 'popular' choices are $J = 5$ or $J = 10$. Obviously, the computational effort for evaluation increases with J, due to both the higher number of classifiers to be trained and evaluated, and the increased amount of training data available in each cycle. Conversely, for simple statistical reasons the performance estimate across folds gets more and more reliable the higher the value of J used. An extreme case of cross-validation is to leave a single instance out at each cycle – this is known as leave-one-out (LOO). A variant of LOO is to group instances by a certain attribute (such as speaker ID) and perform a 'leave-one-group-out' ('leave-one-speaker-out') cross-validation.

11.3.2 Balancing

Related to stratification, a highly relevant issue for computational paralinguistics evaluation is the balancing of class label distributions. Sometimes instances of a certain class of interest occur rarely – for instance, instances of intoxicated speech are rather hard to obtain, while sober speech is easy to acquire in large quantities. As a result, data sets will be highly imbalanced across classes. A similar issue for regression tasks is that continuous labels often resemble a Gaussian distribution with a high concentration of instances around the centre of the scale (Weninger *et al.* 2012). Because the error measures used in training often involve some kind of average deviation from the ground truth (such as mean squared error in case of neural network training), such imbalance can lead to classifier bias in predicting the most frequently occurring class ('majority class'), or predicting the mean of the training label distribution in the case of a regressor. In extreme cases, it can lead to underrepresented classes or parts of the label distribution being completely ignored. Examples of classifiers sensitive to class imbalance include SVMs, DTs and neural networks. In contrast, ML-based classification (e.g., by HMMs) does not suffer from this problem.

Strategies to alleviate this problem can be divided in two rough categories: training set resampling and adjustment of the training objective (Schuller *et al.* 2009a,b,c). The former is rather straightforward to apply for categorical class labels: either training instances of the majority class are deleted ('down-sampling'), or instances of the minority class(es) are duplicated ('up-sampling'). For continuous class labels, this is less straightforward. An advantage of training set resampling is that the classifier training procedure can be left unmodified as a 'black box', which is often desirable in practical system development. Except for very large databases, up-sampling is usually preferred over down-sampling since it does not remove potentially valuable information from the training set. If 'pure' up-sampling is not possible due to space and memory requirements, mixed up- and down-sampling can be applied. In order to focus re-sampling on those instances of particular relevance, one can use more advanced approaches such as the 'synthetic minority over-sampling technique' (Chawla *et al.* 2002). In most cases, test set instances are not balanced, in order to model the natural distribution of instances, as encountered during data collection, in system evaluation.

Adjustment of the classifier training objective for imbalanced training sets is mostly implemented by assigning different weights to different kinds of class confusions ('cost-sensitive classification') or deviations from the ground truth, in order to introduce heavier penalties for errors on the underrepresented instances. This can be particularly suited to sequential classification tasks – for example, in time continuous emotion prediction – where it is not straightforward to resample the training set.

11.3.3 Performance Measures

Having discussed appropriate partitionings into training, development, and test set, let us now discuss actual criteria for the evaluation on the development and test sets (see also Kroschel *et al.* 2011). We will consider the case of classification first, that is, define measures taking into account the correctness of discrete class attributions. Then, we will move on to criteria for regressors, evaluating by continuous-valued error functions.

Classification Measures

Let us assume the task of classifying into $C \geq 2$ distinct classes – one speaks of C-way classification. We can then consider the result of the classification as a mapping $\hat{y} : \mathcal{X} \to \mathcal{C}$, $\mathcal{C} = \{1, \ldots, C\}$. In particular, this implies that each test instance is mapped to exactly one class $i \in \{1, \ldots, C\}$. Thus, the class prediction for a pattern vector $\underline{x} \in \mathcal{T}$ can be denoted as $\hat{y}(\underline{x})$. Let us further denote by $\mathcal{T}_i$ the set of test instances belonging to class i, and by $T_i = |\mathcal{T}_i|$ the corresponding number of instances of class i. Thus we have

$$\mathcal{T} = \bigcup_{i=1}^{C} \mathcal{T}_i = \bigcup_{i=1}^{C} \{\underline{x}_{i,n} \mid n = 1, \ldots, T_i\}. \tag{11.78}$$

It follows that the test set is of cardinality $|\mathcal{T}| = \sum_{i=1}^{C} T_i$. One can also model the assignment of multiple classes to a single instance – this can be of value, for example, for emotion recognition, to model mixtures of classes such as 'happily surprised' or 'angrily surprised' (Mower *et al.* 2011). Mathematically, one then obtains a mapping to the power set of the set of classes, $\hat{y} : \mathcal{X} \to 2^{\mathcal{C}}$. However, for the sake of simplicity this will not be considered further in this chapter.

The most common measure to evaluate the quality of the mapping $\hat{y}$ is the probability of correct classification – usually termed the accuracy or recognition rate:

$$\text{Accuracy} = \frac{\text{\# correctly classified test instances}}{\text{\# test instances}} = \frac{\sum_{i=1}^{C} |\{\underline{x} \in \mathcal{T}_i \mid \hat{y}(\underline{x}) = i\}|}{|\mathcal{T}|}. \tag{11.79}$$

By considering the recognition rate only for a specific class i, one obtains the *recall* measure RE_i:

$$RE_i = \frac{\left|\{\underline{x} \in \mathcal{T}_i \mid \hat{y}(\underline{x}) = i\}\right|}{T_i}. \tag{11.80}$$

With $p_i = T_i/|\mathcal{T}|$ as the prior probability of class i in the test set, we further have

$$\text{Accuracy} = \sum_{i=1}^{C} p_i RE_i. \tag{11.81}$$

The weighting by p_i in equation (11.81) leads to the name weighted accuracy (WA) or weighted average recall (WAR). Obviously, if the distribution of class labels among test instances is highly non-uniform, this leads to a bias of the measure towards the recall of the most frequent class. In particular, if the classifier always chooses the most frequent class i^*, then Accuracy $= p_{i^*}$ – thus yielding seemingly good performance with a 'dummy' classification rule. For such unbalanced problems, one might thus prefer a constant weight $1/C$ in the calculation of accuracy. This weighting yields the unweighted average recall (UAR), sometimes referred to as unweighted accuracy (UA):

$$UA = \frac{\sum_{i=1}^{C} RE_i}{C}. \tag{11.82}$$

Instead of expressing the classifier behaviour in a single measure, one can also assemble the 'confusion matrix' $\underline{C} = (c_{i,j})$ whose entries correspond to the number of confusions of a given pair of classes:

$$c_{i,j} = \left|\{\underline{x} \in \mathcal{T}_i \mid \hat{y}(\underline{x}) = j\}\right|. \tag{11.83}$$

Normalising each row of the confusion matrix to sum to 1, one obtains conditional probabilities $p_{i,j}^c = P(\hat{y}(\underline{x}) = j | \underline{x} \in \mathcal{T}_i)$ that can be useful in the fusion of multiple classifiers (Wöllmer *et al.* 2011a).

Let us now consider the important special case of binary classification ($C = 2$). This plays a major role in detection tasks, where conceptually one class, the 'positive' class, contains the instances of interest (e.g., emotional, intoxicated or sleepy speech) to be found in a large pool of data, while the other class models the 'negative' class comprising all other data. This class is often also referred to as 'background', 'garbage' or 'rejection' class. Without loss of generality, we will assume the set of classes to be $\mathcal{C} = \{+1, -1\}$, containing the positive and the negative class, and $\mathcal{T} = \mathcal{T}_{+1} \cup \mathcal{T}_{-1}$ accordingly. By this definition we obtain the number of true positives (TP) similar to the numerator of equation (11.80):

$$TP = \left|\{\underline{x} \in \mathcal{T}_{+1} \mid \hat{y}(\underline{x}) = +1\}\right|. \tag{11.84}$$

Conversely, the number of false positives (FP) is given by

$$FP = \left|\{\underline{x} \in \mathcal{T}_{-1} \mid \hat{y}(\underline{x}) = +1\}\right|. \tag{11.85}$$

With *TP* and *FP* we can define the precision (PR):

$$PR = \frac{TP}{TP + FP},$$ (11.86)

intuitively corresponding to the conditional probability that a positively classified instance is indeed a positive one. Considering decisions for the negative class -1, one can analogously introduce true negatives (TN) and false negatives (FN):

$$TN = \left|\left\{\underline{x} \in \mathcal{T}_{-1} \mid \hat{y}(\underline{x}) = -1\right\}\right|,$$ (11.87)

$$FN = \left|\left\{\underline{x} \in \mathcal{T}_{+1} \mid \hat{y}(\underline{x}) = -1\right\}\right|.$$ (11.88)

Using the above definitions, recall for a binary problem is usually defined as

$$RE = \frac{TP}{TP + FN}$$ (11.89)

– it is easy to see that this is equivalent to RE_{+1} in terms of (11.80). Further, one defines specificity (SP) and negative predictive value (NPV),

$$SP = \frac{TN}{TN + FP},$$ (11.90)

$$NPV = \frac{TN}{TN + FN}.$$ (11.91)

TP, TP, FN, and TN correspond to the entries of a 2×2 confusion matrix,

$$\underline{C} = \begin{pmatrix} TP & FN \\ FP & TN \end{pmatrix}.$$ (11.92)

From (11.92) it is obvious that the sum of TP, FN, FP and TN is constant – equivalent to the number of instances in the test set. Thus, a trade-off between these measures has to be found. This is often done by uniting precision and recall in the so-called F_1 measure, which is defined as the harmonic mean of both:

$$F_1 = 2\frac{RE \cdot PR}{RE + PR}.$$ (11.93)

The subscript 1 indicates equal weighting of precision and recall. If one wants to put more emphasis on one or the other, one can consider alternative weighting, such as doubling up the weight of one. In general, this yields the F_β measure,

$$F_\beta = (1 + \beta^2)\frac{RE \cdot PR}{RE + \beta^2 PR}.$$ (11.94)

Thus, for example, F_2 would put more emphasis on recall than precision. In terms of TP, FN and FP, we have that

$$F_\beta = \frac{(1 + \beta^2)\text{TP}}{(1 + \beta^2)\text{TP} + \beta^2\text{FN} + FP} \tag{11.95}$$

– thus, the measure F does not take into account the number of true negatives.

An alternative method is to determine the *receiver operating characteristic* (ROC) of the classifier. This is possible if the classifier outputs a probability of the positive class. For example, in neural networks, the activations of the output units indicate posterior class probabilities if they are normalised to sum to 1. One can then define a threshold θ such that all instances with a probability of the positive class above θ are accepted, and compute the true positive ratio (TPR) and false positive ratio (FPR). TPR is equivalent to recall (11.89) in the case of binary classification, while

$$\text{FPR} = \frac{\text{FP}}{\text{TN} + \text{FP}}. \tag{11.96}$$

From (11.89) and (11.96), it follows that FPR and TPR both increase monotonically with more and more instances being classified as positive. Thus, by monotonically increasing θ, a monotonic function FPR(TPR) can be obtained. By interpolation (e.g., using the rectangle rule), one can integrate this function over the range [0, 1] and obtain the *area under curve* (AUC) measure, in the range [0, 1]. A similar measure is obtained by considering the detection error trade-off (DET) curve, which is the function FNR(FPR) which shows the false negative rate (FNR) in terms of the false positive rate and is defined by analogy with the above. One can then calculate the intersection (E, E) of the DET curve with the identity line and obtains the *equal error rate* (EER) E. This corresponds to the optimal classifier configuration if one considers FNR and FPR as equally important.

Note that all the above measures for the two-class case can be generalised to the multi-class scenario if one considers a 'one-versus-all' scenario, picking a class $i \in \{1, \dots, C\}$ as the 'positive' class and considering everything else as negative.

Regression Measures

To conclude the discussion of evaluation criteria, let us now briefly introduce the most common measures for evaluation of regression functions, that is, prediction of continuous-valued quantities instead of discrete class labels. By analogy with the above, we denote the prediction of the regressor by $\hat{y}$, which is now a mapping $\mathcal{X} \to \mathbb{R}$. Just as for classification, we only consider prediction of a single value (univariate regression), not vector-valued quantities (multivariate regression).

Recall that the test set is denoted by $\mathcal{T}$. Let us further denote the actual label of instance $\underline{x}_n \in \mathcal{T}$, $n = 1, \dots, |\mathcal{T}|$, by y_n. Probably the most straightforward measure is the expected deviation of $\hat{y}$ from the actual labels, the mean linear error (MLE)

$$\text{MLE} = \frac{1}{|\mathcal{T}|} \sum_{n=1}^{|\mathcal{T}|} |\hat{y}(\underline{x}_n) - y_n|. \tag{11.97}$$

Note that the absolute value is needed here since otherwise overshooting and underestimation would cancel each other out. Hence the above measure is also called mean absolute error (MAE). An alternative is to take the square instead of the absolute value, yielding the mean square error (MSE). Often also the root of the MSE is considered (root mean squared error (RMSE)). MSE and RMSE are closely related to the *variance* and *bias* of the classifier (Wackerly and Scheaffer 2008). Sometimes the *relative* error is considered instead of the absolute error, resulting, for example, in the mean absolute percentage error (MAPE):

$$\text{MAPE} = \frac{1}{|T|} \sum_{n=1}^{|T|} \left| \frac{\hat{y}(\underline{x}_n) - y_n}{y_n} \right|. \tag{11.98}$$

However, this leads to numerical problems for ground truth labels close or equal to zero and can thus become hard to use in practice.

All of the above measures are scale variant, that is, their value depends on the range of the ground truth labels. For example, if a regressor is trained for arousal values that are determined from observer ratings on a five-point scale (-2 to 2) converted to a quasi-continuum by averaging (see Chapter 6), and the same type of regressor is trained on ratings in $[-1, 1]$, one would expect half the MAE in the latter case, without any meaningful difference in actual performance. A scale-invariant measure for regression is the correlation coefficient (CC) of prediction and ground truth,

$$\text{CC} = \frac{\sum_{n=1}^{|T|} \left(\hat{y}(\underline{x}_n) - \overline{\hat{y}} \right) (y_n - \overline{y})}{\sqrt{\sum_{n=1}^{|T|} \left(\hat{y}(\underline{x}_n) - \overline{\hat{y}} \right)^2 \cdot \sum_{n=1}^{|T|} (y_n - \overline{y})^2}}, \tag{11.99}$$

with the average prediction

$$\overline{\hat{y}} = \frac{1}{|T|} \sum_{n=1}^{|T|} \hat{y}(\underline{x}_n), \tag{11.100}$$

and the average test set label

$$\overline{y} = \frac{1}{|T|} \sum_{n=1}^{|T|} y_n. \tag{11.101}$$

The CC can be understood as the covariance of the predictions and the ground truth labels, normalised to the range $[-1, +1]$ – thus, positive values indicate agreement of the two, zero corresponds to chance, and negative values indicate systematic deviations.

To sum up, MAE and similar measures are often useful as an intuitive measure – for example, in height estimation, the MAE measures by how many centimetres the regressor in question would be mistaken on average. However, as its range varies by task it is of little use for cross-task comparisons (e.g., to determine if age or height can be estimated more precisely).

11.3.4   Result Interpretation

It is obvious that the value of an evaluation measure depends on the choice of the test instances. Results might be skewed by choosing a test set biased towards 'hard' or 'easy' cases, which can be very hard to detect in advance. To remedy this problem, one can introduce an additional level of evaluation, which might be termed 'meta-evaluation': evaluating the reliability of evaluation. We have introduced such 'meta-evaluation' on-the-fly in the context of cross-validation, where the mean and standard deviation of the evaluation measures are usually reported across multiple different partitionings of the same data in order to assess the stability of the evaluation measure depending on the choice of training and test instances. Besides, it is often of interest to assess whether observed performance differences, obtained on a fixed partitioning of the database by multiple classifiers or regressors, are caused by random fluctuations or structural differences.

To address these questions, one typically uses *significance tests*. Let us start by considering the issue of comparing the accuracies of two classification systems on a specific test set, before turning to regressors, and performance comparison across different partitionings. As a first example, we introduce the McNemar test (Dietterich 1998; Gillick and Cox 1989; McNemar 1947). This is based on summarising the results of the two classification systems **A** and **B** in a *contingency table*:

A	**B**	
	correct	incorrect
correct	n_{00}	n_{01}
incorrect	n_{10}	n_{11}

Its elements are defined as follows:

- n_{00} (n_{11}) are the numbers of instances in the test set $\mathcal{T}$ which are classified correctly (incorrectly) by both systems;
- n_{10} is the number of instances classified correctly by **B** but incorrectly by **A**;
- n_{01} is the number of instances classified correctly by **A** but incorrectly by **B**.

Obviously, the sum of the entries in the contingency table, $n_{00} + n_{01} + n_{10} + n_{11}$, equals the number of instances in the test set, $|\mathcal{T}|$. Furthermore, the accuracies of **A** and **B** can be written in terms of the above quantities as $\frac{n_{00}+n_{01}}{|\mathcal{T}|}$ and $\frac{n_{00}+n_{10}}{|\mathcal{T}|}$. Thus, the deviation in classification performance is given by the quantities n_{01} and n_{10}. A significance test aims to answer the question whether this deviation can be explained by a random process with sufficient probability.

To this end, one assumes that the quantities in the contingency table are observations of one random variable, each. In particular, the random variable corresponding to n_{10} shall be denoted by N_{10}. Thus one can formulate a *null hypothesis* H_0 about its distribution, allowing the probability p of the observation n_{10} to be calculated. If p is small, one concludes in turn that the observed difference between the systems is not caused by a random process described by N_{10}. One also speaks of a *significant* difference between **A** and **B** at a *significance level* α if $p \le \alpha$. The exact threshold α depends on the problem statement, particularly on the required confidence in the decision. For automatic classifiers and regressors, one typically uses $\alpha = 0.05$ (5% level) or $\alpha = 0.01$ (1% level).

The null hypothesis of the McNemar test is that N_{10} follows a binomial distribution with success probability $p = 1/2$ and sample size $k := n_{10} + n_{01}$, that is,

$$P(N_{10} = n_{10}) = \binom{k}{n_{10}} \left(\frac{1}{2}\right)^k. \tag{11.102}$$

The underlying idea is that those k elements of the test set which are classified correctly only by one system are distributed between the observed quantities n_{10} and n_{01} with equal probability. One writes

$$H_0 : N_{10} \sim \mathrm{B}\left(k, \frac{1}{2}\right). \tag{11.103}$$

The above null hypothesis is verified by means of a random variable $M \sim \mathrm{B}(k, \frac{1}{2})$. M has expected value $E\{M\} = k/2$. One now considers the probability α that M deviates from its expected value as is described by n_{10} in the contingency table. Because of the symmetry of the binomial distribution, one obtains:

$$p = \begin{cases} 2 \cdot P(n_{10} \leq M \leq k) & \text{if } n_{10} > k/2 \\ 2 \cdot P(0 \leq M \leq n_{10}) & \text{if } n_{10} < k/2 \\ 1 & \text{if } n_{10} = k/2. \end{cases} \tag{11.104}$$

Using the formula for the cumulative distribution function of the binomial distribution, one can determine these probabilities analytically:

$$P(n_{10} \leq M \leq k) = \sum_{m=n_{10}}^{k} \binom{k}{m} \left(\frac{1}{2}\right)^k,$$

$$P(0 \leq M \leq n_{10}) = \sum_{m=0}^{n_{10}} \binom{k}{m} \left(\frac{1}{2}\right)^k. \tag{11.105}$$

The probability p is the probability of error if H_0 is true but is rejected because of the observation of n_{10} – this is called *Type I error*. One has to bear in mind that the value of p does not say whether **A** or **B** is better – this has to be determined by examining the contingency table.

Often, instead of exactly computing α via (11.105), one uses an approximation by a normal distribution. According to the central limit theorem, for large k the random variable

$$N_{10}^* = \frac{N_{10} - \frac{k}{2} - \frac{1}{2}}{\sqrt{\frac{k}{4}}} \tag{11.106}$$

approximately follows a Gaussian distribution with $\mu = 0$ und $\sigma = 1$. To increase the quality of the approximation, one additionally performs a *continuity correction* by the term $-\frac{1}{2}$. With

(11.106) and the cumulative distribution function of the normal distribution (for which no closed-form solution exists; it has to be computed numerically) one can approximate the probabilities from (11.105). The above approximation is particularly useful in practice for large k, for which an exact computation would be difficult for numerical reasons. As a rule of thumb, one should use the approximation (11.106) only if $k > 50$ and n_{10} is not close to zero or k (Gillick and Cox 1989).

A disadvantage of the McNemar test is that one needs the contingency table, which makes it unsuitable for performance comparisons where only the accuracies are known, not the predictions themselves. As an alternative that is easier to compute, one can use a simpler variant of the binomial test. One assumes that the probabilities of correct classification of the systems, p_A and p_B, are known. Without loss of generality we assume that $p_B > p_A$. In practice, one has to estimate these probabilities by the accuracy values obtained. Analogously to the McNemar test, one then supposes that the observed performance difference is due to random fluctuations from the average probability of correct classification, $p_{AB} = (p_A + p_B)/2$, and disproves this hypothesis at a chosen level of significance.

One considers the number of correct classifications on the test set as a random variable N_c, for which one can assume a binomial distribution given the statistical independence of the prediction errors from each other. The null hypothesis is that N_c follows a binomial distribution with success probability p_{AB}:

$$H_0 : N_c \sim B(|T|, p_{AB}). \tag{11.107}$$

One then computes the probability of observing the improved accuracy of **B** as

$$P(N_c > p_B \cdot |T|) = 1 - P(N_c \le p_B \cdot |T|). \tag{11.108}$$

By analogy with the above, the binomial distribution is usually approximated by a normal distribution, leading to the random variable N_c^* given by

$$N_c^* = \frac{N_c - p_{AB} \cdot |T|}{\sqrt{|T|p_{AB}(1 - p_{AB})}} = \frac{N_c/|T| - p_{AB}}{\sqrt{p_{AB}(1 - p_{AB})}} \sqrt{|T|} \tag{11.109}$$

which follows a standard normal distribution (the continuity correction is omitted for the sake of simplicity). As a test quantity, one computes the realisation of N_c^* corresponding to the observed accuracy of **B**,

$$n_{c,\mathbf{B}}^* = \frac{p_B - p_{AB}}{\sqrt{p_{AB}(1 - p_{AB})}} \sqrt{|T|}. \tag{11.110}$$

Then, if

$$1 - P(N_c^* \le n_{c,\mathbf{B}}^*) = 1 - F(n_{c,\mathbf{B}}^*) < \alpha, \tag{11.111}$$

one can reject H_0 at significance level α. The above is a 'one-sided' test, testing if '**B** is better than **A**' – for symmetry reasons in the above definition, this is equivalent to testing if **A** is worse than **B**. One can also define a two-sided test analogously to the McNemar test above, but this is rarely used. Because normally distributed quantities are often denoted by the symbol z, the above test is called a z-*test*.

For a given significance level α, one can now consider the 'line of significance'

$$1 - F(n^*_{c,\mathbf{B}}) = \alpha. \tag{11.112}$$

Solving for p_B, one obtains a function $p_B(p_A)$ indicating how good **B** has to be to surpass the baseline system **A** at significance level α. Equivalently, one can regard the accuracy difference $p_B - p_A$ required for reaching significance at a level α as a function of the baseline accuracy. An example of this function is given in Figure 11.14 for classification experiments with sample sizes of $N = 1000$ and $N = 10\,000$, and α chosen as 0.001, 0.01, or 0.05. The required accuracy improvement is given by the intersection of the vertical line corresponding to the baseline accuracy and the curve corresponding to the level of significance. The vertical line in Figure 11.14 indicates chance level accuracy ($p_A = 0.5$) for a (balanced) binary classification task, as widely encountered in computational paralinguistics.

From Figure 11.14, it is obvious that the required accuracy difference to reach significance becomes smaller and smaller with better baseline accuracy. This is rather intuitive, as one expects small absolute improvements in systems with already good performance to be 'more significant' than the same absolute improvement for lower performing systems.

An advantage of the z-test is that it only requires the accuracies of both systems and the size of the test set. However, the assumption of independence of errors is not necessarily made in practice, for example in the case of sequence labelling by neural networks, where the prediction on one frame strongly depends on the previous predictions. Furthermore, it is assumed that

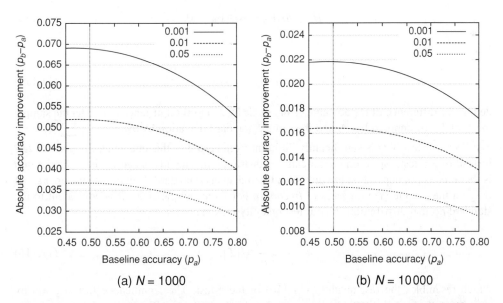

(a) $N = 1000$ (b) $N = 10\,000$

Figure 11.14 Lines of significant absolute accuracy improvements for different levels of significance ($\alpha = 0.001$, 0.01, or 0.05) and sample sizes $N = 1000$ (a) and $N = 10\,000$ (b). Vertical line: chance level accuracy (0.5)

the accuracies of both systems can be estimated reliably, which requires sufficiently large test sets. Despite these theoretical limitations, the z-test is among the most widely used (Dieterich 1998).

Let us now briefly discuss significance testing for regressors, which can be derived from similar considerations to those above. Recall that the correct or incorrect decision of a classifier for each instance can be modelled as a Bernoulli experiment with success probability corresponding to the classifier's (true) accuracy. The modelling of the classifier's accuracy as a normally distributed random variable resulted from the approximation of the binomial distribution (modelling sums of Bernoulli distributed random variables), which is justified by the central limit theorem. More precisely, this approximation holds if the error on each instance is statistically independent of the other errors, and follows the same distribution – one speaks of independent identically distributed (i.i.d.) random variables. If we now consider the absolute deviation of a regressor's prediction on instance n of the test set from the ground truth, $|\hat{y}(\underline{x}_n) - y_n|$, as a random variable, the central limit theorem tells us that the sum of this random variable taken across the test set will be approximately normally distributed for a sufficiently large test set. Obviously, the same holds for the MAE (11.97), which is the above named sum scaled by the size of the test set.

Suppose we have two regressors **A** and **B** delivering predictions on the same test set. Without loss of generality, we assume that **B** has a lower MAE than **A**. We have two random variables $\text{MAE}_\mathbf{A}$ and $\text{MAE}_\mathbf{B}$. A one-tailed significance test is based on the null hypothesis

$$H_0 : \text{MAE}_\mathbf{A} > \text{MAE}_\mathbf{B}, \tag{11.113}$$

or equivalently,

$$H_0 : \text{MAE}_\mathbf{A} - \text{MAE}_\mathbf{B} > 0. \tag{11.114}$$

Let

$$D_n = |\hat{y}_\mathbf{A}(\underline{x}_n) - y_n| - |\hat{y}_\mathbf{B}(\underline{x}_n) - y_n| \tag{11.115}$$

denote the difference of the absolute errors made by **A** and **B** on instance n, and let the sample mean of D_n be denoted by $\overline{D}$, and the sample standard deviation by $\sigma_D = \frac{1}{|\mathcal{T}|-1}\sum_{n=1}^{|\mathcal{T}|}(D_n - \overline{D})^2$. We now test if $\overline{D}$ is significantly different from zero. For this one generally uses a *t-test* instead of a z-test, based on the so-called t-distribution. The advantage of the t-distribution is that it does not depend on the unknown true mean and standard deviation of the errors, but only on the sample size (*degrees of freedom*). For large sample sizes, it is equivalent to the standard normal distribution. As our test quantity we obtain

$$t = \frac{\overline{D}}{\sigma_D}\sqrt{|\mathcal{T}|}, \tag{11.116}$$

which resembles the test quantity (11.110) of the z-test. The significance test is given by comparing the probability that t is observed under H_0 to the significance level α,

$$1 - F_t^{|\mathcal{T}|}(t) < \alpha, \tag{11.117}$$

where $F_t^{|T|}$ is the cumulative distribution function of the t-distribution with $|T|$ degrees of freedom. If this probability is smaller than α, one can reject the null hypothesis of equal performance of **A** and **B** in terms of MAE. By setting the degrees of freedom to the size of the test set, one uses a *paired* t-test, respecting that the sampling of the errors of **A** and **B** is not independent – they are carried out on the same test set.

So far, the choice of the test set has been assumed fixed – thus, the above methods still do not allow us to assess whether some observed performance differences are caused by particular choices of training and test data, rather than fundamental properties of the classifier. Thus, let us now consider the case of repeated evaluation using various partitionings of the data set. Let J denote the number of such partitionings – such as in J-fold cross-validation. Furthermore, let $E_{\mathbf{A}}^{(1)}, \dots, E_{\mathbf{A}}^{(J)}, E_{\mathbf{B}}^{(1)}, \dots, E_{\mathbf{B}}^{(J)}$ denote the evaluation measure reached by classifiers (or regressors) **A** and **B** in partitionings $1, \dots, J$. E can represent any of the measures discussed in this chapter, such as accuracy, UAR, MAE, or RMSE. We can then define the performance difference of **A** and **B** on partitioning j, $j = 1, \dots, J$, as

$$E_D^{(j)} = E_{\mathbf{A}}^{(j)} - E_{\mathbf{B}}^{(j)}. \tag{11.118}$$

Then, we proceed to test whether this difference is significantly different from zero, similar to the MAE significance test above. In particular, a t-test can be used with test statistic

$$t' = \frac{\overline{E_D}}{\sigma_{E_D}} \sqrt{J}, \tag{11.119}$$

where $\overline{E_D}$ and σ_{E_D} are the sample mean and standard deviation of the observed performance differences. Analogously to (11.117), the one-tailed significance test is given by evaluating the inequality

$$1 - F_t^J(t') < \alpha. \tag{11.120}$$

For a more in-depth discussion of evaluation by resampled training and test sets, see Dieterich (1998).

Besides evaluating performance differences across multiple partitionings, the t-test described above is also useful as a replacement for the binomial test for accuracy differences where the assumption of independence across test instances is not met (e.g., if emotion is determined by a neural network or HMM at the frame level). Here, one can compute the evaluation measure of interest (such as the MAE) in a larger unit of analysis that is analysed independently of the others (e.g., a user turn), and then compare the performance measurements obtained analogously to the comparison across partitionings. This has been proposed for word accuracy significance testing in automatic speech recognition by (Gillick and Cox 1989).

References

Baum, L. E., Petrie, T., Soules, G., and Weiss, N. (1970). A maximization technique occurring in the statistical analysis of probabilistic functions of Markov chains. *Annals of Mathematical Statistics*, **41**, 164–171.

Chawla, N. V., Bowyer, K. W., Hall, L. O., and Kegelmeyer, W. P. (2002). Synthetic minority over-sampling technique. *Journal of Artificial Intelligence Research*, **16**, 321–357.

Cortes, C. and Vapnik, V. (1995). Support-vector networks. *Machine Learning*, **20**, 273–297.

Cristianini, N. and Shawe-Taylor, J. (2000). *An Introduction to Support Vector Machines and Other Kernel-Based Learning Methods*. Cambridge University Press, Cambridge.

Deller, J., Proakis, J., and Hansen, J. (1993). *Discrete-Time Processing of Speech Signals*. Macmillan, New York.

Dietterich, T. G. (1998). Approximate statistical tests for comparing supervised classification learning algorithms. *Neural Computation*, **10**, 1895–1923.

Evgeniou, T., Micchelli, C. A., and Pontil, M. (2005). Learning multiple tasks with kernel methods. *Journal of Machine Learning Research*, **6**, 615–637.

Furui, S. (1996). *Digital Speech Processing: Synthesis, and Recognition*. Signal Processing and Communications. Marcel Dekker, New York, 2nd edition.

Gers, F., Schmidhuber, J., and Cummins, F. (2000). Learning to forget: Continual prediction with LSTM. *Neural Computation*, **12**, 2451–2471.

Gillick, L. and Cox, S. J. (1989). Some statistical issues in the comparison of speech recognition algorithms. In *Proc. of ICASSP*, pp. 23–26, Glasgow, UK.

Graves, A. (2008). *Supervised sequence labelling with recurrent neural networks*. Ph.D. thesis, Technische Universität München.

Graves, A. and Schmidhuber, J. (2005). Framewise phoneme classification with bidirectional LSTM and other neural network architectures. *Neural Networks*, **18**, 602–610.

Ho, T. K. (1998). The random subspace method for constructing decision forests. *IEEE Transactions on Pattern Analysis and Machine Intelligence*, **20**, 832–844.

Hochreiter, S. and Schmidhuber, J. (1997). Long short-term memory. *Neural Computation*, **9**, 1735–1780.

Hochreiter, S., Bengio, Y., Frasconi, P., and Schmidhuber, J. (2001). Gradient flow in recurrent nets: the difficulty of learning long-term dependencies. In S. C. Kremer and J. F. Kolen (eds), *A Field Guide to Dynamical Recurrent Neural Networks*, pp. 1–15. IEEE Press, New York.

Jelinek, F. (1997). *Statistical Methods for Speech recognition*. MIT Press, Cambridge, MA.

Joachims, T. (1998). Text categorization with support vector machines: learning with many relevant features. In C. Nédellec and C. Rouveirol (eds), *Proc. of ECML-98, 10th European Conference on Machine Learning*, pp. 137–142, Chemnitz. Springer, Berlin.

Kroschel, K., Rigoll, G., and Schuller, B. (2011). *Statistische Informationstechnik*. Springer, Berlin, 5th edition.

Lacoste, A. and Eck, D. (2005). Onset detection with artificial neural networks. In *Extended abstract of the 1 st Annual Music Information Retrieval Evaluation eXchange (MIREX), held in conjunction with ISMIR*.

Landwehr, N., Hall, M., and Frank, E. (2005). Logistic model trees. *Machine Learning*, **59**(1–2), 161–205.

Lodhi, H., Saunders, C., Shawe-Taylor, J., Cristianini, N., and Watkins, C. (2002). Text classification using string kernels. *Journal of Machine Learning Research*, **2**, 419–444.

Lowerre, B. (1976). *The Harpy Speech Recognition System*. Ph.D. thesis, Carnegie Mellon University, Pittsburgh, USA.

McCulloch, W. and Pitts, W. (1943). A logical calculus of the ideas immanent in nervous activity. *Bulletin of Mathematical Biophysics*, **9**, 115–133.

McNemar, Q. (1947). Note on the sampling error of the difference between correlated proportions or percentages. *Psychometrika*, **12**, 153–157.

Mower, E., Mataric, M. J., and Narayanan, S. S. (2011). A framework for automatic human emotion classification using emotion profiles. *IEEE Transactions on Audio, Speech and Language Processing*, **19**, 1057–1070.

Niemann, H. (2003). *Klassifikation von Mustern*. published online, 2nd, revised and extended edition. http://www5.informatik.uni-erlangen.de/Personen/niemann/klassifikation-von-mustern/m00links.html, last visited 09/20/2007.

O'Shaughnessy, D. (1990). *Speech Communication*. Adison-Wesley, Reading, MA, 2nd edition.

Pachet, F. and Roy, P. (2009). Analytical features: A knowledge-based approach to audio feature generation. *EURASIP Journal on Audio, Speech, and Music Processing*.

Platt, J. (1998). Sequential minimal optimization: A fast algorithm for training support vector machines. Technical Report MSR-98-14, Microsoft Research.

Quinlan, J. (1983). Learning efficient classification procedures and their application to chess end games. In *Machine Learning: An Artificial Intelligence Approach*, pp. 106–121. Tioga Publishing, Palo Alto, CA.

Quinlan, J. (1987). Simplifying decision trees. *International Journal of Man-Machine Studies*, **27**, 221–234.

Quinlan, J. (1996). Bagging, boosting and C4.5. In *Proc. of the 14th National Conference on AI*, volume 5, pp. 725–730, Menlo Park, CA.

Quinlan, J. R. (1993). *C4.5: Programs for Machine Learning*. Morgan Kaufmann.

Rabiner, L. R. (1989). A tutorial on hidden Markov models and selected applications in speech recognition. *Proceedings of the IEEE*, **77**, 257–286.

Riedmiller, M. and Braun, H. (1992). Rprop – a fast adaptive learning algorithm. In *Proc. of the International Symposium on Computer and Information Science*, volume VII.

Rigoll, G. (1994). *Neuronale Netze*. Expert-Verlag, Renningen, Germany.

Rumelhart, D., Hinton, G., and Williams, R. (1987). Learning internal representations by error propagation. In *Parallel Distributed Processing: Explorations in the Microstructure of Cognition*, volume 1, pp. 318–362. MIT Press, Boston, MA.

Ruske, G. (1993). *Automatische Spracherkennung. Methoden der Klassifikation und Merkmalsextraktion*. Oldenbourg, Munich, 2nd edition.

Schalkoff, R. (1994). *Artificial Neural Networks*. McGraw-Hill, New York.

Schölkopf, B. and Smola, A. (2002). *Learning with Kernels: Support Vector Machines, Regularization, Optimization, and Beyond (Adaptive Computation and Machine Learning)*. MIT Press, Cambridge, MA, USA.

Schuller, B. (2006). *Automatische Emotionserkennung aus sprachlicher und manueller Interaktion*. Doctoral thesis, Technische Universität München, Munich.

Schuller, B., Arsić, D., Wallhoff, F., Lang, M., and Rigoll, G. (2005). Bioanalog acoustic emotion recognition by genetic feature generation based on low-level-descriptors. In *Proc. of the International Conference on Computer as a Tool, EUROCON*, pp. 1292–1295, Belgrade.

Schuller, B., Reiter, S., and Rigoll, G. (2006a). Evolutionary feature generation in speech emotion recognition. In *Proc. of the 7th IEEE International Conference on Multimedia and Expo, ICME*, pp. 5–8, Toronto.

Schuller, B., Wallhoff, F., Arsić, D., and Rigoll, G. (2006b). Musical signal type discrimination based on large open feature sets. In *Proc. of the 7th IEEE International Conference on Multimedia and Expo, ICME*, pp. 1089–1092, Toronto.

Schuller, B., Müller, R., Eyben, F., Gast, J., Hörnler, B., Wöllmer, M., Rigoll, G., Höthker, A., and Konosu, H. (2009a). Being bored? Recognising natural interest by extensive audiovisual integration for real-life application. *Image and Vision Computing*, **27**, 1760–1774.

Schuller, B., Schenk, J., Rigoll, G., and Knaup, T. (2009b). "The Godfather" vs. "Chaos": Comparing linguistic analysis based on online knowledge sources and bags-of-N-grams for movie review valence estimation. In *Proc. of the 10th International Conference on Document Analysis and Recognition, ICDAR*, pp. 858–862, Barcelona, Spain.

Schuller, B., Steidl, S., and Batliner, A. (2009c). The Interspeech 2009 Emotion Challenge. In *Proc. of Interspeech*, pp. 312–315, Brighton, UK.

Schuller, B., Batliner, A., Steidl, S., and Seppi, D. (2011). Recognising realistic emotions and affect in speech: State of the art and lessons learnt from the first challenge. *Speech Communication*, **53**, 1062–1087.

Schuller, B., Steidl, S., Batliner, A., Schiel, F., Krajewski, J., Weninger, F., and Eyben, F. (2013a). Medium-term speaker states – a review on intoxication, sleepiness and the first challenge. *Computer Speech and Language, Special Issue on Broadening the View on Speaker Analysis*.

Schuller, B., Steidl, S., Batliner, A., Burkhardt, F., Devillers, L., Müller, C., and Narayanan, S. (2013b). Paralinguistics in speech and language – state-of-the-art and the challenge. *Computer Speech and Language*, **27**, 4–39.

Smola, A. and Schölkopf, B. (2004). A tutorial on support vector regression. *Statistics and Computing*, **14**, 199–222.

Steidl, S., Batliner, A., Seppi, D., and Schuller, B. (2010). On the impact of children's emotional speech on acoustic and language models. *EURASIP Journal on Audio, Speech, and Music Processing, Special Issue on Atypical Speech*, **2010**.

Wackerly, D. and Scheaffer, W. (2008). *Mathematical Statistics with Applications*. Thomson Higher Education, Belmont, CA, 7th edition.

Weninger, F., Krajewski, J., Batliner, A., and Schuller, B. (2012). The voice of leadership: Models and performances of automatic analysis in online speeches. *IEEE Transactions on Affective Computing*, **3**, 496–508.

Werbos, P. (1990). Backpropagation through time: What it does and how to do it. *Proc. of the IEEE*, **78**, 1550–1560.

White, C. M., Rastrow, A., Khudanpur, S., and Jelinek, F. (2009). Unsupervised estimation of the language model scaling factor. In *Proc. of Interspeech*, pp. 1195–1198, Brighton.

Widmer, C. and Rätsch, G. (2012). Multitask learning in computational biology. *Journal of Machine Learning Research*, **25**, 207–216.

Witten, I. H. and Frank, E. (2005). *Data Mining: Practical Machine Learning Tools and Techniques*. Morgan Kaufmann, San Francisco, 2nd edition.

Wöllmer, M., Schuller, B., Eyben, F., and Rigoll, G. (2010). Combining long short-term memory and dynamic Bayesian networks for incremental emotion-sensitive artificial listening. *IEEE Journal of Selected Topics in Signal Processing*, **4**(5), 867–881.

Wöllmer, M., Eyben, F., Schuller, B., and Rigoll, G. (2011a). A multi-stream ASR framework for BLSTM modeling of conversational speech. In *Proc. of ICASSP*, pp. 4860–4863, Prague, Czech Republic.

Wöllmer, M., Blaschke, C., Schindl, T., Schuller, B., Färber, B., Mayer, S., and Trefflich, B. (2011b). On-line driver distraction detection using long short-term memory. *IEEE Transactions on Intelligent Transportation Systems*, **12**(2), 574–582.

Wöllmer, M., Schuller, B., Batliner, A., Steidl, S., and Seppi, D. (2011c). Tandem decoding of children's speech for keyword detection in a child-robot interaction scenario. *ACM Transactions on Speech and Language Processing, Special Issue on Speech and Language Processing of Children's Speech for Child-machine Interaction Applications*, **7**(4). Article 12.

Yang, H., Xu, Z., Ye, J., King, I., and Lyu, M. (2011). Efficient sparse generalized multiple kernel learning. *IEEE Transactions on Neural Networks*, **22**(3), 433–446.

12

System Integration and Application

The stability of the whole is guaranteed by the instability of its parts.

(Karin Meissenburg)

Having discussed the theoretical foundations of computational modelling of paralinguistics, we now dig deeper into the engineering side of things: how to implement systems for actual use in real-life conditions. This may include practical aspects such as using standards as EmotionML for the output encoding or even encoding of the features used (see Section A.2 in the Appendix). As not all these considerations can be covered here, three aspects were selected: distributed processing, autonomous and collaborative learning, and confidence measures. Many other issues exist, which are also partly covered in the literature, such as noise and reverberation robustness, distant talking, speech coding artefacts, and gating, that is, how soon after speech onset a reliable estimate can be made on the paralinguistic target task. Pointers to suitable literature can be found in recent surveys (Schuller *et al.* 2011, 2013a,b). Further, the techniques often strongly resemble those used in related speech and speaker recognition tasks.

12.1 Distributed Processing

Looking at the related field of automatic speech recognition (ASR), we observe that not many developments have had as much influence as the deployment of the technology in mobile services, which have turned out to be a 'killer application'. This comes as the use of ASR in daily life has increased the availability of realistic data for research and development of improved recognition technology, leading to a virtuous circle of better usability and more widespread usage. In particular, self-improving ASR systems have been implemented for mobile services (Gilbert *et al.* 2011). In this respect, ASR may have paved the way for future usage of speech technology in mobile applications. To the best of our knowledge, there does not exist a ready-to-use service for mobile computational paralinguistics. Thus, this section contains a bit of 'tea-leaf reading'. Yet, the area of possible applications is widespread, including 'hearing glasses' for individuals with difficulties in understanding, for example, affective cues, such as persons on the autism spectrum or cochlear-implanted patients, entertainment applications, and gaming. Just as for ASR, implementing such applications

Computational Paralinguistics: Emotion, Affect and Personality in Speech and Language Processing, First Edition.
Björn W. Schuller and Anton M. Batliner. © 2014 John Wiley & Sons, Ltd. Published 2014 by John Wiley & Sons, Ltd.

seems to offer some promise, as the ever present scarcity of labelled, realistic data could be remedied by relying on input from mobile devices such as phones to gradually build paralinguistic models based on real-life speech.

In the following, we will assume that such mobile services require a suitable architecture for distributed processing, where parts of the chain of processing are done on the mobile client while model building and classification/regression are carried out on a server. The design of such architectures will be the focus of this section.

Moving from a stand-alone architecture to a distributed architecture, the questions of transmission and storage of audio features become crucial. Obviously, for the 'low-level' transmission protocols, plenty of standards exist. However, one has to keep in mind the required transmission bandwidth, and the storage cost – in a distributed architecture, we assume that a database of labelled audio features is maintained by the server. To illustrate the required bandwidth, let us consider the baseline feature set for the Interspeech 2009 Emotion Challenge (see Section A.1), where the dimension of the features per speech sample is 384 (Schuller *et al.* 2011), computed as supra-segmental functionals spanning time periods of approximately 1 second. Recall that more recent studies in affect recognition (e.g., Zhang *et al.* 2011) employ thousands of features. Thus, it makes sense to employ feature compression. This can be done, for example, by the split vector quantisation (SVQ) algorithm. In contrast to simple vector quantisation (VQ), it uses several codebooks for different 'parts' of the vectors, that is, subspaces of the feature space. Let us denote the number of codebooks by L – using codebooks of M-bit words, we have

$$L = \left\lceil \frac{N}{2^M} \right\rceil.$$

(12.1)

Formally, a codebook is a mapping $c_l : \{1, \ldots, K\} \to \mathcal{X}$, where $K = 2^M$ is the codebook size. The range of this mapping (the code vectors) is determined by application of a clustering algorithm such as K-means. In the feature compression step for a vector $\underline{x}$ to be transmitted, for $l = 1, \ldots, L$ the index $i_l^* \in \{1, \ldots, K\}$ of the closest codebook entry is determined,

$$i_l^* = \arg \min_{i_l} d(c_l(i_l), \underline{x}).$$

(12.2)

The (weighted) Euclidean distance metric is usually used for $d(\cdot)$. Then, as a compressed feature vector, the vector $(i_1^*, \ldots, i_L^*)^T$ is transmitted. It follows that the transmission size per feature vector is $L \cdot M$ bits. Note that with $L \cdot M$ bits, we can either encode $2^{L \cdot M}$ reference vectors $\in \mathcal{X}$, as would be done in standard VQ, or $(2^M)^L$ combinations of 2^M reference sub-vectors in the L subspaces of $\mathcal{X}$ corresponding to the codebooks. Thus, given a fixed transmission bandwidth, SVQ is better suited to large feature spaces than VQ, as it improves generalisation and drastically reduces the computational complexity of the encoding step (12.2). In the feature decompression step, a feature vector in the original space $\mathcal{X}$ is reconstructed by table look-up.

Assuming $N \cdot 32$ bits for transmission of an uncompressed N-dimensional vector (single precision floating point numbers), the compression rate (CR) obtained by using SVQ is

$$\text{CR} = \frac{32 \cdot N}{L \cdot M}.$$

(12.3)

In Han *et al.* (2012), the trade-off between CR and unweighted average recall (UAR) for affect recognition is analysed, indicating that CRs of 20–40 are feasible without expecting significant

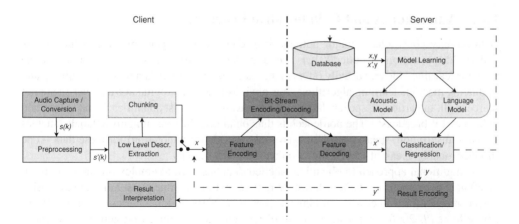

Figure 12.1 Example flow-chart of a distributed system for computational paralinguistics. Components typically contained in a stand-alone system are shown in light grey, while dark grey nodes indicate components in a distributed architecture similar to distributed speech recognition. Network sinks are shown in medium grey

decreases in UAR when applying SVQ. As a positive side effect, the quantisation according to a server-side codebook ensures that speech samples are stored only as references to samples of speakers in the reference database used to create the codebook – thus, no 'personal' features of the speakers are transmitted, in comparison to traditional feature space reduction or feature selection methods.

In Figure 12.1 we observe an important feedback loop from classification to feature extraction, referring to the fact that the distributed architecture allows relying on existing mobile services for ASR, generating lexical features and sending them to a server for recognition.

Ideally, end user systems performing paralinguistic analysis should be free in their choice of a server-side recognition engine, enabling easy integration of new resources and closing the gap between analysis and synthesis applications. To this end, one needs to rely on standardised interfaces. Examples of standards that can be used in distributed paralinguistic analysis are those for distributed speech recognition such as the (European Telecommunications Standards Institute) ES.202.050, and generic communication protocols such as web services. In the area of feature extraction for computational paralinguistics, the CEICES feature coding scheme (Batliner *et al.* 2011) provides a standard at an abstract level (see also Section A.2) and open source feature extraction engines (see Chapter 13 and Eyben *et al.* 2010) do so at the implementation level. Moreover, one needs a standardised interface to communicate recognition results to the client side for interpretation, for example, by a dialogue management or speech synthesis component. In this context, one might mention the Extensible MultiModal Annotation (EMMA) markup language (Baggia *et al.* 2007) or the Emotion Markup Language (EmotionML) proposed in Schröder *et al.* (2007) as examples from the area of affect recognition. Yet, at the time of writing, a generic standard for encoding paralinguistic information still has to emerge.

As has been touched upon in the above discussion, acquiring data on the server, be it in the form of compressed or uncompressed feature vectors, can be used for autonomous learning whose theoretical foundations will be discussed subsequently.

12.2   Autonomous and Collaborative Learning

Data scarcity constitutes a major bottleneck in computational paralinguistics which limits real-world application (Schuller 2012). The term 'data scarcity' (or 'sparsity') stems from the machine learning field and usually refers to the shortage of labelled data points for training and evaluating machine learning algorithms. It is obvious that in the digital age data *per se* are not scarce at all. In particular, speech data as such are available in vast quantities from the Internet, radio and TV broadcasts. The bottleneck is thus obtaining the (para)linguistic labels (such as lexical transcription, speaker attributes, arousal and valence annotation) for these data – one should probably speak of label scarcity rather than data scarcity.

The traditional approach to obtaining labelled data has been to employ human experts for labelling. Such manual annotation, however, remains extremely time-consuming and expensive despite recent efforts in crowd-sourcing (Raykar *et al.* 2010). Semi-supervised learning (SSL: Chapelle *et al.* 2006) seems to be a promising alternative. Here, a system that has been trained on a small amount of human labelled data $\mathcal{L} = \{(\underline{x}_1, y_1), \ldots, (\underline{x}_L, y_L)\}$ automatically annotates a corpus of unlabelled data $\mathcal{U}$. This is usually done iteratively through the following SSL meta-algorithm, stopping after a predefined number I of iterations:

- Initialisation: Set $\mathcal{L}^{(0)} := \mathcal{L}, \mathcal{U}^{(0)} := \mathcal{U}$
- Iteration: Repeat for $i = 1, \ldots, I$:
  1. Train classifier $\hat{y}^{(i-1)}$ on $\mathcal{L}^{(i-1)}$
  2. Use $\hat{y}^{(i-1)}$ to label $\mathcal{U}$
  3. Pick N instances $\underline{x}_1^{(i)} \ldots, \underline{x}_N^{(i)} \in \mathcal{U}$ with highest classifier confidence
  4. Set $\mathcal{L}^{(i)} := \mathcal{L}^{(i-1)} \cup \left\{ \left(\underline{x}_1^{(i)}, \hat{y}(\underline{x}_1^{(i)}) \right), \ldots, \left(\underline{x}_N^{(i)}, \hat{y}(\underline{x}_N^{(i)}) \right) \right\}$
  5. Set $\mathcal{U}^{(i)} := \mathcal{U}^{(i-1)} \setminus \left\{ \underline{x}_1^{(i)}, \ldots, \underline{x}_N^{(i)} \right\}$

Instead of running I iterations, a convergence criterion such as the relative change in the weight vector $\underline{w}$ of a support vector machine (SVM) with respect to the previous iteration step can be used. It remains to clarify how to determine the N instances in step 3 of the SSL algorithm. This requires a measure of the confidence of the classifier in its decision. Often, a simple measure derived from the classification rule can be used. For instance, in SVM classification, the distance from the hyperplane is a rough approximation of classifier confidence – the further away from the hyperplane, the more certain the decision is. The computation of confidence measures will be addressed in more detail below.

SSL is popular in automatic speech recognition (Gilbert *et al.* 2011; Wessel and Ney 2005; Yu *et al.* 2010), and initial studies in affect recognition have been carried out (Zhang *et al.* 2011) – at the time of writing, it remains to be applied to other domains within the computational paralinguistics field.

An extension of the basic SSL algorithm is co-training (Blum and Mitchell 1998). This seems to be especially suited to 'multimodal' tasks, that is, tasks where multiple *views* can be exploited. Co-training relies on two assumptions: *compatibility* and *independence* (Blum and Mitchell 1998; Du *et al.* 2011). Compatibility requires that each view is sufficient to train a good classifier. The assumption of independence demands that the classification of an instance by one view is conditionally independent of the classification by the other. For example, it can be used for acoustic-linguistic recognition of paralinguistic aspects. If only acoustic information is used, the feature vectors can be split in a meaningful way, such as partitioning by energy-related, spectral, and cepstral descriptors.

Suppose that we have two 'modalities' or 'views' A and B on the same classification task. In co-training one iteratively trains a sequence of classifiers on these views, such as feature set subspaces. Thus, the co-training algorithm can be written as follows:

- Initialisation: Set $\mathcal{L}^{(0)} := \mathcal{L}, \mathcal{U}^{(0)} := \mathcal{U}$
- Iteration: Repeat for $i = 1, \ldots, I$:
  1. Train classifiers $\hat{y}_A^{(i-1)}, \hat{y}_B^{(i-1)}$ on $\mathcal{L}^{(i-1)}$
  2. Use $\hat{y}_A^{(i-1)}, \hat{y}_B^{(i-1)}$ to label $\mathcal{U}$
  3. Pick N instances $\underline{x}_1^{(i,A)} \ldots, \underline{x}_N^{(i,A)} \in \mathcal{U}$ with highest confidence of classifier A
  4. Pick N instances $\underline{x}_1^{(i,B)} \ldots, \underline{x}_N^{(i,B)} \in \mathcal{U}$ with highest confidence of classifier B
  5. Set $\mathcal{L}^{(i)} := \mathcal{L}^{(i-1)} \cup \left\{ \left(\underline{x}_1^{(i,A)}, \hat{y}_A(\underline{x}_1^{(i,A)}) \right), \ldots, \left(\underline{x}_N^{(i,A)}, \hat{y}_A(\underline{x}_N^{(i,A)}) \right), \right.$
     $\left. \left(\underline{x}_1^{(i,B)}, \hat{y}_B(\underline{x}_1^{(i,B)}) \right), \ldots, \left(\underline{x}_N^{(i,B)}, \hat{y}_B(\underline{x}_N^{(i,B)}) \right) \right\}$
  6. Set $\mathcal{U}^{(i)} := \mathcal{U}^{(i-1)} \setminus \left\{ \underline{x}_1^{(i,A)}, \ldots, \underline{x}_N^{(i,A)}, \underline{x}_1^{(i,B)}, \ldots, \underline{x}_N^{(i,B)} \right\}$

Note that one can introduce a balancing step into steps 3 and 4 to focus on underrepresented classes, as proposed by Blum and Mitchell (1998). A related modification of the above algorithm is training set resampling, picking a random subset of $\mathcal{U}$ for labelling – as pointed out by Blum and Mitchell (1998), this results in better coverage of the data distribution underlying $\mathcal{U}$, rather than picking only those instances that are 'easy to classify'. Furthermore, the algorithm can be extended to more than two views (Zhou and Li 2005). Finally, it is noteworthy that the independence assumption of the views generally leads to 'contradictory' training sets, where an instance can be added with different labels.

Let us now depart from fully automatic annotation by classification systems and address *collaborative* learning. The aim here is efficient exploitation of the human workforce in labelling, by concentrating efforts on those instances that are 'hard to classify', and whose manual labelling is thus assumed to be of particular interest for system improvement. The active learning (AL) meta-algorithm implements this paradigm by exploiting classifier confidence similarly to SSL – yet focusing, conversely, on the instances with low confidence, and adding them to the training set along with their human annotation. This results in the following AL meta-algorithm:

- Initialisation: Set $\mathcal{L}^{(0)} := \mathcal{L}, \mathcal{U}^{(0)} := \mathcal{U}$
- Iteration: Repeat for $i = 1, \ldots, I$:
  1. Train classifier $\hat{y}^{(i-1)}$ on $\mathcal{L}^{(i-1)}$
  2. Use $\hat{y}^{(i-1)}$ to label $\mathcal{U}$
  3. Pick N instances $\underline{x}_1^{(i)} \ldots, \underline{x}_N^{(i)} \in \mathcal{U}$ with *lowest* classifier confidence
  4. Obtain manual annotation $y_1^{(i)}, \ldots, y_N^{(i)}$ of these instances
  5. Set $\mathcal{L}^{(i)} := \mathcal{L}^{(i-1)} \cup \left\{ \left(\underline{x}_1^{(i)}, y_1^{(i)} \right), \ldots, \left(\underline{x}_N^{(i)}, y_N^{(i)} \right) \right\}$
  6. Set $\mathcal{U}^{(i)} := \mathcal{U}^{(i-1)} \setminus \left\{ \underline{x}_1^{(i)}, \ldots, \underline{x}_N^{(i)} \right\}$

Similarly to SSL, a 'balancing' step can be introduced to focus more on the instances of the minority class (Zhang and Schuller 2012). To select the 'most informative' instances for human labelling, a variety of criteria can be employed in AL besides choosing those with the lowest confidence of the classifier as is exemplified in the above, similar to the method proposed by Riccardi and Hakkani-Tur (2005) for ASR. Alternatives include the expected error reduction method, which aims to measure how much the generalisation of the model would be improved by adding instances (Roy and McCallum 2001), and the expected model change based method, which chooses the instances that impact the current model most greatly (Settles and Craven 2008). Instead of focusing the choice of instances on a particular classifier, one can also utilise multiple classifiers and consider those samples with the lowest classifier agreement as most informative (Liere 2000; McCallum and Nigam 1998). In the area of paralinguistics, AL shows good performance in emotion recognition (Wu and Parsons 2011; Zhang and Schuller 2012); as for SSL, other areas of applications are still to be explored at the time of this writing.

12.3 Confidence Measures

When a computational paralinguistics system (or almost any automatic recognition system) is employed in a real-world application, it is not reasonable to expect perfect accuracy. This may be due to environmental influences (e.g., ambient noise), but also to divergence of the data used to train the system and the application scenario (e.g., training an affect recognition model on acted emotions, but applying it in monitoring of real conversations). Still, whenever accurate recognition is not possible, one would ideally expect the model to signal that its prediction should not be relied upon. In this case, appropriate actions can be taken, for example, a dialogue manager would not change strategy based on unreliable detection of anger.

We have introduced several intuitive concepts of classifier confidence so far. An intuitive notion of classifier confidence is closely related to posterior class probabilities, that is, the system's estimate of the probabilities $p(i|\underline{x})$ that a pattern vector $\underline{x}$ belongs to class $i \in \{1, \ldots, C\}$. For example, in the calculation of the receiver operating characteristic (ROC: see Section 11.3.3) we adjusted a 'confidence' threshold based on the posterior probability of the positive class. It is easy to see how this is intuitively related to confidence – a high posterior probability of the positive class means that the posterior probability of the negative class is close to zero, indicating high confidence; conversely, a posterior probability close to $\frac{1}{2}$ indicates low confidence.

It remains to describe how class posteriors can be obtained from a classifier in the general case. Let us start with the example of a linear classifier, such as an SVM with a linear kernel, whose classification rule is based on computing the distance of the pattern vector $\underline{x}$ to be classified, from the hyperplane given by normal vector $\underline{w}$ and bias b:

$$d(\underline{x}) = \underline{w}^T \underline{x} + b. \tag{12.4}$$

Recall that by the definition of the linear classifier, $d > 0$ indicates a decision for the positive class while $d < 0$ indicates the opposite. Intuitively, the larger the value of d, the higher the posterior probability of the positive class, and vice versa. To obtain a posterior 'pseudo'-probability of the class $+1$ in the range $[0, 1]$, we can thus 'convert' d into a strictly monotonic function that has as domain the real numbers ($\mathbb{R}$) and the range [0, 1]. The prototype of such a function is the sigmoid function, or logistic function, where a and b are free parameters:

$$\pi(u) = \frac{1}{1 + e^{-(a+bu)}} = \frac{e^{a+bu}}{e^{a+bu} + 1}. \tag{12.5}$$

With the distance function (12.4) we obtain the desired mapping from a pattern vector to the posterior pseudo-probability p of the positive class as

$$p(+1|\underline{x}) = \pi(d(\underline{x})). \tag{12.6}$$

The parameters a and b are usually determined by standard iterative curve fitting methods, such as Newton's method (Menard 2002). More specifically, the learning problem in the optimisation is to predict the binary labels of the training data as a Bernoulli trial (outcome 0 for the negative class, or 1 for the positive class) from the values of d on the training data.

A generalisation of the logistic function to multiple outcomes (classes) is the softmax function (11.35), introduced in its basic form in Section 11.2.1, in the context of neural networks for multi-class discrimination. In its general form, it is given by

$$\Pi_i(\underline{u}) = \frac{e^{\underline{b}_i^T \underline{u}}}{\sum_{k=1}^{C} e^{\underline{b}_k^T \underline{u}}} \tag{12.7}$$

For example, the components of the vector $\underline{u}$ can correspond to activations of the output layer of a neural network; in this case, $\underline{u}$ is C-dimensional, and a pseudo-probability equivalent to (11.35) is obtained by choosing $\underline{b}_i$ as the ith unit vector. Determining the coefficients b_i, $i = 1, \ldots, C$, from training data, such as by maximum a posteriori estimation, yields a so-called maximum entropy classifier, which is suitable for deriving posterior probabilities from features such as acoustic or language model likelihoods (White *et al.* 2007), resembling confidences in an intuitive sense. Generalising this idea further, one can think of any discriminative classifier (such as SVMs or DTs) to map a vector consisting of features (such as acoustic stability, classifier likelihoods, or classifier agreement) to class posteriors.

All in all, we conclude that while confidence measures are state-of-the-art in ASR – a survey can be found in (Jiang 2005) – surprisingly little attention has been paid to confidence measures in computational paralinguistics. Furthermore, when we consider tasks with uncertain ground truth such as training an emotion classifier on observer ratings, it can be argued that confidence measures should also take into account the rater (dis)agreement – at the time of writing, this has been addressed only in preliminary studies (Deng *et al.* 2012).

References

Baggia, P., Burnett, D. C., Carter, J., Dahl, D. A., McCobb, G., and Raggett, D. (2007). *EMMA: Extensible MultiModal Annotation markup language*. Manual. http://www.w3.org/TR/emma/.

Batliner, A., Steidl, S., Schuller, B., Seppi, D., Vogt, T., Wagner, J., Devillers, L., Vidrascu, L., Aharonson, V., and Amir, N. (2011). Whodunnit: Searching for the most important feature types signalling emotional user states in speech. *Computer Speech and Language*, **25**, 4–28.

Blum, A. and Mitchell, T. (1998). Combining labeled and unlabeled data with co-training. In *Proc. of the 11th annual conference on Computational Learning Theory*, pp. 92–100, Madison, WI.

Chapelle, O., Schölkopf, B., and Zien, A. (eds) (2006). *Semi-supervised Learning*. MIT Press, Cambridge, MA.

Deng, J., Han, W., and Schuller, B. (2012). Confidence measures for speech emotion recognition: a start. In T. Fingscheidt and W. Kellermann (eds), *Proc. of the 10th ITG Conference on Speech Communication*, pp. 1–4, Braunschweig, Germany.

Du, J., Ling, C. X., and Zhou, Z. (2011). When does co-training work in real data? *IEEE Transactions on Knowledge Discovery and Data Mining*, **23**, 788–799.

Eyben, F., Wöllmer, M., and Schuller, B. (2010). openSMILE – The Munich Versatile and Fast Open-Source Audio Feature Extractor. In *Proc. of the 9th ACM International Conference on Multimedia, MM*, pp. 1459–1462, Florence.

Gilbert, M., Arizmendi, I., Bocchieri, E., Caseiro, D., Goffin, V., Ljolje, A., Phillips, M., Wang, C., and Wilpon, J. (2011). Your mobile virtual assistant just got smarter! In *Proc. of Interspeech*, pp. 1101–1104, Florence.

Han, W., Zhang, Z., Deng, J., Wöllmer, M., Weninger, F., and Schuller, B. (2012). Towards distributed recognition of emotion in speech. In *Proc. of the 5th International Symposium on Communications, Control, and Signal Processing, ISCCSP*, Rome.

Jiang, H. (2005). Confidence measures for speech recognition: a survey. *Speech Communication*, **45**, 455–470.

Liere, R. (2000). *Active Learning with Committees: An Approach to Efficient Learning in Text Categorization Using Linear Threshold Algorithms*. Ph.D. thesis, Oregon State University, Portland, OR.

McCallum, A. and Nigam, K. (1998). Employing EM in pool-based active learning for text classification. In *Proc. of International Conference on Machine Learning (ICML)*, pp. 359–367, Madison, WI.

Menard, S. W. (2002). *Applied Logistic Regression*. Sage Publications, Thousand Oaks, CA, 2nd edition.

Raykar, V. C., Yu, S., Zhao, L. H., Valadez, G. H., Florin, C., Bogoni, L., and Moy, L. (2010). Learning from crowds. *Journal of Machine Learning Research*, **11**, 1297–1322.

Riccardi, G. and Hakkani-Tur, D. (2005). Active learning: theory and applications to automatic speech recognition. *IEEE Transactions on Speech and Audio Processing*, **13**, 504–511.

Roy, N. and McCallum, A. (2001). Toward optimal active learning through sampling estimation of error reduction. In *Proc. of International Conference on Machine Learning (ICML)*, pp. 441–448, Williamstown, MA.

Schröder, M., Devillers, L., Karpouzis, K., Martin, J.-C., Pelachaud, C., Peter, C., Pirker, H., Schuller, B., Tao, J., and Wilson, I. (2007). What should a generic emotion markup language be able to represent? In A. Paiva, R. Prada, and R. W. Picard (eds), *Affective Computing and Intelligent Interaction*, pp. 440–451, Berlin. Springer.

Schuller, B. (2012). The Computational Paralinguistics Challenge. *IEEE Signal Processing Magazine*, **29**(4).

Schuller, B., Batliner, A., Steidl, S., and Seppi, D. (2011). Recognising realistic emotions and affect in speech: State of the art and lessons learnt from the first challenge. *Speech Communication*, **53**, 1062–1087.

Schuller, B., Steidl, S., Batliner, A., Schiel, F., Krajewski, J., Weninger, F., and Eyben, F. (2013a). Medium-term speaker states – a review on intoxication, sleepiness and the first challenge. *Computer Speech and Language, Special Issue on Broadening the View on Speaker Analysis*.

Schuller, B., Steidl, S., Batliner, A., Burkhardt, F., Devillers, L., Müller, C., and Narayanan, S. (2013b). Paralinguistics in speech and language – state-of-the-art and the challenge. *Computer Speech and Language*, **27**, 4–39.

Settles, B. and Craven, M. (2008). An analysis of active learning strategies for sequence labeling tasks. In *Proc. of Empirical Methods in Natural Language Processing (EMNLP)*, pp. 1070–1079, Honolulu.

Wessel, F. and Ney, H. (2005). Unsupervised training of acoustic models for large vocabulary continuous speech recognition. *IEEE Transactions on Speech and Audio Processing*, **13**, 23–31.

White, C., Droppo, J., Acero, A., and Odell, J. (2007). Maximum entropy confidence estimation for speech recognition. In *Proc. of ICASSP*, pp. 809–812.

Wu, D. and Parsons, T. (2011). Active class selection for arousal classification. In *Proc. of Affective Computing and Intelligent Interaction (ACII)*, pp. 132–141, Memphis, TN.

Yu, K., Gales, M., Wang, L., and Woodland, P. (2010). Unsupervised training and directed manual transcription for LVCSR. *Speech Communication*, **52**, 652–663.

Zhang, Z. and Schuller, B. (2012). Active learning by sparse instance tracking and classifier confidence in acoustic emotion recognition. In *Proc. of Interspeech*, pp. 362–365, Portland, OR.

Zhang, Z., Weninger, F., Wöllmer, M., and Schuller, B. (2011). Unsupervised learning in cross-corpus acoustic emotion recognition. In *Proceedings 12th Biannual IEEE Automatic Speech Recognition and Understanding Workshop, ASRU 2011*, pp. 523–528, Big Island, HI.

Zhou, Z. and Li, M. (2005). Tri-training: Exploiting unlabeled data using three classifiers. *IEEE Transactions on Knowledge and Data Engineering*, **17**, 1529–1541.

13

'Hands-On': Existing Toolkits and Practical Tutorial

Our goal is to show that you can develop a robust, safe manned space program and do it at an extremely low cost.

(Elbert Leander 'Burt' Rutan)

An apprentice carpenter may want only a hammer and saw, but a master craftsman employs many precision tools. Computer programming likewise requires sophisticated tools to cope with the complexity of real applications, and only practice with these tools will build skill in their use.

(Robert L. Kruse)

This chapter provides practical examples of paralinguistic feature extraction and live emotion recognition with the openSMILE (open-source Speech and Music Interpretation by Large-space Extraction) toolkit.[1] The toolkit has a strong focus on real-time, incremental processing. Moreover, openSMILE provides a simple console application where modular feature extraction components can be freely configured and connected via a single configuration file. Both incremental on-line processing for live applications and off-line batch processing are supported.

In this chapter we will first give an overview of related toolkits which can also be used to extract some features, or can be used together with openSMILE to perform machine learning tasks and model learning. Then we briefly introduce and describe the openSMILE toolkit and its architecture and finally present a brief hands-on 'computational paralinguistics how-to' in Section 13.3.

13.1 Related Toolkits

Related, popular feature extraction tools used for speech research include the *Hidden Markov Model Toolkit* (HTK: Young *et al.* 2006), the *PRAAT* software (Boersma and Weenink 2005),

---

[1] http://opensmile.sourceforge.net/

Computational Paralinguistics: Emotion, Affect and Personality in Speech and Language Processing, First Edition.
Björn W. Schuller and Anton M. Batliner. © 2014 John Wiley & Sons, Ltd. Published 2014 by John Wiley & Sons, Ltd.

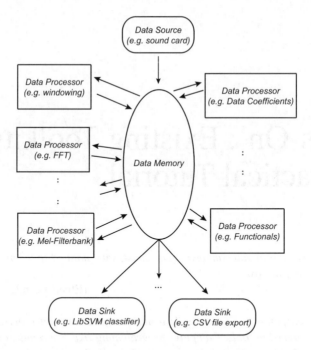

Figure 13.1   Modular architecture of the openSMILE toolkit

the *Speech Filing System*[2] (SFS), the Auditory Toolbox,[3] a MATLAB™ toolbox[4] (Fernandez 2004), the Tracter framework (Garner *et al.* 2009), and the *SNACK*[5] package for the Tcl scripting language. However, not all of these tools are distributed under a permissive open-source licence (e.g., HTK and SFS).

Another comprehensive tool for emotion recognition experiments and dialogue systems is the EmoVoice framework (Vogt *et al.* 2008).

13.2   openSMILE

This section introduces openSMILE's architecture as illustrated in Figure 13.1.[6]

To provide comprehensive and standardised cross-domain feature sets, flexibility and extensibility, and incremental processing support, a number of requirements had to be met. First, incremental processing demands the ability to push audio data sample-wise from arbitrary input streams such as files or the sound card through the chain of processing (cf. Figure 13.2). Then, a ring-buffer memory for features is needed to provide memory efficient temporal context

---

[2]http://www.phon.ucl.ac.uk/resource/sfs/
[3]http://cobweb.ecn.purdue.edu/ malcolm/interval/1998-010/
[4]http://affect.media.mit.edu/publications.php
[5]http://www.speech.kth.se/snack/
[6]A more detailed description can be found in the openSMILE documentation available in the download package at http://sourceforge.net/projects/opensmile/.

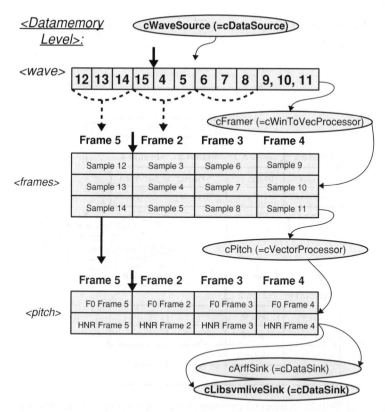

Figure 13.2   openSMILE: incremental data-flow at the low-level descriptor (LLD) level in the ring-buffer memories. Short thick unbroken arrow: current write pointer (Eyben *et al.* 2010)

and buffering of dynamic length. Next, for an efficient design, reusability of data is required to avoid duplicate computation of commonly used parameters such as fast Fourier transform (FFT) spectra (see Figure 13.2). Moreover, fast and 'lightweight' algorithms were favoured and the core functionality was implemented in C and C++ without third-party dependencies where possible, in order to make the toolkit portable and easy to install and use.

A modular design enables an arbitrary combination of features and invites the research community to add new feature extractor components to the code through an application programming interface (API) and a run-time plug-in interface. To handle asynchronous feature streams, universal timing information is available for processing of multiple streams.

Let us now look at openSMILE's modular architecture enabling incremental processing, and the features which are currently implemented. Figure 13.1 shows the overall data-flow architecture of openSMILE. The *data memory* is the central link between all other components – *data sources* (writing from external sources to the memory), *data processors* (reading from the memory, modifying the data, and writing back to the memory), and *data sinks* (reading from the memory and writing to external devices or displaying a result).

The principle of ring buffer based incremental processing can be seen in Figure 13.2 by an example with three levels: wave, frames, and pitch. The 'cWaveSource' component writes

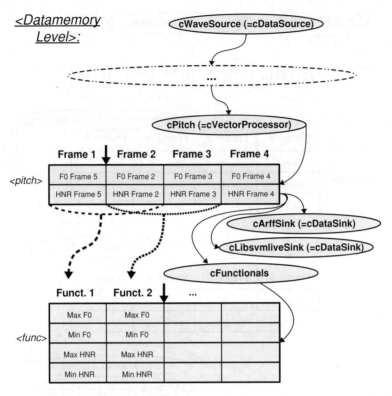

Figure 13.3 openSMILE: incremental data-flow at the functional level in the ring-buffer memories. Short thick unbroken arrow: current write pointer (Eyben *et al.* 2010)

audio samples (from a file or audio device) to the 'wave' level. The current write position pointers are indicated by vertical arrows. The 'cFramer' component produces non-overlapping frames of size 3 from these wave samples. It writes the frames to the 'frames' level. Finally, the 'cPitch' component (simplified in this example) calculates a pitch descriptor for each of the frames. It saves this descriptor to the 'pitch' level. Since all boxes in the plot contain values (i.e., data), the buffers have been filled with previous data, and the write pointers have been warped.

Figure 13.3 shows the processing done at higher levels such as the application of functionals in order to project a time series to single feature values. Two functionals, 'max' and 'min', are shown. These are calculated over two overlapping frames from the pitch parameter and saved to the level 'func'. The buffer size is matched to the block size of the reader and writer. In the pitch functionals example the read block size of the functionals component thus would be two frames because two pitch frames are required for processing at once. openSMILE supports multi-threading for increased computational performance. For utmost parallelisation on multi-core computers, components with computationally demanding operations can be run in separate threads. The individual components can be freely instantiated, configured, and connected to the *data memory* via a simple, text-based configuration file.

Table 13.1 openSMILE: supported low-level descriptors (LLDs). Abbreviations are explained in the text and in the list of abbreviations at the front of the book

Group	LLDs
Waveform	ZCR, extremes, direct component
Signal energy	RMS & logarithmic
Loudness	Intensity & approximated loudness
FFT spectrum	Phase, magnitude (lin., dB, dBA)
ACF, cepstrum	ACF, cepstrum
Mel/Bark spectrum	Bands $0-N_{mel}$
Semitone spectrum	FFT based and filter based
Cepstral	Cepstral features, e.g., MFCC, PLP-CC
Pitch	F_0 via ACF and SHS methods, probability of voicing
Voice quality	HNR, jitter, shimmer
LPC	LPCC, reflection coefficients, residual, LSP
Auditory	Auditory spectra and PLP coefficients
Formants	Centre frequencies and bandwidths
Spectral	Energy in N user-defined bands, multiple roll-off points, centroid, entropy, flux, and relative positions of extrema
Tonal	CHROMA, CENS, CHROMA based features

13.2.1 Available Feature Extractors

openSMILE provides a large number of low-level descriptors (LLDs) for automatic extraction and the application of several filters, functionals, and transformations to these (see Table 13.1). Mel spectra, MFCCs, and PLPs can be computed numerically, compatible with the popular automatic speech recognition (ASR) toolkit HTK (Young *et al.* 2006). PLP computation can be carried out as in the original paper by Hermansky (1990) or in custom modifications by selecting individual steps only.

LLDs can be processed frame by frame with these filters: weighted differential and raised-cosine lowpass as in Schuller *et al.* (2007), first-order infinite impulse response (IIR) lowpass/highpass, comb-filter bank with arbitrary number of filters, moving average smoothing filter, and regression (delta) coefficients (x^W) of arbitrary order W (see Section 8.4.11). These are computed from any feature contour $x(n)$ again in HTK style (Young *et al.* 2006) with the parameter W (order):

$$d(n) = \frac{\sum_{i=1}^{W} i \cdot (x(n+i) - x(n-i))}{2 \sum_{i=1}^{W} i^2}. \tag{13.1}$$

Additional arithmetic operations include add, multiply, and power, for creation of custom features by combining existing operations.

Functionals comprise statistical descriptors, polynomial regression coefficients, and transformations as found in Table 13.2. They can be applied to any data (LLDs or functionals) hierarchically with unlimited depth as described in Schuller *et al.* (2008). Their choice follows

Table 13.2   openSMILE: supported functionals

Group	Functionals
Extremes	Extreme values, positions, ranges
Means	Arithmetic, quadratic, geometric
Moments	Standard deviation, variance, kurtosis, skewness
Percentiles	Percentiles and percentile ranges
Regression	Linear and quadratic approximation coefficients, regression error, and centroid
Peaks	Number of peaks, mean peak distance, mean peak amplitude
Segments	Number of segments by delta thresholding, mean segment length
Sample values	Values of the contour at configurable relative positions
Times/durations	Up- and down-level times, rise/fall times, duration
Onsets	Number of onsets, relative position of first/last on-/offset
DCT	Discrete cosine transformation coefficients
Zero crossings	ZCR, mean crossing rate

the CEICES standard (see Section A.2) developed by seven leading sites in the field (Batliner *et al.* 2006, 2011).

The modular architecture allows the use of any implemented processing functionality in arbitrary combinations, for example, one may use a mel-band filter bank as functionals. This enables brute-forcing of large, unrestricted feature spaces of several thousand features. The idea is not to compute more features than data points, but rather to provide a broad basis of features from which to select those highly relevant to a given task, similar to the work described by Pachet and Roy (2009). For exchange with other popular software modules, supported file formats include Attribute Relation File Format (ARFF: Witten and Frank 2005) from the WEKA data mining software, the LibSVM format, Comma Separated Value (CSV) tables, HTK (Young *et al.* 2006) binary parameter files, and raw binary files. Further features include a built-in audio and speech activity detection which can be used for audio stream chunking in real time.

To conclude, openSMILE was introduced as an example of a feature extractor tailored to be an efficient, on-line or batch scriptable, open-source, cross-platform, and extensible tool implemented in C++. It is increasingly turning into a standard toolkit – in particular in the field of computational paralinguistics.[7] Moreover, the openEAR (open-source Emotion and Affect Recognition) project builds on openSMILE and extends it with integrated classification algorithms and pre-trained models for affect recognition tasks (Eyben *et al.* 2009).

13.3   Practical Computational Paralinguistics How-to

In this section we present a small tutorial which covers how to get started with openSMILE. First, we explain how to obtain and install openSMILE. If you already have a working

---

[7] openSMILE was awarded third place in the Association for Computing Machinery (ACM) Multimedia 2010 Open-Source Software competition. Furthermore, it has been employed as the standard feature extractor for baseline feature computation and used by participants in six research challenges in the field since 2009.

installation of openSMILE you can skip directly to Section 13.3.2, where we explain how to use openSMILE for your first feature extraction task.

13.3.1   Obtaining and Installing openSMILE

The latest stable open-source release of openSMILE can be found at http://opensmile. sourceforge.net/. It is distributed under the GNU General Public Licence (GPL), which allows for private, research, and commercial use, but requires any modifications and dependent code to be made available to the community under the same or a compatible licence.[8]

This tutorial is based on the 1.0.1 open-source release from http://opensmile.sourceforge. net/. Check this website for updates and the latest compilation and installation instructions in the openSMILE book. These are not included here because they are subject to constant change. The release versions include pre-compiled Linux and Windows binaries for 32-bit and 64-bit Intel architecture based computer systems. The source can be built without any third party libraries on Linux systems with autoconf and automake and on Windows systems with Visual Studio. If live audio recording or playback is needed, a version of openSMILE linked to the *PortAudio* library is required.

The following material assumes that you have a working installation of openSMILE on your system and that you are familiar with the command line interface in either Windows or Linux. We will further assume that you are using Linux, that is, the commands and filenames given are for the Linux shell. The syntax for the Windows shell is similar, and we assume that if you are an advanced Windows user, you will know how to run the commands properly in Windows. In general, we recommend running openSMILE under Linux, as it delivers better performance, and batch scripting is easier to carry out due to a broad on-board availability of shells and scripting languages (especially Perl and Python). Always bear in mind that the open-source version of openSMILE is and will always be a *tool* and not a fully featured product or software package. It can be used for data and signal processing, and feature extraction. Conversions of various input data formats, batch processing, handling of data labels and meta-data, and post-processing or conversion of output data formats must be performed with external scripts.

13.3.2   Extracting Features

All functions of openSMILE are available through the main binary *SMILExtract*. To check if you can run SMILExtract, type:

```
SMILExtract -h
```

If you see the usage information and version number of openSMILE, then everything is set up correctly. You will see some lines starting with (MSG) at the end of the output, which is debug output and which you can ignore for the moment.

---

[8]Proprietary licensing and commercial versions of openSMILE and openEAR can be obtained through audEER-ING UG (limited) (http://www.audeering.com/). The company also provides professional consulting and support to researchers and companies who wish to use openSMILE in commercial product development.

Note that you may have to prefix a "./" on Unix-like systems, if SMILExtract is not in your system path but instead in the current directory.

We will now start using SMILExtract to extract basic audio features from an uncompressed wave audio file. You can use your own wave files if you like, or use some of the files provided in the `wav-samples` directory. openSMILE does not support any formats other than uncompressed pulse code modulation (PCM) wave at the time of writing, so if your audio material is in another format you will have to convert it to wave using tools such as ffmpeg or mplayer, which both support any existing audio format. Note that some tools produce an extended wave header format which openSMILE cannot read. When reading such a wave file you will get an error message saying that it is not a supported wave file. The problem can be solved by using the command line tool 'sox' to convert your wave file to a single channel wave file ('sox input.wav -c 1 output.wav') – in this case sox creates an output wave file with the old, standard Resource Interchange File Format (RIFF) header (sometimes it is sufficient to just use 'sox input.wav output.wav').

For a quick start, we use an example configuration file provided with the openSMILE distribution. Type the following command in the top-level directory of the openSMILE package:

```
SMILExtract -C config/demo/demo1\_energy.conf -I wav\_samples/
    speech01.wav -O speech01.energy.csv
```

If you see only (MSG) and (WARN) type messages, and Processing finished! in the last output line, then openSMILE ran successfully. If something fails, you will get an (ERROR) message.

Note for Windows users: Due to sub-optimal exception handling and memory management in Windows, if an exception indicating an error is thrown up in the Dynamic Link Library (DLL) and caught in the main executable, Windows will display a program crash dialogue. In most cases openSMILE will have displayed the error message beforehand, so you can just close the dialogue. In some cases however, Windows kills the program before it can display the error message. If this is the case, please use Linux, or contact the authors and provide some details on your problem.

Now, open the file `speech01.energy.csv` in a text editor to see the result. You can also plot the result graphically using gnuplot[9] or MATLAB for example.

Next, we will generate the configuration file for the above example step by step, to learn how openSMILE configuration files are written. openSMILE can generate configuration templates for a given list of components that you want to have in your configuration. The configuration file that we will now create will be capable of reading a wave file, computing per-frame signal energy, and saving the output to a CSV file. First, create a directory `myconfig` which will hold your new configuration files. Then type the following (without newlines) to generate your first configuration file (demo1.conf):

```
SMILExtract -cfgFileTemplate -configDflt cWaveSource,cFramer,
    cEnergy,cCsvSink -l 1 2> myconfig/demo1.conf
```

The `-cfgFileTemplate` option causes openSMILE to generate a configuration file template. The `-configDflt` takes a comma separated list of components which will be part of

---

[9]See http://www.gnuplot.info/

the generated configuration. In this example: cWaveSource to read a wave file, cFramer to perform windowing on the audio signal, cEnergy to compute the signal RMS energy, and cCsvSink to write the computed energy values to CSV text file. The -1 0 option sets the log level to 1 in order to suppress any debug messages, which should not be in the configuration file (you will still get ERROR messages at log level 1, for example, messages informing you that components you have specified do not exist). The template text is printed to the standard error output, thus we use 2> to dump it to the file myconfig/demo1.conf. If you want to automatically add comments which describe the individual option lines in the generated configuration file, add the option -cfgFileDescriptions to the above command line.

The newly generated configuration file consists of two logical parts. The first part looks like this (note that comments in the examples are initiated by ; or // and may only start at the beginning of a line):

```
;= component manager configuration (= list of enabled
   components!) =
[componentInstances:cComponentManager]
 // this line configures the default data memory:
instance[dataMemory].type = cDataMemory
instance[waveSource].type = cWaveSource
instance[framer].type = cFramer
instance[energy].type = cEnergy
instance[csvSink].type = cCsvSink
// Here you can control the amount of detail displayed for the
// data memory level configuration. 0 is no information at all,
// 5 is maximum detail.
printLevelStats = 1
 // You can set the number of parallel threads (experimental):
nThreads = 1
```

This part contains the configuration of the component manager, which determines what components are instantiated when SMILExtract is run. There always has to be one cDataMemory component, followed by other components. The name given in [] after the instance variable specifies the name of the component instance, which must be unique within one configuration. The value assigned to the instance[].type variable determines the type of the component to be instantiated (e.g., cFramer).

The next part of the file contains the component configuration sections, each of which begins with a section header as illustrated in the following example:

```
[waveSource:cWaveSource]
...
[framer:cFramer]
...
[energy:cEnergy]
...
[csvSink:cCsvSink]
...
```

The section header follows the format [instanceName:componentType], where instan-ceName must be the same name as in the componentInstances section. The template component configuration sections are generated with all available values explicitly set at their default values. This functionality is currently experimental; some values might override other values or have a different meaning if explicitly specified. Therefore, you are advised to carefully check all the available options and list only those in the configuration file which you require. Nonetheless, it is considered good practice to explicitly specify as many options (even if they are identical to the default values) as possible, in case the default value changes in a later release. Specifying values explicitly will ensure compatibility of your configuration files with future versions. Moreover, this will increase the readability of your configuration files because all parameters can be viewed in one place without looking up the defaults in the manual. To see up-to-date documentation (the documentation in the openSMILE book is only updated for major releases) of all available configuration options for a particular component, type the command

```
SMILExtract -H cComponent
```

and replace cComponent with the name of the component in question. The component name in this case is always evaluated as if it had a wildcard at the end. For example, cM as component name would match cMfcc and cMelSpec and thus would show the online help for both components.

Once the template configuration file is ready, you have to configure the component connections. To do this, assign so-called data memory 'levels' to the dataReader and dataWriter components by replacing the XXXX placeholders in each source, sink, or processing component in the respective reader.dmLevel and writer.dmLevel lines. You can choose arbitrary alphanumeric names for the writer levels, since the dataWriters automatically register and create the level you specify as writer.dmLevel with the data memory. The component-to-component connection is achieved by assigning the level as read level to reader.dmLevel of another component. Here, the following rules apply. For each level only *one* writer may exist, that is, only one component can write to a level; however, there is no limit to the number of components that read from a level (enabling reusability of data by multiple components). Further, one component can read from more than one level if you specify multiple data memory level names separated by a ; such as reader.dmLevel = energy;loudness to read data from the levels energy and loudness. Data are thus concatenated column-wise.

In our example we want the cFramer component to read from the input PCM stream, which is provided by the cWaveSource component, then create overlapping frames of 25 ms length every 10 ms. Finally, we want to save these frames to a new level which we call "energy". Thus, we change

```
[waveSource:cWaveSource]
writer.dmLevel = <<XXXX>>
```

to

```
[waveSource:cWaveSource]
writer.dmLevel = wave
```

and in the framer section

```
[framer:cFramer]
reader.dmLevel = <<XXXX>>
writer.dmLevel = <<XXXX>>
...
```

to

```
[framer:cFramer]
reader.dmLevel = wave
writer.dmLevel = waveframes
copyInputName = 1
frameMode = fixed
frameSize = 0.025000
frameStep = 0.010000
frameCenterSpecial = left
noPostEOIprocessing = 1
```

(Note that we removed a few unnecessary frameSize* options, set the frameSize to 25 ms (0.025 seconds), and changed the rate at which frames are sampled (frameStep) to 0.010 seconds.)

Next in the processing chain is the cEnergy component, which reads the audio frames created by the cFramer component and computes the logarithmic signal energy for each frame. The cCsvSink finally writes the energy values to a CSV format text file called myenergy.csv. To achieve this, we change the corresponding lines to

```
[energy:cEnergy]
reader.dmLevel = waveframes
writer.dmLevel = energy
...
rms = 0   ; disables root-mean-square energy
log = 1   ; enables logarithmic energy
...
[csvSink:cCsvSink]
reader.dmLevel = energy
filename = myenergy.csv
...
```

We are now ready to run our own configuration file in SMILExtract:

```
SMILExtract -C myconfig/demo1.conf
```

This will open the file "input.wav" in the current directory (be sure to copy a suitable wave file and rename it to "input.wav"), then compute frame-based signal energy, and save the result to the file "myenergy.csv". The output should be the same as with the example configuration file which we have used before.

To use a file other than "input.wav" as input, change the `filename=` line in the waveSource section. If you want to be able to pass the input file name and the output file name to openSMILE on the SMILExtract command line, you have to add a command to the configuration file to define a custom command line option. To do this, change the filename lines of the wave source and the CSV sink to

```
[waveSource:cWaveSource]
...
filename = \cm[inputfile(I):file name of the input wave file]
...
[csvSink:cCsvSink]
...
filename = \cm[outputfile(O):file name of the output CSV file]
...
```

You can now run

```
SMILExtract -C myconfig/demo1.conf -I wav\_samples/speech01.wav
   -O speech01.energy.csv
```

This concludes our brief introduction to configuring and running openSMILE. For the most recent documentation, refer to the openSMILE book and the on-line help contained in the binary as well as reading the source. The configuration file created in the above example is of limited use in practice. openSMILE is released with much more powerful configuration files than this example.

Predefined configurations for HTK compatible computation of MFCCs 0–12 with delta and acceleration coefficients and without/with cepstral mean removal are contained in the files

```
MFCC12_0_D_A.conf
MFCC12_0_D_A_Z.conf
```

respectively. Instead of the zeroth cepstral coefficient, logarithmic energy can be used by using these configuration files:

```
MFCC12_E_D_A.conf
MFCC12_E_D_A_Z.conf
```

The same files are available for HTK compatible PLP coefficients:

```
PLP_0_D_A.conf
PLP_0_D_A_Z.conf
PLP_E_D_A.conf
PLP_E_D_A_Z.conf
```

A set of basic prosodic descriptors (pitch and intensity) is contained in

```
prosodyShs.conf
```

for pitch extracted with the SHS algorithm and in

```
prosodyAcf.conf
```

for pitch extracted with a cepstrum and ACF based method. In newer releases `smileF0.conf` contains the configuration to run the most recent and best speech pitch extraction algorithm available in openSMILE. At the moment this is SHS followed by a Viterbi smoothing stage.

Further, all the baseline feature sets of the Interspeech Emotion and Paralinguistics Challenges are available as openSMILE configuration files. The feature set of the Interspeech 2009 Emotion Challenge is available in the file `emo_IS09.conf` and `IS09.conf` in newer releases. The feature set of the Interspeech 2010 Paralinguistics Challenge is available in the file `paraling_IS10.conf` and `IS10.conf` in newer releases. Only in newer releases are the features sets for the Interspeech 2011 Speaker State and 2012 Speaker Trait Challenge, the Audio Visual Emotion Challenges (AVEC) 2011, 2012, and 2013, as well as the 2013 Computational Paralinguistics Evaluation (ComParE) Challenge contained. The features contained in these sets (excluding the most recent sets of 2013) are summarised in Tables A.1 and A.2 in the Appendix.

The general command to extract features with a Challenge configuration file is (example for IS10):

```
SMILExtract -C paraling_IS10.conf -I input.wav -O output.arff -
    instname input1
-label1 class_label_for_input_wav
```

This will extract a single static feature vector for the utterance in the file `input.wav` and create the ARFF file `output.wav` containing a header with the names of all the features and a single feature vector for `input.wav`. If `output.arff` already exists, the feature vector for `input.wav` will be appended to the file.

The Challenge configuration files require (via an include) an additional configuration file named `arff_targets.conf`. This file configures the reference label targets to be included in the ARFF file in order to be able to perform and evaluate classification experiments with the WEKA software on the ARFF files produced. It also defines the option `-label1` (or similar for other Challenge tasks, for example, age, gender). The `-instname` parameter controls which instance name is written (as first string attribute) to the ARFF file. This is useful at a later stage for assigning automatically produced predictions for the task or certain feature values to a particular instance (and associated with it the corresponding input file). The instance name could be the same as the input file name. Actually, this is often the default choice.

If a single wave file contains multiple utterances, and the start/end times of these are known from an external label file, openSMILE can be called once for each utterance on the same input wave file by using `-start` and `-end` command line options, which accept the start and end time of the segment to process in seconds. For example:

```
SMILExtract -C paraling_IS10.conf -I input.wav -O output.arff -
    instname input1 -label1 class_label_for_input_wav -start
    1.2 -end 5.0
```

For these options to work, the configuration file must define them and route them to the appropriate options in the wave source component. This can be done by adding the following two lines to the configuration file in the cWaveSource section:

```
[waveIn:cWaveSource]
...
```

```
start = \cm[start{0}:start in input wave file in seconds]
end = \cm[end{-1}:end in input wave file in seconds]
```

See the `avec2011.conf` file for an example. For other output formats, the configuration file needs to be modified, and the cArffSink component needs to be replaced by a cCsvSink component for CSV text file output, for example. Alternatively, the 'live' classifier sinks can be used to classify the feature vectors directly (e.g., cLibsvmLiveSink) and display the results on the console output.

13.3.3 Classification and Regression

openSMILE is primarily a feature extractor. However, it also has components which implement on-line classifiers. These include LibSVM and recurrent neural networks at the time of writing. The current releases of openEAR[10] (Eyben *et al.* 2009) show examples of how to use these on-line classifiers. Further examples are found in the latest code releases from the SEMAINE (Sustained Emotionally coloured Machine-human Interaction using Non-verbal Expression) EU project.[11]

Before openSMILE can be used in live recognition mode, it is required to build good models for your task. General models, for example, those shipped with openEAR, are trained on small, prototypical databases and do not generalise well to real input data, where reverberation, background noise, microphone transfer functions, etc. pose challenging conditions.

Therefore, to build your own personal toy computational paralinguistics recogniser, record some data, for example, from yourself and ideally also from a few friends with your laptop computer's microphone, or through your mobile phone. Recordings should be a couple of phonetically balanced short sentences (4–6 seconds in length). Including a small number of distinct emotion, affective state, or speaker attribute classes (e.g., angry, neutral, sad, happy, or young, old, male, female) will give you the best results for a start.

Once you have recorded all your data, put the files into sub-directories (representing the emotion classes) of a single directory (named after your database), for example:

```
mydatabase/angry/recording1_angry.wav
mydatabase/happy/recording1_happy.wav
mydatabase/happy/recording2_happy.wav
```

For the above directory structure openSMILE provides a Perl script in `scripts/modeltrain` to automate the process of feature extraction. From that directory, run:

```
perl stddirectory_smilextract.pl /path/to/mydatabase emo_IS09.
   conf mydatabase_emo_IS09.arff
```

You can also use a different configuration file, but you must adapt the options for the ARFF instance targets in either the cArffSink configuration section of the `stddirectory_smilextract.pl` Perl script.

---

[10]http://www.openaudio.eu
[11]http://www.semaine-project.eu/

The above command produces the ARFF file `mydatabase_emo_IS09.arff` in the current directory. You can load this file with the WEKA software using the WEKA Explorer Graphical User Interface (GUI) and try a wide range of classifiers and classifier options there. For details we refer to the WEKA documentation available at http://www.cs.waikato.ac.nz/ml/weka/.

LibSVM models which can be used in a live recogniser with SMILExtract can be built with the `buildmodel.pl` script in the directory `scripts/modeltrain`. The script requires the programs `svm-train` and `svm-scale` to be pre-compiled. See the script's source code for additional documentation.

References

Batliner, A., Steidl, S., Schuller, B., Seppi, D., Laskowski, K., Vogt, T., Devillers, L., Vidrascu, L., Amir, N., Kessous, L., and Aharonson, V. (2006). Combining efforts for improving automatic classification of emotional user states. In *Proc. of IS-LTC 2006*, pp. 240–245, Ljubljana.

Batliner, A., Steidl, S., Schuller, B., Seppi, D., Vogt, T., Wagner, J., Devillers, L., Vidrascu, L., Aharonson, V., and Amir, N. (2011). Whodunnit: Searching for the most important feature types signalling emotional user states in speech. *Computer Speech and Language*, **25**, 4–28.

Boersma, P. and Weenink, D. (2005). Praat: doing phonetics by computer (Version 4.3.14). http://www.praat.org/.

Eyben, F., Wöllmer, M., and Schuller, B. (2009). openEAR – Introducing the Munich Open-Source Emotion and Affect Recognition Toolkit. In *Proc. of Affective Computing and Intelligent Interaction (ACII)*, pp. 576–581, Amsterdam.

Eyben, F., Wöllmer, M., and Schuller, B. (2010). openSMILE – The Munich Versatile and Fast Open-Source Audio Feature Extractor. In *Proc. of the 9th ACM International Conference on Multimedia, MM*, pp. 1459–1462, Florence.

Fernandez, R. (2004). *A Computational Model for the Automatic Recognition of Affect in Speech*. Ph.D. thesis, MIT Media Arts and Science.

Garner, P. N., Dines, J., Hain, T., El Hannani, A., Karafiat, M., Korchagin, D., Lincoln, M., Wan, V., and Zhang, L. (2009). Real-time ASR from meetings. In *Proc. of Interspeech*, pp. 2119–2122, Brighton.

Hermansky, H. (1990). Perceptual linear predictive (plp) analysis for speech. *Journal of the Acoustical Society of America*, **87**, 1738–1752.

Pachet, F. and Roy, P. (2009). Analytical features: A knowledge-based approach to audio feature generation. *EURASIP Journal on Audio, Speech, and Music Processing*.

Schuller, B., Eyben, F., and Rigoll, G. (2007). Fast and robust meter and tempo recognition for the automatic discrimination of ballroom dance styles. In *Proc. of ICASSP*, pp. 217–220, Honolulu.

Schuller, B., Wimmer, M., Mösenlechner, L., Kern, C., Arsić, D., and Rigoll, G. (2008). Brute-forcing hierarchical functionals for paralinguistics: A waste of feature space? In *Proc. of ICASSP*, pp. 4501–4504, Las Vegas, NV.

Vogt, T., André, E., and Bee, N. (2008). EmoVoice – a framework for online recognition of emotions from voice. In *Proc. of an IEEE Tutorial and Research Workshop on Perception and Interactive Technologies for Speech-Based Systems (PIT 2008)*, volume 5078 of *Lecture Notes in Computer Science*, pp. 188–199, Kloster Irsee. Springer, Berlin.

Witten, I. H. and Frank, E. (2005). *Data Mining: Practical Machine Learning Tools and Techniques*. Morgan Kaufmann, San Francisco, 2nd edition.

Young, S., Evermann, G., Gales, M., Hain, T., Kershaw, D., Liu, X., Moore, G., Odell, J., Ollason, D., Povey, D., Valtchev, V., and Woodland, P. (2006). *The HTK Book*. Cambridge University Engineering Department. For HTK version 3.4.

14

Epilogue

I think we're on the road to coming up with answers that I don't think any of us in total feel we have the answers to.

(Kim Anderson, Mayor of Naples, Florida)

In this book we have presented an overview of computational paralinguistics: its history, topics, methodologies, and the state of the art in Part I; and computational modelling – features, machine learning procedures, and system integration, together with a hands-on, in Part II. Throughout, we have referred to lacunas and desiderata. It might be informative to contrast the state of the art in computational paralinguistics with that in automatic speech processing, especially in automatic speech recognition (ASR): more than 40 years of research in ASR versus some 15 years of research in computational paralinguistics. ASR is about words, their sequence in the spoken word chain, and the meaning (semantics) encoded in this sequence. Computational paralinguistics is about a multitude of different phenomena signalled in between or by words or with modulations onto words. Parallel are the developments from (strictly) controlled to more realistic data, from small to larger databases, from simple to complex modelling and feature spaces, and from the modelling of single phenomena towards more complex phenomena. Different is the type of reference (in the sense of Section 2.4): a word is a word is a word – even if there is almost an indefinite number of combinations of words, and even if the targeted sequence of words can be obscured by noise, slurring, disfluencies, foreign accents, and so on. However, in principle, the target is not as evasive as it can be in computational paralinguistics where there often is no stable ground truth (again in the sense of Section 2.4), and where data and especially single classes can be very sparse and processed with more or less competing annotations and models. These difficulties result in a performance which is often sub-optimal, compared to the state of the art in ASR. It has been – and still is – one of the 'cross-sub-cultural' tasks to make it plausible that some 75% correct for a two-class problem can be as 'good' (meaning: representing the state of the art in the field) or even 'better' than some 75% correct word recognition with a lexicon of some 1000 words.

Especially with the advent of easily obtainable toolboxes, all the means necessary for employing a multiplicity of machine learning approaches are now available for studies within computational paralinguistics, such as (the fusion of) large feature vectors modelling acoustic,

Computational Paralinguistics: Emotion, Affect and Personality in Speech and Language Processing, First Edition.
Björn W. Schuller and Anton M. Batliner. © 2014 John Wiley & Sons, Ltd. Published 2014 by John Wiley & Sons, Ltd.

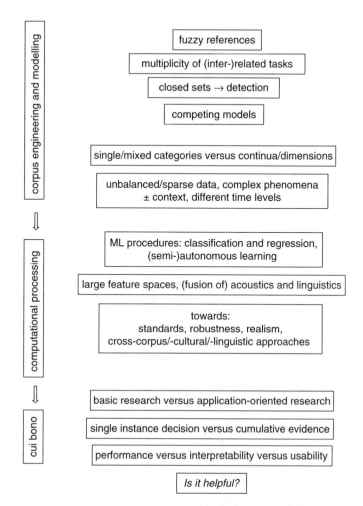

Figure 14.1   The state of the art of computational paralinguistics – and beyond

linguistic, or other types of information. However, there is as yet no smooth cooperation between different fields such as phonetics, linguistics, psychology, sociology, and medicine on the one hand, and engineering approaches on the other hand, nor a satisfying and fruitful cross-disciplinary understanding and cooperation. It will take some time to match together the different standards that we have mentioned throughout this book, especially in Chapters 2 and 3. Some of the key topics in this respect are the different performance-based metrics, the different criteria such as classification performance versus interpretability, or 'closed-world' performance versus (different kinds of) usability measures. Eventually, we are aiming at answers for the simple question *Is it helpful?* – for basic research and especially within applications.

Figure 14.1 closes this epilogue and attempts to summarise the present state of the art of computational paralinguistics and the future directions in which it might go.

Appendix

A.1 openSMILE Feature Sets Used at Interspeech Challenges

Tables A.1 and A.2 show the low-level descriptors (LLDs) and functionals and their frequency across the six openSMILE toolkit standard feature sets which were used for the Interspeech and Audio/Visual Emotion Challenge (AVEC) baselines. The challenges are: the Interspeech 2009 Emotion Challenge (Schuller *et al.* 2009), the Interspeech 2010 Paralinguistics Challenge (Schuller *et al.* 2010), the Interspeech 2011 Speaker State Challenge (Schuller *et al.* 2011b), the Interspeech 2012 Speaker Trait Challenge (Schuller *et al.* 2012b), and the first and second Audio/Visual Emotion Challenge (AVEC 2011 (Schuller *et al.* 2011a) and AVEC 2012 (Schuller *et al.* 2012a)).

All six feature sets contain supra-segmental features. That is, the acoustic LLDs such as energy and pitch (which are sampled at a fixed rate – typically 5 or 10 ms), are summarised over a given segment (of variable length) into a single feature vector of fixed length. This is achieved by applying statistical functionals to the LLD. Each functional maps each LLD signal to a single value for the given segment. Examples of functionals are mean, standard deviation, higher-order statistical moments, and quartiles. The openSMILE toolkit (Eyben *et al.* 2010) is used for feature extraction.

All LLDs are computed for short, overlapping windows of the original audio signal. The windows are typically 20–60 ms long and are shifted at a rate of 5 or 10 ms (cf. above).

LLDs are filtered over time by a simple moving average low-pass filter in order to remove artefacts introduced by this windowing. First-order delta coefficients (resembling the first derivative) are computed for each LLD. The total number of features is – in principle – obtained by multiplying the number of LLDs by 2 (considering the delta coefficients), and then multiplying by the number of functionals.

For the Interspeech 2009 Emotion Challenge feature set (EC), for example, 16 LLDs and 16 delta LLDs by 12 functionals yields 384 features. However, for the other feature sets exceptions hold from this strict brute-forcing rule. These exceptions are indicated in footnotes in Tables A.1 and A.2 which are explained in the captions. The exceptions eliminate features

Computational Paralinguistics: Emotion, Affect and Personality in Speech and Language Processing, First Edition.
Björn W. Schuller and Anton M. Batliner. © 2014 John Wiley & Sons, Ltd. Published 2014 by John Wiley & Sons, Ltd.

Table A.1 Low-level descriptors (LLDs) in openSMILE standard features sets. Interspeech 2009 Emotion Challenge (EC), Interspeech 2010 Paralinguistic Challenge (PC), Interspeech 2011 Speaker State Challenge (SSC), Interspeech 2012 Speaker Trait Challenge (STC), Audio/Visual Emotion Challenges (AVEC) 2011/2012 (A'11, A'12). The PC and SSC sets also include the number of voiced segments (F_0 onsets)

	EC	PC	SSC	STC	A'11	A'12
	Number of descriptors					
# LLDs	16	38	59	64	31	31
# Functionals	12	21	39	40	42	38
# Features	384	1 582	4 368	6 125	1 941	1 841
LLDs:						
RMS energy	✓		✓	✓	✓	✓
sum of auditory spec. (loudness)		✓¹	✓	✓	✓	✓
sum of RASTA filt. aud. spec.			✓	✓		
zero crossing rate	✓		✓	✓	✓	✓
energy in 250–650 Hz, 1–4 kHz			✓	✓	✓	✓
spectral roll-off pts. 25, 50, 75, 90%			✓	✓	✓	✓
spectral flux			✓	✓	✓	✓
spectral entropy			✓	✓	✓	✓
spectral variance			✓	✓	✓	✓
spectral skewness			✓	✓	✓	✓
spectral kurtosis			✓	✓	✓	✓
spectral slope			✓	✓		
psychoacoustic sharpness				✓	✓	✓
harmonicity				✓	✓	✓
MFCC 0		✓				
MFCC 1–10	✓	✓	✓	✓	✓	✓
MFCC 11–12	✓	✓	✓	✓		
MFCC 13–14		✓		✓		
log mel frequency band 0–7		✓¹				
LSP frequency 0–7		✓				
26 RASTA spec. bands (0–8 kHz)			✓	✓		
F_0 (ACF based)	✓					
F_0 (SHS based)		✓				
F_0 (SHS based, Viterbi smoothing)			✓	✓	✓	✓
F_0 envelope		✓				
probability of voicing	✓	✓	✓	✓	✓	✓
jitter		✓	✓	✓	✓	✓
jitter (delta: 'jitter of jitter')		✓	✓	✓	✓	✓
shimmer		✓	✓	✓	✓	✓
logarithmic HNR				✓	✓	✓

Description of feature name abbreviations: RMS, root mean square; MFCC, Mel-frequency cepstral coefficients; RASTA refers to a technique of band-pass filtering in the log-spectral domain as used in PLP feature extraction; PLP, perceptual linear predictive coding; LSP, line spectral pair; F_0, fundamental frequency; ACF, autocorrelation function; SHS, sub-harmonic summation; HNR, harmonics-to-noise ratio.

¹Only used for the TUM AVIC baseline (PC).

Table A.2 Functionals in openSMILE standard features sets. Interspeech 2009 Emotion Challenge (EC), Interspeech 2010 Paralinguistic Challenge (PC), Interspeech 2011 Speaker State Challenge (SSC), Interspeech 2012 Speaker Trait Challenge (STC), Audio/Visual Emotion Challenges (AVEC) 2011/2012 (A'11, A'12)

Functional:	EC	PC	SSC	STC	A'11	A'12
positive arithmetic mean				√[4]	√[4]	√[4]
arithmetic mean	√	√	√	√[4]	√[4]	√[4]
root quadratic mean				√	√	√
contour centroid			√	√		
standard deviation (std. dev.)	√	√	√	√	√	√
flatness				√	√	√
skewness	√	√	√	√	√	√
kurtosis	√	√	√	√	√	√
quartiles 1, 2, 3		√[1]	√	√	√	√
inter-quartile ranges 2–1, 3–2, 3–1		√[1]	√	√	√	√
percentile 1%, 99%		√	√	√	√	√
percentile range 1%–99%		√	√	√	√	√
% signal above min. + .25, .5 range				√	√	√
% signal above min. + .75 range		√[1]		√		
% signal above min. + .9 range		√[1]		√	√	√
% signal below min. + .25 range			√	√		
% signal below min. + .5, .75, .9 range				√		
% frames signal rising			√	√	√	√
% frames signal falling			√	√		
% frames signal has left/right curv.			√[6]	√[6]		
% frames that are non-zero			√[2]	√[2]		
linear regression offset	√	√[1]		√[3]		
linear regression slope	√	√[1]	√	√[3]	√[3]	√[3]
linear regression error (lin.)		√[1]		√[3]	√[3]	√[3]
linear regression error (quad.)	√	√[1]	√			
quadratic regression coefficient a			√	√[3]	√[3]	√[3]
quadratic regression coefficient b			√	√[3]		
quadratic regression error (lin.)					√[3]	√[3]
quadratic regression error (quad.)			√	√[3]		
maximum, minimum	√					
maximum − minimum (range)	√					
rising/falling slopes mean & std. dev.				√[3]	√[3]	√[3]
inter peak distances mean & std. dev.				√[3]	√[3]	√[3]
amplitude mean of maxima				√[3]	√[3]	√[3]
amplitude mean of minima				√[3]	√[3]	√[3]
amplitude range of maxima				√[3]	√[3]	√[3]
relative position of max, min	√	√[1]		√		
linear predictive coding gain			√	√	√[3,5]	√[3,5]
linear predictive coding coeff. 1–5			√	√	√[3,5]	√[3,5]
peak distances mean			√	√[3]		
peak distances standard deviation			√	√[3]		
peak value mean			√	√[3]		
peak value mean – arithmetic mean			√	√[3]		
segm. length mean, max, min, std. dev.			√[2]	√	√[5]	
input duration in seconds		√[2]	√[2]			

[1]Only used for the TUM AVIC baseline (PC). [2]Only applied to F_0. [3]Not applied to delta coefficient contours. [4]For delta coefficients the mean of only positive values is applied, otherwise the arithmetic mean is applied. [5]Not applied to voicing related LLDs. [6]Only applied to voicing related LLDs.

which do not contribute any meaningful information, for example, because they are always constant.

A.2 Feature Encoding Scheme

The feature sets shown in the previous section provide a first step towards unification of features. However, if one wishes to encode the type of features in a proprietary feature set, a standard encoding scheme seems desirable. There is no agreed-upon taxonomy of feature types yet, let alone a machine-readable feature encoding scheme, except the one based on Batliner *et al.* (2011) which will be shown here. A common understanding may be guaranteed within a single research group; this is, however, not possible across groups and different research traditions. Just to give one simple example: features representing the temporal alignment of pitch or energy extrema – which are measured in milliseconds – can be conceived still as representing pitch or energy. This is not incorrect because pitch or energy extrema without such an alignment are not very meaningful. However, when classifying prosodic boundaries, accentuation, or emotion, one realises that there is a high correlation between these parameters and genuine duration parameters such as overall duration of phonetic/linguistic entities. Thus, we define such features as duration features – but of course, one still wants to keep in mind that they are modelling pitch or energy contours. Another example is 'mixed' parameters: there has to be agreement on whether to attribute them to the one or the other category, or to a 'mixed' or 'not-attributable' category. All these questions are part of a general taxonomy of acoustic or linguistic parameters.

If a database has been thoroughly processed and annotated, another problem pertains to size of units, manual versus automatic processing, etc. To give another simple example: pitch values can be obtained from different pitch detection algorithms, or from fully manual procedures, or they can be based on manual corrections of automatic extraction results. These questions are part of a partly general, partly specific taxonomy of types of processing.

The feature encoding scheme developed within the CEICES initiative is one step towards a standardisation, comparable to first attempts to define standards such as SAMPA for phonetic coding or EARL (Schröder *et al.* 2007) for emotion coding.

The scheme is realised in 8-bit ASCII coding; fields are delimited by full stops. Each position (column) has its specific semantics. Each line fully describes one feature.

Table A.3 provides an overview on the entire scheme, Table A.4 shows the details for linguistic feature types, and Table A.5 displays the codes of functionals. The tables include a large number of abbreviations to keep the scheme to three one-page tables. Note that linguistic descriptors (L) are encoded as 6-tuples, acoustic descriptors (A) as 20-tuples, and functionals (F) as 6-tuples. Full stops within one field are only delimiters, without their own semantics. Each column within a field is used to encode specific types. The six L positions are mutually exclusive. In the A field, sub-fields (two columns each) delimited by full stops denote one acoustic type; mixed types denoted in two different sub-fields are possible.

To give an example of a feature code, the minimum of words' pitch could have the following code:

S1I00010M1D4R5111L000000A00.00.10.00.00.00.00.00.00.00C0000110000F00.01.01
N00X0000000000T- - - - - - - - -PWoPitchMin.

Table A.3 Feature coding scheme: overview. The alpha-numeric number assigns the position of the respective column within that field. l, r: left, right; ctxt: context

Identifier	S1
Site Id	**I12345**
Mic quality	**M1** 1 close-talk, 2 reverberated, 3 room, 9 other
Domain	**D1** 1 fixed frame-rate, 2 subword, 3 word, 4 chunk, 5 turn, 6 speaker, 9 other
Reference	**R1 domain segm. I: basic** 0 none, 1 acoust., 2 synt., 3 semantic, 5 synt. + acoust., 9 other **2 domain segm. II: strategy** 0 implicit, 1 manual + heuristics, 2–3 dummy manual, 4 other manual, 5–7 dummy autom., 8 other autom., 9 other **3 subdomain segm.** 0 none, 1 manual, 2 autom., 9 other **4 LLD/orthography/non-verbals** 1 manual, 2 automatic, 3 mixed, 9 other
Linguistic	**L123456**
Acoustics	**A12. duration (1 interval, 2 position)** **34. energy** **56. F0** **78. spectral** **90. formant (9 number, 0 type)** **ab. cepstral (number)** **cd. voice quality** **ef. wavelets (number)** **gh. pause** **ij dummy**
Comp. context	**C1 voiced, 2 unvoiced, 3 fixed frame, 4 sub word, 5 word,** **6 chunk, 7 turn, 8 speaker, 9 whole db, 0 outside sub-corpus** 0 not appl., 1 applicable, 2 incl. l local ctxt, 3 incl. r local ctxt, 4 incl. l+r local ctxt, 6 only l local ctxt, 7 only r local ctxt, 8 only l+r local ctxt
Functionals	**F12.34.56**
Norm. context	**N12**
etXetera	**X1234567890**
Text	**T1234567890** string
Previous name	**P1234567890.....** alphanumeric

Table A.4 Linguistic descriptors in detail

---

1. column: Bag-of-Words

---

0 non-applicable
1 stopping
2 stemming
3 stopping and stemming
4 truncated words
5 character *N*-Gram
6 class-based clustering
7 dummy
8 dummy
9 other

---

2. column: POS

---

0 non-applicable
1 API
2 APN
3 AUX
4 NOUN
5 PAJ
6 VERB
9 other

---

3. column: Higher Semantics

---

0 none
1 vocative
2 positive-valence
3 negative-valence
4 commands-directions
5 interjections
6 dummy
7 dummy
8 dummy
9 other

---

4. column: Non-Verbals

---

0 none
1 breathing
2 laughter
3 cough
4 interjection
5 (human) noise
6 dummy
7 dummy
8 dummy
9 other

---

5. column: Disfluencies

---

0 not applicable
1 filler pauses
2 syllable length
4 hesitation (nasal)
5 hesitation (vocal)

(continued)

Table A.4 *(Continued)*

6 hesitation (vocal, nasal)
7 dummy
8 dummy
9 other

6. column: reserved for later use

Table A.5 Functionals in detail by type and code. Abbreviations: cf combination functional, ivl interval, lin linear, pos position, quad quadratic, rvl reversal, std dev standard deviation, zcr zero crossing rate

00 non-applicable

extreme values

01 min, 02 max, 03 min pos, 04 max pos,
05 range, 06 mean min dist, 07 mean max dist, 08 min slope,
09 max slope, 50 on-pos, 51 off-pos, 59 other pos

mean

10 arithmetic, 11 quadratic, 12 geometric, 13 harmonic,
14 absolute, 15 conf. ivl both, 16 conf. ivl upper, 17 conf. ivl lower, 18 centroid

percentiles

20 quartile 1, 21 quartile 2 (median), 22 quartile 3, 23 quartile range 21,
24 quartile range 32, 25 quartile range 31 (iqr), 29 percentile other

higher statistical moments

30 std dev, 31 variance, 32 skewness, 33 kurtosis,
34 length, 35 sum, 36 zcr, 37 most frequent value (mode)

specific functionals/regression

38 up level time, 39 down level time, 40 micro variation, 41 #segments,
42 #rvl points, 43 #peaks, 44 mean dist rvl points, 45 std dev dist of rvl points,
46 mean peak distance, 47 ratio, 48 error, 49 other stat functional,
60 reg error, 61 lin reg coeff 1, 62 lin reg coeff 2, 63 quad reg coeff 1,
64 quad reg coeff 2, 65 quad reg coeff 3, 66 #ivls,
67 mean ivl length, 68 #positive ivls, 69 #negative ivls
70 DCT coeff 1, 71 DCT coeff 2, 72 DCT coeff 3, 73 DCT coeff 4, 74 DCT coeff 5,
79 other spectral coefficient

genetic functions

80 cf absolute value, 81 cf signum, 82 cf log, 83 cf reciprocal value, 84 cf power,
85 cf add, 86 cf minus, 87 cf mult, 88 cf div, 89 cf other

linguistic functionals

90 boolean TF, 91 word count TF, 92 log word count TF, 99 other TF functional

References

Batliner, A., Steidl, S., Schuller, B., Seppi, D., Vogt, T., Wagner, J., Devillers, L., Vidrascu, L., Aharonson, V., and Amir, N. (2011). Whodunnit: Searching for the most important feature types signalling emotional user states in speech. *Computer Speech and Language*, **25**, 4–28.

Eyben, F., Wöllmer, M., and Schuller, B. (2010). openSMILE – The Munich Versatile and Fast Open-Source Audio Feature Extractor. In *Proc. of the 9th ACM International Conference on Multimedia, MM*, pp. 1459–1462, Florence.

Schröder, M., Devillers, L., Karpouzis, K., Martin, J.-C., Pelachaud, C., Peter, C., Pirker, H., Schuller, B., Tao, J., and Wilson, I. (2007). What should a generic emotion markup language be able to represent? In A. Paiva, R. Prada, and R. W. Picard (eds) *Affective Computing and Intelligent Interaction*, pp. 440–451, Berlin. Springer.

Schuller, B., Steidl, S., and Batliner, A. (2009). The Interspeech 2009 Emotion Challenge. In *Proc. of Interspeech*, pp. 312–315, Brighton, UK.

Schuller, B., Steidl, S., Batliner, A., Burkhardt, F., Devillers, L., Müller, C., and Narayanan, S. (2010). The Interspeech 2010 Paralinguistic Challenge. In *Proc. of Interspeech*, pp. 2794–2797, Makuhari, Japan.

Schuller, B., Valstar, M., Eyben, F., McKeown, G., Cowie, R., and Pantic, M. (2011a). AVEC 2011 – The First International Audio/Visual Emotion Challenge. In B. Schuller, M. Valstar, R. Cowie, and M. Pantic (eds) *Proceedings of the First International Audio/Visual Emotion Challenge and Workshop, AVEC 2011, held in conjunction with the International HUMAINE Association Conference on Affective Computing and Intelligent Interaction 2011, ACII 2011, Memphis, TN*, volume II, pp. 415–424. Springer, Berlin.

Schuller, B., Batliner, A., Steidl, S., Schiel, F., and Krajewski, J. (2011b). The Interspeech 2011 Speaker State Challenge. In *Proc. of Interspeech*, pp. 3201–3204, Florence.

Schuller, B., Valstar, M., Cowie, R., and Pantic, M. (2012a). AVEC 2012 – The Continuous Audio/Visual Emotion Challenge. In *Proc. of the 2nd International Audio/Visual Emotion Challenge and Workshop, AVEC, Grand Challenge and Satellite of ACM ICMI 2012*, Santa Monica, CA.

Schuller, B., Steidl, S., Batliner, A., Nöth, E., Vinciarelli, A., Burkhardt, F., Son, R. v., Weninger, F., Eyben, F., Bocklet, T., Mohammadi, G., and Weiss, B. (2012b). The Interspeech 2012 Speaker Trait Challenge. In *Proc. of Interspeech*, pp. 254–257, Portland, OR.

Index

Computational Paralinguistics: Emotion, Affect and Personality in Speech and Language Processing, First Edition.
Björn W. Schuller and Anton M. Batliner. © 2014 John Wiley & Sons, Ltd. Published 2014 by John Wiley & Sons, Ltd.